METHODS IN MOLECULAR BIOLOGY™

AF182498

Series Editor
John M. Walker
School of Life Sciences
University of Hertfordshire
Hatfield, Hertfordshire, AL10 9AB, UK

For other titles published in this series, go to
www.springer.com/series/7651

G-Quadruplex DNA

Methods and Protocols

Edited by

Peter Baumann

Howard Hughes Medical Institute
Stowers Institute for Medical Research, Kansas City, MO, USA
Department of Molecular and Integrative Physiology, University of Kansas Medical Center, Kansas City, KS, USA

Humana Press

Editor
Peter Baumann
Howard Hughes Medical Institute
Stowers Institute for Medical Research
Kansas City, MO
USA
and
Department of Molecular and Integrative Physiology
University of Kansas Medical Center
Kansas City, KS
USA
peb@stowers.org

ISSN 1064-3745 e-ISSN 1940-6029
ISBN 978-1-4939-5639-5 ISBN 978-1-59745-363-9 (eBook)
DOI 10.1007/978-1-59745-363-9
Springer New York Dordrecht Heidelberg London

© Humana Press, a part of Springer Science+Business Media, LLC 2010
Softcover re-print of the Hardcover 1st edition 2010
All rights reserved. This work may not be translated or copied in whole or in part without the written permission of the publisher (Humana Press, c/o Springer Science+Business Media, LLC, 233 Spring Street, New York, NY 10013, USA), except for brief excerpts in connection with reviews or scholarly analysis. Use in connection with any form of information storage and retrieval, electronic adaptation, computer software, or by similar or dissimilar methodology now known or hereafter developed is forbidden.
The use in this publication of trade names, trademarks, service marks, and similar terms, even if they are not identified as such, is not to be taken as an expression of opinion as to whether or not they are subject to proprietary rights.
While the advice and information in this book are believed to be true and accurate at the date of going to press, neither the authors nor the editors nor the publisher can accept any legal responsibility for any errors or omissions that may be made. The publisher makes no warranty, express or implied, with respect to the material contained herein.

Printed on acid-free paper

Humana Press, a part of Springer Science+Business Media (www.springer.com)

Preface

A square planar arrangement of four guanine bases was first proposed to explain the unusual property of guanosine to form gels. This G-quartet structure may have easily remained an odd curiosity, if it wasn't for the intriguing possibility that such interactions of guanosine bases have functions in biology. Chromosome termini in most eukaryotes are comprised of repetitive, G-rich DNA sequences that can form remarkably stable stacks of G-quartets, often referred to as G-quadruplexes. The observations that G-quadruplex structures form readily in vitro under physiological conditions and that suitable sequences are present at the ends of chromosomes of most eukaryotes have prompted much interest in the role of G-quartets in biology. Recent reports have provided experimental support for physiological functions of G-quartets not just as telomeres, but also in the control of gene expression and in mRNA maturation. The realization that the human genome harbors literally hundreds of thousands of potentially G-quartet-forming sequences has raised the exciting possibility that many biological functions of these structures remain to be discovered.

Recent work revealed that stabilizing G-quadruplexes in telomeric DNA inhibits telomerase activity, providing impetus for the development of G-quartet-interacting drugs. The therapeutic potential of G-quartets, however, goes far beyond telomerase inhibitors. G-quartet-containing oligonucleotides have been recognized as a potent class of aptamers effective against STAT3 and other transcription factors implicated in oncogenesis. Outside the realms of biology and therapeutics, G-quartets provide insights into molecular self-assembly and supramolecular chemistry and have recently found applications as sensors in nano-technology.

This book aims to present a collection of detailed methods and protocols for studying G-quartet formation, dynamics, and molecular recognition. We believe that this volume will be a useful resource for those familiar with G-quartets, as well as an easy entry point for those researchers from diverse fields who are just developing an interest in G-quadruplex DNA.

Peter Baumann

Acknowledgements

Many people have contributed to our collective knowledge on G-quartets and each of them has my profound gratitude. I am especially thankful to Dr. Martin Gellert, who first sparked my interest in G-quadruplex DNA. I am greatly indebted to Dr. Rachel Helston, whose editorial contributions were critical for the completion of this volume. I also thank Carla Anderson for her administrative help as well as the many people at the Stowers Institute who have provided support. Finally, I owe thanks to Dr. John Walker and Humana Press for giving me the opportunity to share this volume with you.

Contents

Contributors

PETER BAUMANN • *Howard Hughes Medical Institute, Stowers Institute for Medical Research, Kansas City, MO, USA*
Department of Molecular and Integrative Physiology, University of Kansas Medical Center, Kansas City, KS, USA
TRACY M. BRYAN • *Children's Medical Research Institute and the University of Sydney, Sydney, Australia*
ROBERT BUSCAGLIA • *James Graham Brown Cancer Center, University of Louisville, Louisville, KY, USA*
JONATHAN B. CHAIRES • *James Graham Brown Cancer Center, University of Louisville, Louisville, KY, USA*
CHENG-CHUNG CHANG • *Department of Chemistry, National Chung-Hsing University, Taichung, Taiwan, Republic of China*
TA-CHAU CHANG • *Institute of Atomic and Molecular Sciences, and Genomic Research Center, Academia Sinica, Taipei, Taiwan, Republic of China*
Department of Chemistry, National Taiwan University, Taipei, Taiwan, Republic of China; Institute of Biophotonics Engineering, National Yang-Ming University, Taipei, Taiwan, Republic of China
PABLO ENGLEBIENNE • *Department of Chemistry, McGill University, Montreal, QC, Canada*
NICHOLA C. GARBETT • *James Graham Brown Cancer Center, University of Louisville, Louisville, KY, USA*
TIMUR I. GAYNUTDINOV • *Department of Nuclear Medicine, Warren G. Magnuson Clinical Center, National Institutes of Health, Bethesda, MD, USA*
ROBERT D. GRAY • *James Graham Brown Cancer Center, University of Louisville, Louisville, KY, USA*
TAEKJIP HA • *Department of Physics, University of Illinois at Urbana Champaign, Urbana, IL, USA*
Howard Hughes Medical Institute, Urbana, IL, USA
SHOZEB HAIDER • *The Cancer Research UK Biomolecular Structure Group, The School of Pharmacy, University of London, London, UK*
STEVE G. HERSHMAN • *Department of Pathology and Laboratory Medicine, University of Pennsylvania School of Medicine, Philadelphia, PA, USA*
LAURENCE H. HURLEY • *Department of Pharmacology and Toxicology, College of Pharmacy, University of Arizona, Tucson, AZ, USA*
MICHAEL B. JARSTFER • *Division of Medicinal Chemistry and Natural Products, School of Pharmacy, University of North Carolina, Chapel Hill, NC, USA*
F. BRAD JOHNSON • *Department of Pathology and Laboratory Medicine, Institute on Aging, University of Pennsylvania School of Medicine, Philadelphia, PA, USA*
ROXANNE KIELTYKA • *Department of Chemistry, McGill University, Montreal, QC, Canada*

HANS J. LIPPS • *Institute of Cell Biology, University of Witten/Herdecke, Witten, Germany*
LUIS A. MARKY • *Department of Pharmaceutical Sciences, Department of Biochemistry and Molecular Biology, Eppley Institute for Research in Cancer, University of Nebraska Medical Center, Omaha, NE, USA*
CHONGKHAM S. MEKMAYSY • *James Graham Brown Cancer Center, University of Louisville, Louisville, KY, USA*
NICOLAS MOITESSIER • *Department of Chemistry, McGill University, Montreal, QC, Canada*
DAVID MONCHAUD • *Institut Curie, Section Recherche, CNRS UMR 176, Center Universitaire Paris XI, Orsay, France*
IAN K. MOON • *Division of Medicinal Chemistry and Natural Products, School of Pharmacy, University of North Carolina, Chapel Hill, NC, USA*
STEPHEN NEIDLE • *The Cancer Research UK Biomolecular Structure Group, The School of Pharmacy, University of London, London, UK*
RONALD D. NEUMANN • *Department of Nuclear Medicine, Warren G. Magnuson Clinical Center, National Institutes of Health, Bethesda, MD, USA*
BURAK OKUMUS • *Department of Physics, University of Illinois at Urbana Champaign, Urbana, IL, USA*
CHRIS M. OLSEN • *Department of Pharmaceutical Sciences, University of Nebraska Medical Center, Omaha, NE, USA*
KATRIN PAESCHKE • *Institute of Cell Biology, University of Witten/Herdecke, Witten, Germany*
IGOR G. PANYUTIN • *Department of Nuclear Medicine, Warren G. Magnuson Clinical Center, National Institutes of Health, Bethesda, MD, USA*
LUIGI PETTRACONE • *James Graham Brown Cancer Center, University of Louisville, Louisville, KY, USA*
JAN POSTBERG • *Institute of Cell Biology, University of Witten/Herdecke, Witten, Germany*
PAUL RYVKIN • *Department of Pathology and Laboratory Medicine, Penn Center for Bioinformatics, and Graduate Group in Genomics and Computational Biology, University of Pennsylvania School of Medicine, Philadelphia, PA, USA*
CHRISTIANE SCHAFFITZEL • *Institute for Molecular Biology and Biophysics, ETH Zürich, Zürich, Switzerland*
HANADI SLEIMAN • *Department of Chemistry, McGill University, Montreal, QC, Canada*
JASMINE S. SMITH • *Department of Pathology and Laboratory Medicine, Cancer Biology Program, and Institute on Aging, University of Pennsylvania School of Medicine, Philadelphia, PA, USA*
DAEKYU SUN • *Department of Pharmacology and Toxicology, College of Pharmacy, University of Arizona, Tucson, AZ, USA*
MARIE-PAULE TEULADE-FICHOU • *Institut Curie, Section Recherche, CNRS UMR 176, Center Universitaire Paris XI, Orsay, France*
LI-SAN WANG • *Department of Pathology and Laboratory Medicine, Institute on Aging, Penn Center for Bioinformatics, and Graduate Group in Genomics and Computational Biology, University of Pennsylvania School of Medicine, Philadelphia, PA, USA*

Chapter 1

G-Quadruplexes: From Guanine Gels to Chemotherapeutics

Tracy M. Bryan and Peter Baumann

Abstract

G-quartets are square planar arrangements of four guanine bases, which can form extraordinarily stable stacks when present in nucleic acid sequences. Such G-quadruplex structures were long regarded as an in vitro phenomenon, but the widespread presence of suitable sequences in genomes and the identification of proteins that stabilize, modify, or resolve these nucleic acid structures have provided circumstantial evidence for their physiological relevance. The therapeutic potential of small molecules that can stabilize or disrupt G-quadruplex structures has invigorated the field in recent years. Here we review some of the key observations that support biological functions for G-quadruplex DNA as well as the techniques and tools that have enabled researchers to probe these structures and their interactions with proteins and small molecules.

Key words: G-quadruplex, G-quartet, Guanosine, Telomerase, Telomere

1. Introduction

More than four decades before Watson and Crick proposed their structure for DNA, the German chemist Ivar Bang noted that guanylic acid forms gels at high millimolar concentrations (1). This unusual physical property puzzled researchers for the next 50 years until Gellert and colleagues collected fiber x-ray diffraction data on guanylic acid (2), revealing the assembly of tetrameric units into large helical structures that account for the gel-like properties of the aqueous solution. Four molecules of guanylic acid form a square planar arrangement in which each of the four bases is the donor and acceptor of two hydrogen bonds, now referred to as a G-quartet (Fig. 1.1). As the interest in nucleic acids intensified over the following decades, it became clear that guanosine homo-oligomers can adopt the same structure, both in the ribose and deoxyribose forms (3, 4). For years, little consideration

P. Baumann (ed.), *G-Quadruplex DNA: Methods and Protocols*, Methods in Molecular Biology, vol. 608
DOI 10.1007/978-1-59745-363-9_1, © Humana Press, a part of Springer Science+Business Media, LLC 2010

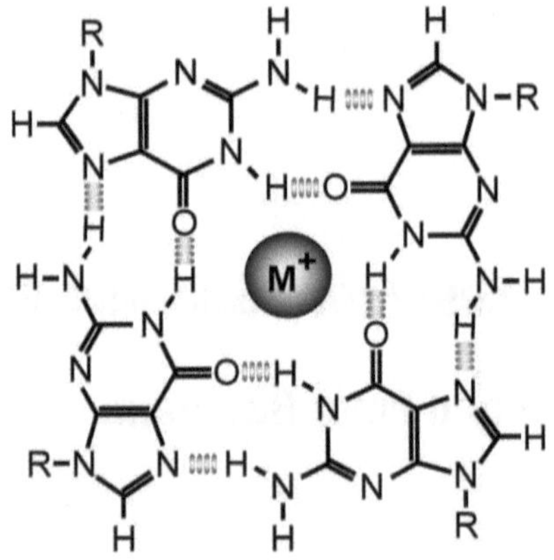

Fig. 1.1. The G-quartet is a square planar arrangement of four guanine bases each of which serves as the donor and acceptor of two hydrogen bonds. The monovalent metal ion shown in the center is critical for stability when stacks of G-quartets form a G-quadruplex.

was given to possible roles for G-quartets in biological systems until Henderson and colleagues made the observation that oligonucleotides corresponding to the G-rich strand of telomeric DNA display unexpectedly high electrophoretic mobility on nondenaturing polyacrylamide gels (5). Structural probing later showed that G-rich sequences found at telomeres and in the immunoglobulin switch region can indeed adopt stable four-stranded structures now known as G-quadruplexes (5–8).

Among the five nucleosides commonly found in DNA and RNA, the property to form stable and extensive self-associations is limited to guanosine because of its unique hydrogen bonding donor and acceptor sites. Cations play a critical role in stabilizing G-quadruplex structures by occupying the central cavity and neutralizing the electrostatic repulsion of inwardly pointing guanine O6 oxygens. It was recognized early on that the ability to stabilize guanosine gels differed greatly between cations (9), suggesting that the ionic radius is important for complex stability. In the alkali series K^+ promotes the most stable G-quadruplexes, followed by Rb^+, Na^+, Cs^+, and Li^+. Electrostatic effects are also likely to affect the relative ability of cations to stabilize G-quadruplexes (10). The hydration energy of monovalent cations is inversely proportional to their ionic radii; hence the larger the cation the less hydrophilic it is, making it more likely to preferentially partition itself at the interior of the G-quartet. The same effect of different monovalent cations was also observed for the stability of structures formed by telomeric oligonucleotides, demonstrating that single-stranded telomeric DNA can fold into G-quadruplex structures under conditions within the physiological range (8).

2. Structural Diversity

From the earliest days of studying G-quadruplexes in vitro, it was apparent that these structures exhibit extensive structural polymorphism. G-quadruplexes may form from one (intramolecular) or two or more (intermolecular) DNA strands; another way of classifying them is whether the DNA strand orientation is antiparallel (Fig. 1.2a), parallel (Fig. 1.2b), or hybrid (Fig. 1.2c). Correspondingly, the nucleotide linkers between G-quartet stacks can adopt a multitude of loop structures (Fig. 1.2). G-quadruplex conformation is influenced by both the DNA sequence and the conditions used in the folding reaction such as the nature of the

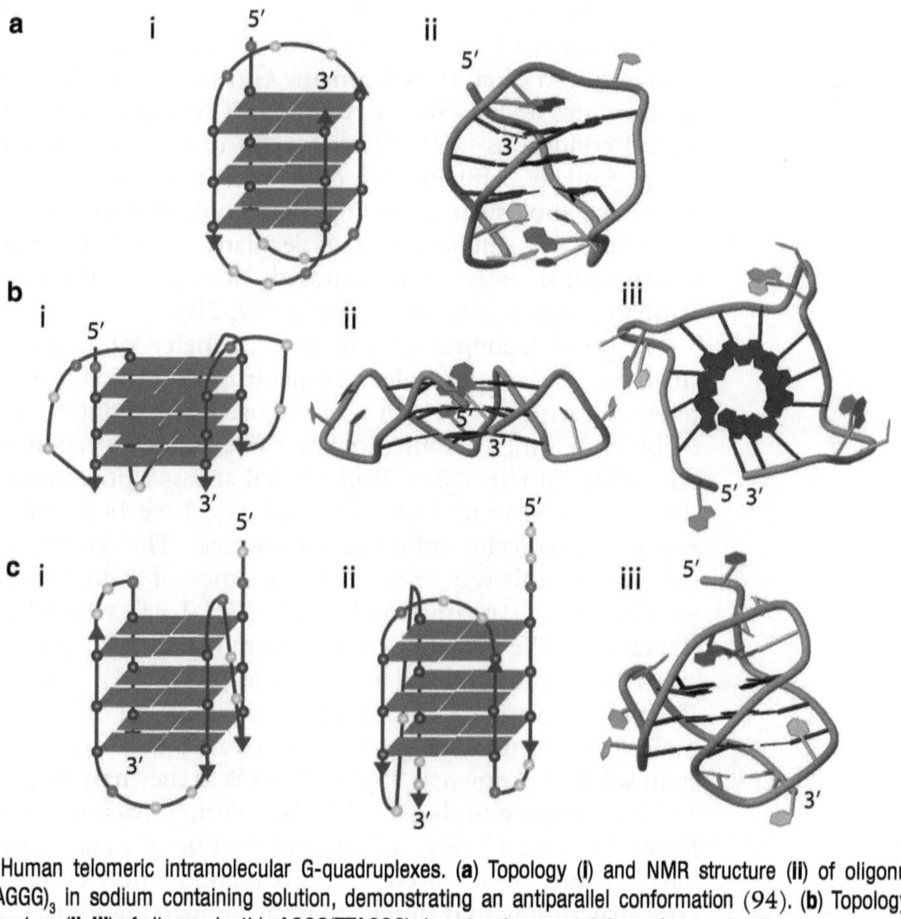

Fig. 1.2. Human telomeric intramolecular G-quadruplexes. (**a**) Topology (**i**) and NMR structure (**ii**) of oligonucleotide AGGG(TTAGGG)$_3$ in sodium containing solution, demonstrating an antiparallel conformation (94). (**b**) Topology (**i**) and crystal structure (**ii**, **iii**) of oligonucleotide AGGG(TTAGGG)$_3$ in potassium containing solution, showing a parallel "propeller" structure (23). The crystal structure is shown as a side view (**ii**) and a top view (**iii**). (**c**) Hybrid conformations in potassium containing solution. Hybrid 1 (**i**) and hybrid 2 (**ii**) topologies illustrate differences in loop structures (28, 29). The NMR structure of hybrid 2 is shown in (**iii**) (28). T Bryan in *Molecular Themes in DNA Replication*, ed. Lynne Cox, Royal Society of Chemistry, Cambridge, 2009, p **264** – Reproduced by permission of The Royal Society of Chemistry.

stabilizing cation. Although some general trends are apparent, e.g. potassium can favor parallel conformations (11), there are always exceptions to these rules; e.g. both antiparallel potassium-stabilized and parallel sodium-stabilized G-quadruplexes exist and can be quite stable (12–15). Thus it is difficult to predict the propensity of a sequence to fold into a particular structure, and each sequence needs to be characterized empirically under different folding conditions. The existence of multiple G-quadruplex conformations in equilibrium in the same solution (12, 16, 17) emphasizes the (often-overlooked) need to purify individual isomers prior to analysis.

The stability of G-quadruplexes also varies widely; it depends not only on the identity of the stabilizing cation, but also on the DNA length and sequence, and the strand stoichiometry and alignment (18). Nevertheless, as a G-quartet contains eight hydrogen bonds in comparison to the two or three present between Watson and Crick base pairs, it might be expected that G-quadruplexes have equal or higher stability than duplex DNA. This is indeed often the case: many G-quadruplexes have melting temperatures well in excess of 60 or 70 °C under otherwise physiological conditions (18). This suggests that G-quadruplex DNA can potentially compete with duplex formation in vivo. In agreement, the molecular crowding agent polyethylene glycol, typically used to simulate the molecularly crowded intracellular environment, was demonstrated to favor formation of G-quadruplexes over duplex DNA (19, 20).

A good example of the heterogeneity of G-quadruplex structures is the intramolecular quadruplex formed from human telomeric sequence, which is of intense interest due to its ability to block telomere elongation by the cancer-associated enzyme telomerase in vitro (21). Both crystal and solution structures of the oligonucleotide $AGGG(TTAGGG)_3$ have been solved and reveal dramatically different topologies. The NMR solution structure of this sequence in the presence of sodium is an antiparallel basket-type quadruplex (22) (Fig. 1.2a), while the crystal structure in the presence of potassium represents a parallel propeller-type intramolecular G-quadruplex (23) (Fig. 1.2b). Recently, two variations of a third conformation of human intramolecular telomeric G-quadruplex have been detected in potassium solution, known as "hybrid" forms as they have both parallel and antiparallel strands (24–27). The solution structures of the two forms ("hybrid 1", Fig. 1.2c(i) and "hybrid 2", Fig. 1.2c(ii, iii)) have recently been solved, and reveal an identical G-quadruplex core structure with differences in the connecting loops (28, 29). The equilibrium between hybrid 1 and hybrid 2 was greatly influenced by the 3′ sequence of the oligonucleotide, with GGG ends favoring formation of hybrid 1 (28, 29). As only ~5% of telomeres in human cells end in GGG (30), this might imply that hybrid 2 predominates in vivo. The in vivo equilibrium may also be affected by temperature, ionic conditions, and the presence of particular proteins.

As potassium levels in mammalian cells are ~150 mM and generally higher than sodium levels (31), one of the potassium structures may be the more physiologically relevant conformation. But which one is it? It has been argued that the parallel conformation seen in the crystal structure is not biologically relevant and may simply represent an artifact of the crowding conditions introduced by the crystalline state (32). However, the presence of 40% polyethylene glycol induced a shift from hybrid to parallel G-quadruplexes and the authors of this study postulate that molecular crowding conditions may in fact more accurately represent the in vivo situation (33).

3. Biological Roles for G-quadruplexes

As many nucleic acid sequences rich in guanosines are capable of forming G-quadruplexes, one wonders how prevalent these structures truly are within cells. Telomeric DNA has received much attention in this regard, in part because chromosomes end in single stranded overhangs of the G-rich strand which may fold into G-quadruplex structures. But extended single-strandedness is not a prerequisite for G-quadruplex formation. Transient destabilization of duplex DNA during transcription, replication, or DNA repair may well be sufficient to allow G-quadruplex DNA formation at many sites in the genome. Bioinformatic analysis has identified 375,000 candidate sequences within the human genome that could form G-quadruplex structures (34, 35). It is possible that not all of these sequences form stable quadruplexes under physiological conditions (36). However, the nonrandom distribution of potentially G-quadruplex-forming sequences across the genome, as well as the nonrandom length and sequence of loop regions, argues that natural selection may be at work. Coding sequences are underrepresented for the transcribed strand suggesting that G-quartet formation in mRNA may be detrimental (34). Despite the underrepresentation of coding sequences, the frequency at which potentially G-quadruplex-forming sequences are found within transcribed regions displays an intriguing correlation with gene function. They are frequently found in proto-oncogenes including c-MYC, VEGF, c-kit, HIF-1a, and BCL2, but are significantly underrepresented in tumor suppressor genes (37).

A role for G-quadruplexes in gene regulation seems likely as putative G-quadruplex-forming sequences are concentrated in promoter regions. Nearly half of all known genes in the human genome harbor such sequences within 1,000 nucleotides upstream of the transcription start site (38). Regions of the human genome that are both within promoters and hypersensitive to nuclease cleavage show the greatest enrichment of potential quadruplex elements. The nuclease sensitivity of these sites indicates that the DNA is not bound by nucleosomes or other proteins and therefore may be more prone to G-quadruplex formation. This bias may at least in part reflect the G-richness of many transcription factor binding sites. Careful examination of individual promoter sequences will be required to dissect the contributions of the different pathways and structures in gene regulation. At least in the case of the c-MYC promoter it has been shown that the G-quadruplex-forming region plays a critical role in regulating expression of this gene. A single point mutation which destabilizes the G-quadruplex resulted in a 3-fold increase in basal transcriptional activity of the c-MYC promoter (39). Conversely, a cationic porphyrin known to stabilize a G-quadruplex structure was able to suppress c-MYC transcriptional activation. These results strongly argue for a regulatory role of this particular G-quadruplex as a repressor of c-MYC transcription.

Despite much circumstantial evidence in favor of the existence of telomeric G-quadruplex structures in human cells, as well as tantalizing hints at potential functions, the actual role(s) of these structures in vivo have remained enigmatic. The different conformations may carry out distinct roles. Intermolecular G-quadruplexes could facilitate telomere–telomere associations; such interactions have been observed in the telomere-rich environment of the macronuclei of ciliated protozoa and there is evidence that they are mediated by G-quadruplexes (40, 41). It was postulated 20 years ago that intermolecular parallel G-quadruplexes may be involved in the alignment of sister chromatids during meiosis (6), but there remains no direct evidence for this intriguing possibility. The clustering of telomeres in a meiotic bouquet arrangement has been observed in almost all organisms (42) and the demonstration that G-quadruplexes are involved in telomere bouquet formation would represent a major advance in understanding this ubiquitous structure. There is indirect evidence for this hypothesis; a component of the meiosis-specific synaptonemal complex in *Saccharomyces cerevisiae*, Hop1, was demonstrated to promote pairing of double-stranded DNA helices via G-quartet formation, implicating intermolecular G-quadruplexes as the vehicles of chromosomal synapsis during meiotic prophase (43). Furthermore, deletion of a G-quadruplex-specific nuclease, *KEM1*, blocks meiosis in yeast, consistent with the hypothesis that G-quadruplex DNA may be involved in homologous chromosome pairing during meiosis (44).

G-quadruplexes may be involved in conferring capping and protective functions to telomeres. They may sequester the telomere from inappropriate elongation by telomerase, or protect it from nucleolytic degradation or end-to-end fusion. Intramolecular antiparallel G-quadruplexes have indeed been shown to be resistant to telomerase elongation in vivo (13, 21, 45), although parallel intermolecular structures are extended by telomerase (13). There is some evidence that G-quadruplexes play a protective role; incubation of duplex DNA with human cell extract elicited a DNA damage response which was alleviated by addition of a 3′ tail capable of forming a G-quadruplex (46). However, the possibility that the protective function was mediated by a telomere-binding protein in the extract was not ruled out.

Finally, G-quadruplexes may play an important role in inhibiting the activation of the alternative lengthening of telomeres (ALT) pathway, a recombination-based mechanism for telomere elongation. As a single-stranded 3′ overhang is essential during the early steps of recombination, sequestration of the single-stranded region at the ends of chromosomes into G-quadruplexes could form an effective barrier against ALT. Several proteins that are potentially involved in the ALT mechanism are also known to unwind G-quadruplexes, including RPA (47) and the RecQ helicases BLM (48) and WRN (49). Unfolding of telomeric G-quadruplexes may allow access to the telomere by recombination proteins and enable initiation of the ALT mechanism of telomere elongation.

There are several reasons to believe that regulatory functions of G-quadruplexes may be more prevalent at the level of RNA than of DNA. Firstly, RNA is single-stranded, at least when first synthesized, and although extensive Watson–Crick base-pairing occurs in some RNAs, a substantial portion of most RNAs remains single-stranded because of the absence of a complementary sequence. Secondly, G-quadruplexes are even more stable in RNA than in DNA and once formed they are highly refractory to unfolding (50). Roles for G-quadruplexes in RNA regulation, splicing, and processing are further supported by the enrichment of candidate sequences in 5′ UTRs (51), first introns (52), and near polyadenylation signals (53). The presence of a G-quadruplex has been experimentally verified in the 5′ UTR of NRAS and a repressive effect on translation has been documented (51). As nearly 3000 mRNAs have potentially G-quadruplex-forming sequences in their 5′ UTR, it is tempting to speculate that G-quadruplex structures may be widely used to control gene expression at the translational level.

The fragile X mental retardation protein (FMRP) associates with polysomes and is thought to regulate mRNA translation. In vitro selection for RNAs that are preferentially bound by FMRP identified RNA ligands which form intramolecular G-quartets indicating that G-quadruplex-containing mRNAs may be the target of FMRP regulation (54). Indeed, when FMRP-containing ribonucleoprotein complexes were immunoprecipitated from mouse brain, nearly 70% of the associated mRNAs contained sequences predicted to form G-quartet structures (55). Such strong correlation argues for a role of this structure in identifying the class of RNAs regulated by FMRP.

Alternative splicing of a number of genes is affected by G-rich sequences in the pre-mRNA and regulatory roles in vivo have been proposed. One of the most interesting examples was discovered in the context of studying the effects of G-quadruplex stabilizing drugs on telomerase activity in cancer cells. Early in vitro experiments had shown that stabilizing G-quartet structures in single-stranded telomeric DNA could inhibit elongation by telomerase (45). It was therefore believed that the G-quartet stabilizing compound 12459 caused telomere shortening and apoptosis in a lung adenocarcinoma cell line by binding to the ends of chromosomes and inhibiting telomerase. However, closer examination revealed that the effect was largely mediated by stabilization of G-quartets in the pre-mRNA of the catalytic subunit of telomerase causing a shift in splicing pattern such that an inactive form of TERT is produced (56). To what extent G-quartet structures are involved in regulating alternative splicing of TERT and other genes in the absence of stabilizing compounds is presently unclear, but the potential for modulating the expression of many genes at this level is attractive.

Another potentially G-quadruplex-forming RNA is telomeric repeat-containing RNA (TERRA)(57). It appears that the C-rich strand of telomeric DNA is actively transcribed from several promoters within subtelomeric DNA and the G-rich RNA product remains associated with telomeric chromatin. As the complementary RNA was not detected, one would expect TERRA to form quadruplex structures unless prevented from doing so by interactions with proteins or telomeric DNA.

Given the abundance of nucleic acid sequences that can form G-quadruplex structures and the evidence supporting their formation under physiological conditions, there is little doubt that such structures form in vivo. There is also accumulating evidence that numerous proteins interact with G-quadruplex DNA and in some cases promote their unfolding. An issue that has been far more difficult to resolve is whether there is a positive regulatory role for G-quadruplex DNA in biology. The presence of a specific nucleic acid structure is inherently difficult to verify in vivo. Intracellular transcription of G-rich DNA in *Escherichia coli* has been shown to produce loops of the non-template strand containing G-quadruplex structures that are detectable by electron microscopy (58). Arguably the most direct evidence for G-quadruplex DNA existing in cells is that antibodies raised against G-quadruplex DNA label the macronuclei of a ciliate (41). A concern often raised about such experiments is that the reagent used for detection may drive the equilibrium towards the folded form, thus creating the very structure it is designed to detect. Nevertheless, Lipps and colleagues have used such antibodies in an intriguing series of experiments aimed at dissecting telomere structure throughout the cell cycle. Their work has led to a model in which telomere end binding proteins TEBP α and β actively stabilize G-quadruplexes for most of the cell cycle. During S-phase, TEBP β is phosphorylated and dissociates from the telomere. At the same time telomerase is recruited and G-quadruplex structures are resolved making the chromosome ends available for extension by telomerase (59, 60).

4. Applications for G-quadruplex Stabilizing or Disrupting Compounds

In 1991, Zahler and colleagues demonstrated for the first time that an intramolecular telomeric G-quadruplex could not be extended by *Oxytricha* telomerase in vitro (45). On the basis of this finding, a substantial effort has been made to identify synthetic and natural compounds that lock telomeric DNA in a G-quadruplex conformation and thus impede telomere elongation in vivo. Given the requirement for telomere maintenance in the indefinite proliferation of cancer cells, such molecules are promising candidates as anticancer drugs. A large number of G-quadruplex-interacting ligands from many chemical classes have been described (61, 62). Those ligands which have been conclusively demonstrated to inhibit telomerase in vitro include the 2,6-diamidoanthraquinone BSU-1051 (63), the perylene diimide PIPER (64), the porphyrin TMPyP4 (65), the trisubstituted acridine BRACO19 (66, 67), bisquinolinium compounds such as 360A, 307A, and the PhenDC series (66, 68, 69), and the natural product telomestatin (66, 70).

Telomestatin is one of the most well-studied G-quadruplex ligands because of its ability to greatly stabilize G-quadruplexes and its high specificity for these structures. Telomestatin induces and specifically recognizes the human intramolecular (71) antiparallel (72) G-quadruplex conformation. Telomestatin initially appeared to be a very potent telomerase inhibitor in vitro with an EC_{50} value of 5 nM (70), although this is now known to be at least one order of magnitude greater (66). Nevertheless, at relatively low doses ($\leq 2\ \mu M$), telomestatin causes gradual telomere shortening and growth arrest or apoptosis in a large number of cancer cell lines (73–78), supporting its use as a telomerase inhibitor in vivo. It has recently become clear, however, that classical telomerase inhibition is only part of the telomeric mechanism of action of telomestatin and related drugs. Higher doses of telomestatin ($\geq 5\ \mu M$) lead to proliferation defects within a time frame that is too short for the effects to be explained by telomere shortening (73, 76). This effect is independent of the telomerase status of the cells, and is likely due to direct uncapping of the chromosome termini in tumor cells. There are now several lines of evidence to support the uncapping mechanism; namely, treatment with telomestatin has been shown to cause degradation of the telomeric 3′ G-overhang (73, 76, 79), rapid dissociation of the telomere capping proteins TRF2 and Pot1 from telomeric termini (76, 79, 80), and an increase in DNA damage signals at the telomeres (79). Other G-quadruplex-stabilizing ligands such as BRACO19 and the pentacyclic acridine RHPS4 also cause disruption of the protective telomere cap structure (81–84).

It was initially envisaged that telomerase inhibition by G-quadruplex stabilizers would be a very specific cancer therapy, because of the absence of active telomerase in most normal tissues. A general effect on telomere structure raises the worrying possibility of toxic effects on normal cells. Nevertheless, several of the afore-mentioned drugs show good selectivity for cancer cell lines over normal cells, for unknown reasons (76–78, 84). This may be due to a different telomere cap structure in normal versus cancer cells, or the existence of intact checkpoint pathways; these possibilities remain to be explored. This raises the exciting possibility that G-quadruplex-stabilizers will constitute a specific cancer therapy that has the capability of overcoming the time-lag required for telomere shortening to occur.

Other considerations when evaluating potential telomere-targeted drugs include their specificity for particular G-quadruplex conformations, given the large number of potential G-quadruplex forming sequences in the human genome. For example, the porphyrin TMPyP4 interacts with telomeric G-quadruplexes with a minimal degree of specificity over its interaction with a G-quadruplex in the promoter of the c-*Myc* oncogene (85, 86). The cellular effects of other ligands, however, are clearly mediated primarily through the telomeres; for example, overexpression of telomere proteins TRF2 and POT1 rendered xenograft tumors resistant to the effects of RHPS4 (84). Furthermore, the implications of the extension of some types of G-quadruplexes by telomerase are also unknown (13). While telomere-targeted G-quadruplex-stabilizing molecules are showing great promise as anti-cancer drugs, their mechanisms of cellular action and the likelihood of adverse effects on healthy, proliferating cells must be further investigated prior to clinical use.

5. Methodologies Used to Study G-quadruplex Structures

There are many simple techniques that can be used to probe aspects of the structure of a G-quadruplex. Native gel electrophoresis revealed early on that G-rich oligonucleotides have an unusual structure that results in aberrant migration on a nondenaturing acrylamide gel (5, 6), and this remains an accessible and straightforward technique to reveal the presence of a G-quadruplex. Intramolecular G-quadruplexes have a compact structure and thus migrate faster through a cation-containing gel than their linear counterparts (8), while intermolecular G-quadruplexes migrate slower due to increased molecular weight (6, 7). Native gel electrophoresis is also invaluable in enabling purification of G-quadruplexes, an important consideration given the heterogeneity of structures that can form from a single oligonucleotide (*see* Moon and Jarstfer, this volume).

Other techniques are required to verify that the aberrantly migrating structures contain G-quartets. Circular dichroism (CD) spectroscopy is a convenient diagnostic tool in this regard, and has the additional advantage of being able to discriminate between G-quadruplex conformations. In this technique, the sample is exposed to circularly polarized light; if there is a chiral species in the solution, it will generally interact asymmetrically with the light, with the asymmetry varying with wavelength. Although it is difficult to predict a CD spectrum from a structure, characteristic spectra corresponding to different G-quadruplex conformations have been determined empirically. Parallel-stranded G-quadruplexes show a peak at 260 nm and a trough at 240 nm, while a peak at 295 nm and a trough at 260 nm are diagnostic of anti-parallel structures (87, 88). The recently-described "hybrid" structures formed from human telomeric oligonucleotides (Fig. 1.2c) show a strong peak at 290 nm with a shoulder out to about 270 nm, and troughs at 235 and 255 nm (24, 27). If carried out over a range of temperatures, CD spectroscopy can also be used to observe melting of a G-quadruplex and hence determine thermodynamic parameters such as T_m, ΔH, and ΔG^0, vital information for comparing stabilities of structures (12, 87); (*see* Olsen and Marky, this volume). G-quadruplexes also show changes in UV absorbance at 295 nm relative to their linear counterparts, so UV spectroscopy may also be used to derive thermodynamic parameters that are reflective of G-quadruplex stability (*see* Olsen and Marky, this volume).

One of the earliest techniques used to verify G-quartet models of telomeric structure was dimethylsulfate (DMS) footprinting (6–8, 89). DMS methylates the N7 position of guanine; subsequent treatment with piperidine breaks the DNA backbone at methylated sites. Gel electrophoresis allows visualization of the length of the cleaved fragments. In a G-quadruplex the N7 is hydrogen bonded and protected from methylation (*see* Fig. 1.1), resulting in little or no cleavage at the guanines involved in G-quartets (*see* Sun and Hurley, this volume).

A powerful recent method for probing G-quadruplex conformation and dynamics is single-molecule fluorescence resonance energy transfer (FRET) (*see* Okumus and Ha, this volume).

In FRET, the oligonucleotide to be folded into a G-quadruplex is labeled with a donor and an acceptor fluorophore. Upon folding of the DNA, the donor fluorophore transfers its energy to the acceptor, with an efficiency that depends on their distance apart and relative orientation. By performing FRET on a dilute solution or a surface-immobilized sample, and capturing the resulting energy emission with a confocal fluorescence microscope, the dynamics of folding of a single molecule can be observed; this removes the need to average the signal from a population of non-synchronously-folding molecules, allowing sensitive dynamic analysis (90). Application of this technique to the human intramolecular telomeric G-quadruplex has revealed that two conformations coexist in solution in both sodium and potassium buffers, and each conformation can be further divided into long-lived (minutes) and short-lived (seconds) species (16, 17).

The above methods provide a wealth of information about G-quadruplex behavior and conformation, but in order to determine precise molecular structures high-resolution techniques such as nuclear magnetic resonance (NMR) and x-ray crystallography are required. NMR first revealed the strand orientations and loop configurations of several telomeric G-quadruplexes and led to high resolution structures (22, 91–95). X-ray crystallography was also successful in generating high resolution structures (96, 97); there are now more than 30 reported structures of G-quadruplexes, some with resolutions less than 1 Å. In some cases, structures of the same molecule solved using both techniques differ either subtly (92, 96) or quite dramatically (23, 94). It is likely that this is a result of the molecular crowding conditions introduced by crystallization. Which technique best represents the in vivo situation is equivocal, and further advances in technology will be needed to determine the true structure of G-quadruplexes within living cells.

References

1. Bang I (1910) Untersuchungen über die Guanylsäure. Biochem Z 26:293–311
2. Gellert M, Lipsett MN, Davies DR (1962) Helix formation by guanylic acid. Proc Natl Acad Sci U S A 48:2013–2018
3. Chantot JF, Guschlbauer W (1972) Mechanism of gel formation by guanine nucleosides. Jerus Symp Quantum Chem Biochem 4:205–214
4. Ralph RK, Connors WJ, Khorana HG (1962) Secondary structure and aggregation in deoxyguanosine oligonucleotides. J Am Chem Soc 84:2265–2266
5. Henderson E, Hardin CC, Walk SK, Tinoco I Jr, Blackburn EH (1987) Telomeric DNA oligonucleotides form novel intramolecular structures containing guanine–guanine base pairs. Cell 51:899–908
6. Sen D, Gilbert W (1988) Formation of parallel four-stranded complexes by guanine-rich motifs in DNA and its implications for meiosis. Nature 334:364–366
7. Sundquist WI, Klug A (1989) Telomeric DNA dimerizes by formation of guanine tetrads between hairpin loops. Nature 342:825–829
8. Williamson JR, Raghuraman MK, Cech TR (1989) Monovalent cation-induced structure of telomeric DNA: the G-quartet model. Cell 59:871–880
9. Chantot J, Guschlbauer W (1969) Physicochemical properties of nucleosides 3. Gel formation by 8-bromoguanosine. FEBS Lett 4:173–176

10. Williamson JR (1994) G-quartet structures in telomeric DNA. Annu Rev Biophys Biomol Struct 23:703–730
11. Miura T, Benevides JM, Thomas GJ Jr (1995) A phase diagram for sodium and potassium ion control of polymorphism in telomeric DNA. J Mol Biol 248:233–238
12. Balagurumoorthy P, Brahmachari SK (1994) Structure and stability of human telomeric sequence. J Biol Chem 269:21858–21869
13. Oganesian L, Moon IK, Bryan TM, Jarstfer MB (2006) Extension of G-quadruplex DNA by ciliate telomerase. EMBO J 25:1148–1159
14. Phan AT, Patel DJ (2003) Two-repeat human telomeric d(TAGGGTTAGGGT) sequence forms interconverting parallel and antiparallel G-quadruplexes in solution: distinct topologies, thermodynamic properties, and folding/unfolding kinetics. J Am Chem Soc 125: 15021–15027
15. Schultze P, Smith FW, Feigon J (1994) Refined solution structure of the dimeric quadruplex formed from the Oxytricha telomeric oligonucleotide d(GGGGTTTTGGGG). Structure 2:221–233
16. Lee JY, Okumus B, Kim DS, Ha T (2005) Extreme conformational diversity in human telomeric DNA. Proc Natl Acad Sci U S A 102:18938–18943
17. Ying L, Green JJ, Li H, Klenerman D, Balasubramanian S (2003) Studies on the structure and dynamics of the human telomeric G quadruplex by single-molecule fluorescence resonance energy transfer. Proc Natl Acad Sci U S A 100:14629–14634
18. Hardin CC, Perry AG, White K (2000) Thermodynamic and kinetic characterization of the dissociation and assembly of quadruplex nucleic acids. Biopolymers 56:147–194
19. Kan ZY, Lin Y, Wang F, Zhuang XY, Zhao Y, Pang DW, Hao YH, Tan Z (2007) G-quadruplex formation in human telomeric (TTAGGG)4 sequence with complementary strand in close vicinity under molecularly crowded condition. Nucleic Acids Res 35:3646–3653
20. Miyoshi D, Matsumura S, Nakano S, Sugimoto N (2004) Duplex dissociation of telomere DNAs induced by molecular crowding. J Am Chem Soc 126:165–169
21. Zaug AJ, Podell ER, Cech TR (2005) Human POT1 disrupts telomeric G-quadruplexes allowing telomerase extension in vitro. Proc Natl Acad Sci U S A 102:10864–10869
22. Wang Y, Patel DJ (1993) Solution structure of a parallel-stranded G-quadruplex DNA. J Mol Biol 234:1171–1183
23. Parkinson GN, Lee MP, Neidle S (2002) Crystal structure of parallel quadruplexes from human telomeric DNA. Nature 417:876–880
24. Ambrus A, Chen D, Dai J, Bialis T, Jones RA, Yang D (2006) Human telomeric sequence forms a hybrid-type intramolecular G-quadruplex structure with mixed parallel/antiparallel strands in potassium solution. Nucleic Acids Res 34:2723–2735
25. Luu KN, Phan AT, Kuryavyi V, Lacroix L, Patel DJ (2006) Structure of the human telomere in K+ solution: an intramolecular (3+1) G-quadruplex scaffold. J Am Chem Soc 128:9963–9970
26. Phan AT, Luu KN, Patel DJ (2006) Different loop arrangements of intramolecular human telomeric (3+1) G-quadruplexes in K+ solution. Nucleic Acids Res 34:5715–5719
27. Xu Y, Noguchi Y, Sugiyama H (2006) The new models of the human telomere d[AGGG(TTAGGG)3] in K+ solution. Bioorg Med Chem 14:5584–5591
28. Dai J, Carver M, Punchihewa C, Jones RA, Yang D (2007) Structure of the Hybrid-2 type intramolecular human telomeric G-quadruplex in K+ solution: insights into structure polymorphism of the human telomeric sequence. Nucleic Acids Res 35:4927–4940
29. Phan AT, Kuryavyi V, Luu KN, Patel DJ (2007) Structure of two intramolecular G-quadruplexes formed by natural human telomere sequences in K+ solution. Nucleic Acids Res 35:6517–6525
30. Sfeir AJ, Chai W, Shay JW, Wright WE (2005) Telomere-end processing the terminal nucleotides of human chromosomes. Mol Cell 18:131–138
31. Orlov SN, Hamet P (2006) Intracellular monovalent ions as second messengers. J Membr Biol 210:161–172
32. Li J, Correia JJ, Wang L, Trent JO, Chaires JB (2005) Not so crystal clear: the structure of the human telomere G-quadruplex in solution differs from that present in a crystal. Nucleic Acids Res 33:4649–4659
33. Xue Y, Kan ZY, Wang Q, Yao Y, Liu J, Hao YH, Tan Z (2007) Human telomeric DNA forms parallel-stranded intramolecular G-quadruplex in K+ solution under molecular crowding condition. J Am Chem Soc 129:11185–11191
34. Huppert JL, Balasubramanian S (2005) Prevalence of quadruplexes in the human genome. Nucleic Acids Res 33:2908–2916
35. Todd AK, Johnston M, Neidle S (2005) Highly prevalent putative quadruplex sequence motifs in human DNA. Nucleic Acids Res 33:2901–2907

36. Risitano A, Fox KR (2004) Influence of loop size on the stability of intramolecular DNA quadruplexes. Nucleic Acids Res 32:2598–2606
37. Eddy J, Maizels N (2006) Gene function correlates with potential for G4 DNA formation in the human genome. Nucleic Acids Res 34:3887–3896
38. Huppert JL, Balasubramanian S (2007) G-quadruplexes in promoters throughout the human genome. Nucleic Acids Res 35:406–413
39. Siddiqui-Jain A, Grand CL, Bearss DJ, Hurley LH (2002) Direct evidence for a G-quadruplex in a promoter region and its targeting with a small molecule to repress c-MYC transcription. Proc Natl Acad Sci U S A 99:11593–11598
40. Lipps HJ (1980) In vitro aggregation of the gene-sized DNA molecules of the ciliate *Stylonychia mytilus*. Proc Natl Acad Sci U S A 77:4104–4107
41. Schaffitzel C, Berger I, Postberg J, Hanes J, Lipps HJ, Pluckthun A (2001) In vitro generated antibodies specific for telomeric guanine-quadruplex DNA react with *Stylonychia lemnae* macronuclei. Proc Natl Acad Sci U S A 98:8572–8577
42. Harper L, Golubovskaya I, Cande WZ (2004) A bouquet of chromosomes. J Cell Sci 117: 4025–4032
43. Anuradha S, Muniyappa K (2004) Meiosis-specific yeast Hop1 protein promotes synapsis of double-stranded DNA helices via the formation of guanine quartets. Nucleic Acids Res 32:2378–2385
44. Liu Z, Gilbert W (1994) The yeast KEM1 gene encodes a nuclease specific for G4 tetraplex DNA: implication of in vivo functions for this novel DNA structure. Cell 77:1083–1092
45. Zahler AM, Williamson JR, Cech TR, Prescott DM (1991) Inhibition of telomerase by G-quartet DNA structures. Nature 350: 718–720
46. Tsai YC, Qi H, Liu LF (2007) Protection of DNA ends by telomeric 3′ G-tail sequences. J Biol Chem 282:18786–18792
47. Salas TR, Petruseva I, Lavrik O, Bourdoncle A, Mergny JL, Favre A, Saintome C (2006) Human replication protein A unfolds telomeric G-quadruplexes. Nucleic Acids Res 34:4857–4865
48. Sun H, Karow JK, Hickson ID, Maizels N (1998) The Bloom's syndrome helicase unwinds G4 DNA. J Biol Chem 273: 27587–27592
49. Mohaghegh P, Karow JK, Brosh RM Jr, Bohr VA, Hickson ID (2001) The Bloom's and Werner's syndrome proteins are DNA structure-specific helicases. Nucleic Acids Res 29:2843–2849
50. Sacca B, Lacroix L, Mergny JL (2005) The effect of chemical modifications on the thermal stability of different G-quadruplex-forming oligonucleotides. Nucleic Acids Res 33:1182–1192
51. Kumari S, Bugaut A, Huppert JL, Balasubramanian S (2007) An RNA G-quadruplex in the 5′ UTR of the NRAS proto-oncogene modulates translation. Nat Chem Biol 3:218–221
52. Eddy J, Maizels N (2008) Conserved elements with potential to form polymorphic G-quadruplex structures in the first intron of human genes. Nucleic Acids Res 36: 1321–1333
53. Kikin O, Zappala Z, D'Antonio L, Bagga PS (2008) GRSDB2 and GRS_UTRdb: databases of quadruplex forming G-rich sequences in pre-mRNAs and mRNAs. Nucleic Acids Res 36:D141–D148
54. Darnell JC, Jensen KB, Jin P, Brown V, Warren ST, Darnell RB (2001) Fragile X mental retardation protein targets G quartet mRNAs important for neuronal function. Cell 107:489–499
55. Brown V, Jin P, Ceman S, Darnell JC, O'Donnell WT, Tenenbaum SA, Jin X, Feng Y, Wilkinson KD, Keene JD, Darnell RB, Warren ST (2001) Microarray identification of FMRP-associated brain mRNAs and altered mRNA translational profiles in fragile X syndrome. Cell 107:477–487
56. Gomez D, Lemarteleur T, Lacroix L, Mailliet P, Mergny JL, Riou JF (2004) Telomerase downregulation induced by the G-quadruplex ligand 12459 in A549 cells is mediated by hTERT RNA alternative splicing. Nucleic Acids Res 32:371–379
57. Azzalin CM, Reichenbach P, Khoriauli L, Giulotto E, Lingner J (2007) Telomeric repeat containing RNA and RNA surveillance factors at mammalian chromosome ends. Science 318:798–801
58. Duquette ML, Handa P, Vincent JA, Taylor AF, Maizels N (2004) Intracellular transcription of G-rich DNAs induces formation of G-loops, novel structures containing G4 DNA. Genes Dev 18:1618–1629
59. Paeschke K, Juranek S, Simonsson T, Hempel A, Rhodes D, Lipps HJ (2008) Telomerase recruitment by the telomere end binding protein-beta facilitates G-quadruplex DNA unfolding in ciliates. Nat Struct Mol Biol 15:598–604

60. Paeschke K, Simonsson T, Postberg J, Rhodes D, Lipps HJ (2005) Telomere end-binding proteins control the formation of G-quadruplex DNA structures in vivo. Nat Struct Mol Biol 12:847–854
61. De Cian A, Lacroix L, Douarre C, Temime-Smaali N, Trentesaux C, Riou JF, Mergny JL (2008) Targeting telomeres and telomerase. Biochimie 90:131–155
62. Monchaud D, Teulade-Fichou MP (2008) A hitchhiker's guide to G-quadruplex ligands. Org Biomol Chem 6:627–636
63. Sun D, Thompson B, Cathers BE, Salazar M, Kerwin SM, Trent JO, Jenkins TC, Neidle S, Hurley LH (1997) Inhibition of human telomerase by a G-quadruplex-interactive compound. J Med Chem 40:2113–2116
64. Fedoroff OY, Salazar M, Han H, Chemeris VV, Kerwin SM, Hurley LH (1998) NMR-Based model of a telomerase-inhibiting compound bound to G-quadruplex DNA. Biochemistry 37:12367–12374
65. Wheelhouse RT, Sun DK, Han HY, Han FX, Hurley LH (1998) Cationic porphyrins as telomerase inhibitors: the interaction of tetra-(*N*-methyl-4-pyridyl)porphine with quadruplex DNA. J Am Chem Soc 120:3261–3262
66. De Cian A, Cristofari G, Reichenbach P, De Lemos E, Monchaud D, Teulade-Fichou MP, Shin-Ya K, Lacroix L, Lingner J, Mergny JL (2007) Reevaluation of telomerase inhibition by quadruplex ligands and their mechanisms of action. Proc Natl Acad Sci U S A 104:17347–17352
67. Read M, Harrison RJ, Romagnoli B, Tanious FA, Gowan SH, Reszka AP, Wilson WD, Kelland LR, Neidle S (2001) Structure-based design of selective and potent G quadruplex-mediated telomerase inhibitors. Proc Natl Acad Sci U S A 98:4844–4849
68. De Cian A, Delemos E, Mergny JL, Teulade-Fichou MP, Monchaud D (2007) Highly efficient G-quadruplex recognition by bisquinolinium compounds. J Am Chem Soc 129:1856–1857
69. Pennarun G, Granotier C, Gauthier LR, Gomez D, Hoffschir F, Mandine E, Riou JF, Mergny JL, Mailliet P, Boussin FD (2005) Apoptosis related to telomere instability and cell cycle alterations in human glioma cells treated by new highly selective G-quadruplex ligands. Oncogene 24:2917–2928
70. Shin-ya K, Wierzba K, Matsuo K, Ohtani T, Yamada Y, Furihata K, Hayakawa Y, Seto H (2001) Telomestatin, a novel telomerase inhibitor from *Streptomyces anulatus*. J Am Chem Soc 123:1262–1263
71. Kim MY, Vankayalapati H, Shin-Ya K, Wierzba K, Hurley LH (2002) Telomestatin, a potent telomerase inhibitor that interacts quite specifically with the human telomeric intramolecular g-quadruplex. J Am Chem Soc 124:2098–2099
72. Rezler EM, Seenisamy J, Bashyam S, Kim MY, White E, Wilson WD, Hurley LH (2005) Telomestatin and diseleno sapphyrin bind selectively to two different forms of the human telomeric G-quadruplex structure. J Am Chem Soc 127:9439–9447
73. Gomez D, Paterski R, Lemarteleur T, Shin-Ya K, Mergny JL, Riou JF (2004) Interaction of telomestatin with the telomeric single-strand overhang. J Biol Chem 279:41487–41494
74. Kim MY, Gleason-Guzman M, Izbicka E, Nishioka D, Hurley LH (2003) The different biological effects of telomestatin and TMPyP4 can be attributed to their selectivity for interaction with intramolecular or intermolecular G-quadruplex structures. Cancer Res 63:3247–3256
75. Shammas MA, Shmookler Reis RJ, Li C, Koley H, Hurley LH, Anderson KC, Munshi NC (2004) Telomerase inhibition and cell growth arrest after telomestatin treatment in multiple myeloma. Clin Cancer Res 10: 770–776
76. Tahara H, Shin-Ya K, Seimiya H, Yamada H, Tsuruo T, Ide T (2006) G-Quadruplex stabilization by telomestatin induces TRF2 protein dissociation from telomeres and anaphase bridge formation accompanied by loss of the 3′ telomeric overhang in cancer cells. Oncogene 25:1955–1966
77. Tauchi T, Shin-Ya K, Sashida G, Sumi M, Nakajima A, Shimamoto T, Ohyashiki JH, Ohyashiki K (2003) Activity of a novel G-quadruplex-interactive telomerase inhibitor, telomestatin (SOT-095), against human leukemia cells: involvement of ATM-dependent DNA damage response pathways. Oncogene 22:5338–5347
78. Tauchi T, Shin-ya K, Sashida G, Sumi M, Okabe S, Ohyashiki JH, Ohyashiki K (2006) Telomerase inhibition with a novel G-quadruplex-interactive agent, telomestatin: in vitro and in vivo studies in acute leukemia. Oncogene 25:5719–5725
79. Gomez D, Wenner T, Brassart B, Douarre C, O'Donohue MF, El Khoury V, Shin-Ya K, Morjani H, Trentesaux C, Riou JF (2006) Telomestatin-induced telomere uncapping is modulated by POT1 through G-overhang extension in HT1080 human tumor cells. J Biol Chem 281:38721–38729

80. Gomez D, O'Donohue MF, Wenner T, Douarre C, Macadre J, Koebel P, Giraud-Panis MJ, Kaplan H, Kolkes A, Shin-ya K, Riou JF (2006) The G-quadruplex ligand telomestatin inhibits POT1 binding to telomeric sequences in vitro and induces GFP-POT1 dissociation from telomeres in human cells. Cancer Res 66:6908–6912
81. Burger AM, Dai F, Schultes CM, Reszka AP, Moore MJ, Double JA, Neidle S (2005) The G-quadruplex-interactive molecule BRACO-19 inhibits tumor growth, consistent with telomere targeting and interference with telomerase function. Cancer Res 65:1489–1496
82. Leonetti C, Amodei S, D'Angelo C, Rizzo A, Benassi B, Antonelli A, Elli R, Stevens MF, D'Incalci M, Zupi G, Biroccio A (2004) Biological activity of the G-quadruplex ligand RHPS4 (3, 11-difluoro-6, 8, 13-trimethyl-8H-quino[4, 3, 2-kl]acridinium methosulfate) is associated with telomere capping alteration. Mol Pharmacol 66:1138–1146
83. Phatak P, Cookson JC, Dai F, Smith V, Gartenhaus RB, Stevens MF, Burger AM (2007) Telomere uncapping by the G-quadruplex ligand RHPS4 inhibits clonogenic tumour cell growth in vitro and in vivo consistent with a cancer stem cell targeting mechanism. Br J Cancer 96:1223–1233
84. Salvati E, Leonetti C, Rizzo A, Scarsella M, Mottolese M, Galati R, Sperduti I, Stevens MF, D'Incalci M, Blasco M, Chiorino G, Bauwens S, Horard B, Gilson E, Stoppacciaro A, Zupi G, Biroccio A (2007) Telomere damage induced by the G-quadruplex ligand RHPS4 has an antitumor effect. J Clin Invest 117:3236–3247
85. Halder K, Chowdhury S (2007) Quadruplex-coupled kinetics distinguishes ligand binding between G4 DNA motifs. Biochemistry 46:14762–14770
86. Lemarteleur T, Gomez D, Paterski R, Mandine E, Mailliet P, Riou JF (2004) Stabilization of the c-myc gene promoter quadruplex by specific ligands' inhibitors of telomerase. Biochem Biophys Res Commun 323:802–808
87. Balagurumoorthy P, Brahmachari SK, Mohanty D, Bansal M, Sasisekharan V (1992) Hairpin and parallel quartet structures for telomeric sequences. Nucleic Acids Res 20:4061–4067
88. Hardin CC, Henderson E, Watson T, Prosser JK (1991) Monovalent cation induced structural transitions in telomeric DNAs: G-DNA folding intermediates. Biochemistry 30:4460–4472
89. Panyutin IG, Kovalsky OI, Budowsky EI, Dickerson RE, Rikhirev ME, Lipanov AA (1990) G-DNA: a twice-folded DNA structure adopted by single-stranded oligo(dG) and its implications for telomeres. Proc Natl Acad Sci U S A 87:867–870
90. Ha T, Enderle T, Ogletree DF, Chemla DS, Selvin PR, Weiss S (1996) Probing the interaction between two single molecules: fluorescence resonance energy transfer between a single donor and a single acceptor. Proc Natl Acad Sci U S A 93:6264–6268
91. Smith FW, Feigon J (1992) Quadruplex structure of Oxytricha telomeric DNA oligonucleotides. Nature 356:164–168
92. Smith FW, Feigon J (1993) Strand orientation in the DNA quadruplex formed from the Oxytricha telomere repeat oligonucleotide d(G4T4G4) in solution. Biochemistry 32:8682–8692
93. Wang Y, Patel DJ (1992) Guanine residues in d(T2AG3) and d(T2G4) form parallel-stranded potassium cation stabilized G-quadruplexes with anti glycosidic torsion angles in solution. Biochemistry 31:8112–8119
94. Wang Y, Patel DJ (1993) Solution structure of the human telomeric repeat d[AG3(T2AG3)3] G-tetraplex. Structure 1:263–282
95. Wang Y, Patel DJ (1994) Solution structure of the Tetrahymena telomeric repeat d(T2G4)4 G-tetraplex. Structure 2:1141–1156
96. Kang C, Zhang X, Ratliff R, Moyzis R, Rich A (1992) Crystal structure of four-stranded Oxytricha telomeric DNA. Nature 356: 126–131
97. Laughlan G, Murchie AI, Norman DG, Moore MH, Moody PC, Lilley DM, Luisi B (1994) The high-resolution crystal structure of a parallel-stranded guanine tetraplex. Science 265:520–524

Chapter 2

Molecular Modeling and Simulation of G-Quadruplexes and Quadruplex-Ligand Complexes

Shozeb Haider and Stephen Neidle

Abstract

Methods for the molecular modeling and simulation of G-quadruplex structures and their drug/ligand complexes are discussed, and a range of protocols is presented for undertaking a variety of tasks including model-building, ligand docking, dynamics simulation, continuum solvent modeling, energetic calculations, principal component analysis, and quantum chemical computations. The scope and limitations of these approaches are discussed.

Key words: G-quadruplex, Molecular modeling, Ligand complexes, Molecular dynamics, Simulations

1. Introduction

Guanine (G)-quadruplexes are built from short lengths of G-tract separated by lengths of general sequence. In the case of intramolecular quadruplexes, at least four G-tracts are required:

$$G_{3-5}\ N_{L1}\ G_{3-5}\ N_{L2}\ G_{3-5}\ N_{L3}\ G_{3-5}$$

In general the G-tracts form the underlying core of quadruplex structures, with sets of four guanines at a time interacting together to form planar hydrogen-bonded G-quartets, which can then stack on top of each other. Quadruplexes can be formed (1, 2) from a single strand (termed unimolecular, or intramolecular quadruplexes), from two strands (bimolecular, or dimeric), or from four separate strands (tetramolecular). All have a requirement for alkali metal ion stabilization with $K^+ > Na^+$; these are coordinated to the O6 guanine atoms at the centre of a G-quartet and form a central ion channel. The N_L sequences link the G-quartets to form loops

P. Baumann (ed.), *G-Quadruplex DNA: Methods and Protocols*, Methods in Molecular Biology, vol. 608
DOI 10.1007/978-1-59745-363-9_2, © Humana Press, a part of Springer Science+Business Media, LLC 2010

and grooves, with variability in the nature of the connections being a major factor in the resultant variety of quadruplex topologies that have been observed (1). There are currently, as of autumn 2009, only 32 crystal structures of quadruplexes in the Protein Data Bank (PDB), and a rather larger number of NMR-derived structures. The former have been recently reviewed (3). There are as yet no general rules governing the folding of these sequences, although a start on their classification has been made (4). Evidence to date indicates that it is not yet possible to reliably predict overall quadruplex topology, although the simple topological rules for short N_L linkers appear to be robust (5). Folding is unpredictable once linkers have > two nucleotides, and especially when they themselves contain guanine residues, as has been shown by the unexpected and unique arrangement formed by a 22-mer sequence from the promoter region of the *c-kit* oncogene (6). The human genome contains over 250,000 distinct nontelomeric putative quadruplex sequences (7, 8) of which those in oncogenic promoter regions have been most studied (9). Quadruplexes formed from human telomeric sequences comprise repeats of the simple sequence d(TTAGGG), whereas non-telomeric sequences generally have no such symmetry.

Small-molecule ligands can promote the formation of quadruplex structures from telomeric DNA, which can then inhibit the telomerase enzyme and destabilize telomere end-capping in cancer cells (10). This finding has led to studies aimed at designing, synthesizing, and evaluating such molecules as anticancer agents [reviewed in, for example, refs 11–14). A large number of quadruplex-binding ligands have been reported [summarized in ref 15), the majority of which share the common structural feature of a planar aromatic chromophore.

There are remarkably few detailed crystal or NMR structures for ligand–quadruplex complexes (16, 17). Those for bimolecular human telomeric quadruplexes all show a single topology, the parallel fold (18–20), as does the porphyrin complex with a *c-myc* oncogene promoter intramolecular quadruplex (21). The topology of human telomeric intramolecular quadruplexes is more varied, with crystallographic studies on a 22-mer showing the all-parallel fold (18), whereas NMR studies on several related sequences with small changes at 5′ and 3′ ends show (3+1) antiparallel folds (22–25).

1.1. Quadruplex Modeling – Challenges and Approaches

The structural polymorphism of many quadruplexes is as yet incompletely understood and presents a challenge for molecular simulation studies that to date has not been met. The problems of modeling individual quadruplex structures are similar to those of nucleic acids generally, but with the added complexity of the central ion channel (26). Given the variety of nucleic acid structures and their complexes combined with the inherent flexibility of nucleic acids, there are many problems to which computational techniques such as molecular dynamics (MD) can contribute. This has been made possible due to the increasingly accurate parameter determination in nucleic acid force fields and algorithmic development (27, 28), inclusion of explicit counter ions and solvent molecules, as well as the use of more complex methods for evaluation of long-range electrostatic effects, which are important in charged systems. The maturity of the field is further indicated by the substantial body of recent literature on application of novel computational methods to a variety of biomolecular systems that contain complex nucleic acid arrangements such as DNA quadruplexes and drug-DNA complexes. Such improvement in methods and more careful comparison with experimental data give us increasing confidence in modeling methods (29).

2. Methodology and Force Fields

The most common modeling method is that of molecular dynamics (MD). It is based on solving Newton's laws of motion for all atoms in the system. The force on each atom is calculated from the derivative of the sum of potential energy terms for Coulombic, van der Waals, bond length, bond angle, and dihedral angle contributions. The acceleration on an atom can be calculated from the force and integrated to calculate velocities, which in turn can be integrated to find atomic position vectors. The time course of these position vectors forms the trajectory. The integration time step is adjusted depending upon the highest frequency vibrations in the system e.g. bond stretching along C–H and O–H bonds. The trajectories usually employ the NPT statistical ensemble that is generated if the number of atoms, pressure, and temperature are kept constant during the simulation. Cheap computational power means that simulations can now be carried out using explicitly solvated systems. In such a system, the solute is immersed in a large box of explicitly solvated water molecules and counter-ions. The box is replicated in all directions to satisfy periodic boundary conditions. The molecules are described by simple pair-additive atomistic potentials known as force fields that treat atoms as Lennard-Jones van der Waals spheres with partial constant point charges localized at the individual atomic centers, linked by harmonic springs supplemented by valence angle and torsion profiles mimicking the covalent structures.

The explicitly solvated simulations employ the particle-mesh Ewald (PME) method (30) or atom-based force shift approaches (31, 32) for taking into account the long-range electrostatic effects in an efficient manner. These effects have been shown to be significant in nucleic acid systems because of the charge on the phosphate backbone and counter-ions and are even more important for quadruplex DNAs with their multi-faceted electrostatics features. Such complications resulted in the expulsion of the cations from the central electronegative channel in the quadruplex core, leading to the collapse of the structure in the first MD simulation of a quadruplex structure (33). Introduction of the atom-based force-shift truncation method using a 12 Å cutoff and PME treatment of electrostatics (34, 35) produced stable and very similar nanosecond MD simulations of nucleic acids. The CPU time requirements are similar for optimized cutoff radius and convergence parameters for PME summation, but the periodic boundary conditions necessary with standard implementations of PME make it slower than a spherical cutoff in a non–periodic geometry adapted to the shape of the system being studied. The pros and cons of the Ewald summation method and the periodicity it imposes on the system have been studied in detail and the results suggest that the artifacts of the method are small for biomolecular systems when comparing to errors arising from sampling and force field limitations (36).

Progress in force field development in recent years has made stable multi-nanosecond molecular simulations routine, although challenges in adeguately simulating loop regions remain to be fully overcome (71). Several force field parameter sets such as in the AMBER package (parm99SB) (http://amber.scripps.edu) (27), CHARMM27 (37), and the latest GROMOS (38) force field have all yielded reasonable results for the simulation of conventional B-DNA conformations. Implementation of CHARMM for the simulation of unusual nucleic acid structures such as quadruplex DNA has not yet been extensively reported. The CHARMM force field contains similar functional forms including bond stretching, angle bending, torsion angle, and nonbonded interaction, but they are all derived differently (37). However, use of CHARMM27 to simulate folded RNAs has resulted in unstable trajectories (39). The GROMOS force field is yet to be tested and published independently for nucleic acids. A recent 10 ns benchmarking simulation using this force field by the authors on quadruplex DNA resulted in a complete loss of four-stranded structure. One should avoid using force fields that have not been explicitly parameterized for nucleic acids and tested for quadruplex structures.

An improved version of the AMBER parm99 force field (parmbsc0) for nucleic acids has been reported recently (40). It emphasizes the correct representation of α/γ concerted rotation in the nucleic acid backbone. The force field has been derived by fitting to high-level quantum mechanical data, verified by comparison with very high-level quantum mechanical calculations, and by a very extensive comparison between simulation and experimental data (40). The total simulation time used to validate the force field includes 1 μs molecular simulations in aqueous solution.

In addition to the improvement of force fields, one of the main computational challenges is to simulate large systems over longer time scales. The time scale of events happening in real biological time is much longer than what can be simulated with the computational power available today. The result is limited sampling of conformational space. Faster computers would improve sampling but at the same time would also result in accumulation of force field deficiencies that can have detrimental effects over time. Enhanced sampling of conformational space can be approached by running multiple simulations using a rational approach of multiple starting structures or by using enhanced sampling methods. It must be kept in mind that the force fields being used to simulate biomolecular systems are over-simplified representations that are unable to accurately capture all energy contributions simultaneously.

The square planar arrangement of guanines in a G-quartet results in the carbonyl oxygens pointing towards the helical axis within the central core of the structure. Repeats of stacked G-quartets result in the formation of a central channel that is lined by carbonyl oxygen atoms, and thus the central channel running along the helical axis is highly electronegative in character. To avoid electrostatic repulsion, quadruplexes are stabilized by cations (preferably monovalent) that are embedded within the channel. Depending upon the size of the cation, they can be sandwiched symmetrically between two planes of the four G-quartets, each forming a square anti-prismatic arrangement in which the square plane of oxygen atoms above the ion is rotated with respect to the plane below, as observed in the crystalline state with K^+ ions (41). Two K^+ ions very rarely occur with a separation of <3.5 Å in order to avoid an unstable electrostatic configuration. However, K^+ ions in principle may occupy adjacent sites and form stable complexes in which the cation is encapsulated, sandwiched, and coordinated, as observed in potassium-coordinated crown ethers (42). By contrast, sodium ions have a smaller ionic radius which can fit into the in-plane site and have been observed to lie close to each quartet plane (43, 44). The channel cations impart stability to the structure and have been observed in all quadruplex crystal structures solved to date, as well as in NMR structures. These cations are mobile and can readily exchange with the bulk solvent and have been observed experimentally on a millisecond timescale (45). However, once the cations are removed the rigidity of the structure is immediately lost (46). The solvent can move freely within the quadruplex core but is unable to stabilize the structure. At no point are the quadruplex G-quartets left vacant by cations and their stability is dependent on the cations associated with them. The present Amber ff99SB and parmbsc0 force field parameter sets are capable of accommodating monovalent cations in the simulations and have been shown to stabilize G-quadruplexes

(47, 48). However, the radii of K^+ and Na^+ ions are over-sized. Studies by Sponer and coworkers have shown that the positions for in-plane Na^+ ion are under-populated and K^+ ions can readily move out of the channel (49). Reducing the cation radii increases the sampling of ions inside the channel (49, 50).

Simulating a sugar-phosphate backbone in nucleic acids has always been a challenge for force fields because of both its flexible nature and the anionic electrostatic potential generated by the phosphate groups. The contributions from the complex electronic structure of the backbone change with solvation and conformational dynamics are not taken into account by nonpolarizable atom–atom pair additive force fields. A new and improved version of the parm94 (51) and parm99 (52) force fields has been introduced (parmbsc0) that is able to accommodate the α/γ angles in the backbone of nucleic acids including G-quadruplexes (40).

3. Methods

3.1. Model Preparation for Molecular Modeling

1. The appropriate coordinates of the structure (x-ray or NMR) can be obtained from the Nucleic Acid Database (NDB) or Protein Data Bank (PDB). This forms the initial coordinates for molecular modeling.
2. The initial starting coordinates are subjected to molecular mechanics energy minimization.
3. The first round of minimization involves 1,000 steps of steepest descent with line searching.
4. The second round of minimization involves 1,000 steps of Polak-Ribiere conjugate gradient with a derivative convergence of 0.05 kj $Å^{-1}$ mol^{-1}.
5. This is followed by a short run of MD (50 ps at 300 K) with a 2.0 fs time step.
6. A final 1,000 steps of molecular mechanics energy minimization (Polak-Ribiere conjugate gradient) of the time-averaged MD structure is employed to obtain a low-energy final model for further studies.

3.2. Ligand-Quadruplex Modeling

Quantitative molecular modeling can be performed in order to visualize and describe in energetic terms how small molecules interact with and stabilize G-quadruplex structures. An appropriate binding site for the ligand can be chosen depending on the model being chosen for the study, for example, within a loop in the human intramolecular G-quadruplex NMR structure (PDB code 143D) or an external binding site in the human intramolecular quadruplex crystal structure (PDB code 1KF1). As the solution and the crystal structure of the intramolecular G-quadruplex formed by four repeats of the human sequence are different, there are at least two plausible ways by which a ligand can bind to these systems. If there is no loop or the loops are positioned laterally as in the crystal structure (PDB code 1KF1), then the molecule can sit externally on the quartets. The other alternative is that the ligand is intercalated between the diagonal loop and the quartets, which is the most plausible model for ligand binding to the Na^+ solution structure of the human 22-mer intramolecular G-quadruplex, or to any quadruplex with a loop positioned above the G-quartets.

To create a pseudo-intercalation ligand binding site in the human intramolecular G-quadruplex NMR structure (PDB code 143D) the following steps are employed:

1. The binding site is to be created between the diagonal TTA loop and the G-quartet segment of the structure (at the 5′ AG step).
2. Two phosphate backbones are broken at the 5′ AG step.
3. The two halves of the structure are separated so that the separation of the A:A base pair and the G-quartet is increased from 3.4 to 6.8 Å.
4. The sugar-phosphate backbones are reconnected.
5. Positional restraints are placed on the structure except for residues contributing to the pseudo-intercalation binding site which includes the reconnected sugar-phosphate backbone.
6. The first round of minimization involves 1,000 steps of steepest descent with line searching.
7. The second round of minimization involves 1,000 steps of Polak–Ribiere conjugate gradient with a derivative convergence of 0.05 kjÅ–1 mol–1.
8. A final 1,000 steps of molecular mechanics energy minimization (Polak–Ribiere conjugate gradient) is employed to relieve any resulting steric distortion while retaining the intercalation geometry between the G-quartet and the loop motifs.

3.3. Modeling Ligands and Docking in G-Quadruplex Structures

1. Molecular models of the ligands are created in the builder package of the Insight suite of programs.
2. The CFF force field is used to parameterize the ligand.
3. The partial charges are added semi-empirically using the MOPAC package.
4. The ligand is minimized using 1,000 steps of steepest descent with line searching.

5. The ligand is then docked in the binding site using a multi-phased docking protocol employing the grid docking method available within the Affinity program in the Insight suite of packages. An advantage of this module is that it can explore molecular orientations while interactively monitoring changes in the electrostatic and van der Waals ligand G-quadruplex interactions.
6. The ligand is randomly oriented 200 times and is centered on the more polar face of the quadruplex.
7. The van der Waals radius is set to 10% of the full value (×0.1).
8. The charges are not considered and the nonbonded cut-off value is set to 8.0 Å.
9. The system is minimized for 300 steps using the conjugate gradient method.
10. The maximum allowable change for succeeding structures (energy tolerance) is set to 10,000 kcal/mol and the energy range is set to 30–40 kcal/mol.
11. The 75 lowest energy structures are used for the second phase of docking.
12. Simulated annealing is used to further refine the initial placement for the 75 structures.
13. During the second phase, the van der Waals radius value is adjusted back to its full value (×1.0).
14. Nonbonded cut-off is set to 12 Å.
15. The system is minimized for 300 steps of conjugate gradient.
16. This is followed by a short burst of MD where the starting temperature is set at 500 K and the system is cooled to 300 K over 7.5–10 ps.
17. The resulting structures are minimized for 800 steps of conjugate gradient.
18. Twenty-five structures with the lowest total energy are used for further evaluation.
19. Many different positions in three-dimensional space are evaluated in this procedure. The selection of low energy conformation is made on the basis of maximization of π-electron overlap between the G-quartet, ligand chromophore, and the nucleobases that sit over the ligand.
20. A typical ligand chromophore may be positioned directly over the centre of the adjacent G-quartet while its substituent side chains protrude out through to the sugar–phosphate backbone.
21. Quantitatively, the docked molecules are selected on the basis of three criteria: (a) total energy of the system, (b) energy of

the ligand and the binding pocket, and (c) the total number of intermolecular hydrogen bonds.

22. The final confirmation of the ligand complex is then subjected to a further 500 steps of molecular mechanics energy minimization employing the conjugate gradient algorithm.

3.4. Simulating Ligand-G-Quadruplex Complexes

1. The ligand is docked onto the G-quadruplex using the docking protocol mentioned above.
2. The docked ligand-G-quadruplex complex is then transferred to the AMBER9.0 or 10.0 package.
3. The force field parameters for the ligand are calculated employing the new general amber force field (GAFF) for small organic molecules in the *antechamber* program.
4. The parameters are then extrapolated onto the ligand in the *leap* program.
5. The system is then set up in the *leap* program using protocols mentioned below.
6. MD simulations are carried out using the *sander* program within AMBER.
7. The trajectory is analyzed using the program *ptraj* within AMBER.

3.5. Multimeric G-Quadruplex Model Building

1. The crystal structure of the 22-mer human telomeric DNA d[$AG_3(T_2AG_3)_3$] (PDB code 1KF1) is taken as a primary unit for the construction of the higher-order model. It consists of three stacks of G-quartets connected by TTA loops.
2. The adenine nucleobase at the extreme 5′ end is removed to generate a 21-mer.
3. Two 21-mer units are taken and positioned end-to-end (3′→5′).
4. The two units are rotated 30° relative to each other. The 30° angle is taken from the original crystallographic analysis and is the twist angle found between two consecutive stacked G-quartets.
5. The rise and rotation of G-quartets between the two units are kept at values observed in the primary unit.
6. The rise is positioned at an optimal separation of 3.5 Å.
7. A TTA loop is extracted from the crystal structure.
8. The two units are joined by the extracted TTA loop.
9. The final 45 nucleotide multimer is subjected to three rounds of molecular mechanics energy minimization and a short burst of MD (described above) to relieve any steric clashes within the model.

Models of higher order (four quadruplex unit repeats) multimeric structures can be generated by taking two 45-mer units and positioning and joining them by a TTA loop in a manner analogous to that described above. The resultant model is then subjected to similar protocols of molecular mechanics energy minimization and dynamics procedures to relieve structural stress.

A pseudo-intercalation ligand binding site can also be generated in the 45-mer human telomeric DNA between the two 21-mer units.

1. The sugar–phosphate backbone is broken between the TTA loop and the joining 21-mer G-quadruplex units.
2. The two 21-mer units are separated and the distance between the two units is increased from 3.5 to 7.0 Å.
3. The sugar–phosphate backbone is reconnected.
4. Positional restraints are placed on the structure except for residues contributing to the pseudo-intercalation binding site, which include the reconnected sugar–phosphate backbone.
5. The first round of minimization involving 1,000 steps of steepest descent with line searching.
6. The second round of minimization involves 1,000 steps of Polak–Ribiere conjugate gradient with a derivative convergence of 0.05 kj $Å^{-1}$ mol^{-1}.
7. A final 1,000 steps of molecular mechanics energy minimization (Polak–Ribiere conjugate gradient) is carried out to relieve any resulting steric clashes arising from the generation of the intercalation site between the two units.

3.6. Molecular Dynamics Simulations

1. The initial starting structure is taken from a structural database (NDB or PDB) and imported in the AMBER package using the *leap* program.
2. The x-ray structures of G-quadruplexes have revealed a vertical alignment of cations along the helical axis within the central core of the structure. The ions are retained in their positions as observed in the crystal structure. However, in the case of NMR derived solution structures, the ions need to be placed at appropriate positions in the structure. The cations are positioned between the G-quartets in case of K^+ ions and within or close to the plane of an individual G-quartet for Na^+ ions.
3. The positions of the water molecules observed in the crystal structure are retained. The structure is then solvated in a periodic box containing explicit TIP3P water. The dimensions of the box are such that its boundaries extend at least 10 Å from any solute atom.

4. Additional positively charged K^+ counter-ions are included in the system to neutralize the charge on the DNA backbone. This also maintains the consistency with the crystallization conditions and prepares the simulation to run in a uniform K^+ ionic environment.
5. The counter-ions can be placed automatically by the *leap* program throughout the water box at grid points of negative Coulombic potential.
6. The final net charge on the system should be zero.
7. MD simulations are carried out using the *sander* module in the AMBER9.0 simulations package.
8. Positional restraints of 500 kcal/mol are placed on the entire G-quadruplex structure (*ntr=1*). This ensures that only the water molecules and the counter-ions move during the first round of minimization.
9. The system is then equilibrated with explicit TIP3P water by 3,000 steps (*maxcyc=3000*) of molecular mechanics energy minimization which employs 500 steps of steepest descent followed by conjugate gradient minimization for the remainder of the steps (*ncyc=500*). The other parameter for the minimization is the use of periodic boundary conditions (ntb=1, constant volume) with a non-bonded Lennard-Jones cutoff of 10Å (*cut=10.0*).
10. The nonbonded pair list is updated every 50 steps (*ntpr=50*).
11. A second round of energy minimization is carried out using the same parameters as in the first round. The only difference is that there are no restraints on the DNA (*ntr=0*).
12. Restrained MD is carried out for twenty picoseconds (*nstlim=10,000*) with positional restraints of 50 kcal/mol being placed on G-quadruplex DNA (*ntr=1*).
13. The time step is set at 2 fs (*dt=0.002*).
14. The coordinates are read in with no initial velocity information (*ntx=1, irest=0*).
15. Temperature scaling is switched on (*ntt=3*) with the temperature being increased from 0 K (*tempi=0*) to 300 K (*temp0=300*) using a Langecin temperature equilibration scheme.
16. The periodic boundary conditions are switched on (*ntb=2*) with constant pressure. The system needs to be equilibrated at constant pressure to get proper density and to avoid box edge effects (*ntb=2, ntp>0*).
17. Constant pressure dynamics is carried out by setting *pres0=1*. This is the reference pressure (units in bar, 1 bar=~1 atm) at which the system is maintained. Pressure regulations only apply when constant pressure periodic boundary conditions are used (*ntb=2*).

18. The constant pressure dynamics flag with isotropic position scaling is used (*ntp = 1*) and the pressure relaxation time is set to 2.0 ps (*taup* =*2*).
19. The SHAKE algorithm is enabled for hydrogen atoms (ntc = 2) with a tolerance of 0.0005 Å and a 2 fs time step. The SHAKE feature constrains the vibrational stretching of hydrogen bond lengths and effectively fixes the bond distance to the equilibrium value.
20. The force evaluation for calculating bind interactions involving hydrogen atoms is omitted (*ntf*=*2*). If the SHAKE algorithm is being used then it is not necessary to calculate forces for constrained bonds.
21. The energy output frequency is set at 500 steps (*ntpr* =*500*) in human readable format in the mdout and mdinfo files.
22. The coordinates in the mdcrd trajectory file are updated every 500 steps (*ntwx* =*500*).
23. The temperature and energy are written to the mden file after every 500 steps (*ntwe* =*500*).
24. The coordinates are written to the restart file after every 1,000 steps (*ntwr* =*500*).
25. The system is further subjected to a second round of MD calculations for 200 ps (*nstlim* = *100,000*) in which the constraints on the G-quadruplex DNA are relaxed to 5.0 kcal/mol.
26. The parameters for MD in the second round are exactly the same as for round one. The only change in parameter *ntx* =*7*. This option allows the coordinates and the velocities to be read in from the last restart file. The box information is also read if *ntb*>*0*.
27. The system is then subjected to ten ns (*nstlim* =*5,000,000*) of production MD in which there are no restraints on the system (*ntr* =*0*).
28. As the calculation needs to be restarted as a continuation of the previous round of dynamics, velocities in the coordinate input file are required (*irest* = *1*).
29. The value for ntx is retained at seven, which allows the coordinates and velocities to be read in from the restart file.
30. The rest of the parameters remain the same as in the previous two MD runs.
31. The Particle Mesh Ewalds (PME) summation term is used for all simulations with the charge grid spacing set at 1.0 Å.
32. The trajectories are analyzed using *ptraj* module available in the AMBER suite of programs.
33. The trajectories can be visualized using the VMD program (53).

3.7. Continuum Solvent Modeling

Sometimes simulating explicitly solvated systems can be too computationally expensive. One approach is to employ continuum solvent methods where explicit solvent is replaced with hybrid explicit/implicit (54) or completely implicit models (55). The solvation energies and solvent-dependent conformational changes can be predicted reliably using the Poisson–Boltzmann (PB) approach; however their computational complexity can hinder their use in a MD simulation. The generalized-Born (GB) method is much faster and can be parameterized to yield reasonable solvation energies. Both PB and GB approaches, combined with structural snapshots from explicit solvent MD simulations, have been used in estimating free energies in nucleic acids including G-quadruplex DNA (48, 50, 56). This is carried out by running conventional explicit solvent MD simulations and then postprocessing the trajectory where the explicit solvent and periodicity are removed. The energies are averaged over a sufficient number of snapshots. The MM-PBSA or MM-GBSA free energy methods allow calculation of free energy changes for processes that are not accessible to conventional free energy algorithms. Applications to G-quadruplex DNA require explicit inclusion of ions in the channel.

Continuum solvent modeling can also be applied to calculating ligand binding energies. This can be carried out using two approaches: (a) single trajectory approach where the ΔGs are derived from a single trajectory of the ligand–quadruplex complex and (b) multiple trajectory approach where the free energy difference is evaluated using three separate trajectories of the complex, receptor, and the ligand. It is generally agreed that the single trajectory approach may be more reliable as it cancels sampling errors in the intramolecular terms. These errors can be very significant in separate trajectories.

1. The MM-PBSA method is used to calculate approximate free energies (using a single trajectory approach in the case of ligand binding energies).
2. A conventional MD simulation is carried out using the *sander* program in the AMBER package. The parameters are described above.
3. Snapshots are collected every 20 ps for energetic analysis.
4. The electrostatic contribution to the solvation free energy is calculated using the Delphi II program (57).
5. The hydrophobic contribution to the solvation free energy is determined with solvent accessible surface area dependent terms.
6. Dielectric constants of 1.0 and 80.0 are assigned to the solute and the solvent respectively.
7. A grid spacing of 0.5 Å is chosen, with the longest linear dimension of the molecule occupying 80% of this grid.

8. The AMBER parm99SB charge set and BONDI radii (58) are used.
9. The three K^+ ions are explicitly included within the quadruplex channel.
10. The radius of K^+ ion was determined to be 2.025 Å, by adjusting it until ($\Delta G_{polar} + \Delta G_{nonpolar}$) was equal to the experimental $\Delta G_{solvation}$ of −80.6 kcal/mol.
11. All other energy terms are calculated with programs distributed with AMBER.
12. The solute entropic contribution is estimated with the *nmode* program, using snapshots collected every 200 ps.
13. Each snapshot is minimized in the gas phase, using a distance-dependent dielectric of $\varepsilon = 4r$ before the vibrational mode frequencies are calculated.

3.8. Enhanced Sampling Methods

Enhanced sampling methods deal with pronounced sampling of small parts of the molecules such as loops as in the case of G-quadruplex structures. The conformation adopted by the loops can differ and thus in theory MD simulations should be able to show the stability and correctness of one structure over another and whether the different conformations are inter-convertible on an affordable time scale. However, it is quite difficult for conversions to occur between two different conformational states when either of the structures is accurate during the course of a conventional MD simulation. Local enhanced sampling (LES) can be applied to the loops. The selected part of the molecule is split into N copies that are simulated independently, while the rest of the molecule is simulated in the standard manner. The energy barrier height is reduced proportionally to the number of copies being used (1/N).

1. The initial model is taken from a database (NDB or PDB).
2. The loop conformational space is searched with simulated annealing procedures in the Discover module of the Insight suite of packages.
3. During the simulated annealing procedures, the G-quartets are kept fixed while the loops are allowed to move.
4. The simulated annealing runs are carried out in implicit solvent using a distance dependent dielectric ($\varepsilon = 4r$) that mimics the solvent.
5. The initial loop conformation is minimized using 1,000 steps of Polak–Ribiere conjugate gradient with a derivative convergence of 0.05 kj $Å^{-1}$ mol^{-1}.
6. During each cycle, the loop is first heated to 1,000 K over 2 ps, simulated at 1,000 K for 2 ps, and eventually cooled to 300 K for 1 ps.

7. The resulting structure is again minimized using 1,000 steps of Polak–Ribiere conjugate gradient with a derivative convergence of 0.05 kj $Å^{-1}$ mol^{-1}.
8. The next loop conformation is generated from heating of the latest minimized conformation.
9. The structures obtained from the simulated annealing runs are clustered into conformational families on the basis of the root mean square deviation (rmsd) analysis between all structure pairs.
10. Pairwise rmsds between all structure pairs are calculated.
11. Clustering is then carried out according to the method used by the NMRCLUST program (59).
12. Selected structures from the clusters are subjected to extensive MD simulations in explicit solvent using the AMBER program.
13. The ions are placed in the structure when these are not present in the experimental template.
14. Additional cations are added in order to neutralize the charge on the system.
15. The system is then solvated in a pre-equilibrated TIP3P water box.
16. The box size depends on the system but is always extended at least 10 Å from the solute in every direction.
17. The equilibration procedure consists of ten steps, beginning with 1,000 steps of molecular mechanics energy minimization and 25 ps of MD where the solvent is only allowed to move.
18. The whole system is then minimized for 1,000 steps followed by 3 ps of dynamics with a restraint of 25 kcal/mol on the DNA.
19. The DNA restraints were lowered by 5 kcal/mol during each of the next five rounds of 1,000 step minimizations.
20. Finally the system is heated to 300 K over 20 ps with no further restraints.
21. The parameters for MD are used as described above in the protocol for MD simulations.
22. The local enhanced sampling (LES) simulations are carried out in a subset of loop conformations which are generated after an initial equilibration period of 1 ns simulations in explicit solvent.
23. Five copies of each loop are generated using the *Addles* module of AMBER9.0 software.
24. Both LES (loops) and non-LES regions (G-quartets) are maintained at 300 K in separate water baths.

25. After LES simulations are finished, the final copies are averaged.
26. Molecular mechanics and the Poisson Boltzmann Solvent Accessibility (MM-PBSA) method are used to calculate the energies and the results are compared to pre-LES energies.
27. The MM-PBSA method to calculate energies is described above.

3.9. Principal Components Analysis (Essential Dynamics)

An important realization in the analysis of a trajectory obtained by MD simulation is that not every aspect of motion is equally important for function. The concept of essential subspace was introduced, which contains large anharmonic motion of atoms and it is these motions that are more biologically relevant than smaller positional fluctuations. The configurational space that contains only a few degrees of freedom in which these anharmonic motions occur can be identified by reducing the dimensionality of the data that is obtained from MD simulations. Principal components analysis (PCA) is a method that takes the trajectory of a MD simulation and extracts the dominant modes in the motion of the molecule. The overall rotation and translation of the structure during the time course of the trajectory are removed by a translation to the average geometrical center of the molecule and by a least squares fit superimposition onto a reference structure. The configurational space is then constructed using a simple linear transformation in Cartesian coordinate space to generate a $3N \times 3N$ covariance matrix. The matrices are summed and averaged over the whole trajectory. The resulting matrix is then diagonalized generating a set of eigenvectors that gives a vectoral description of each component of the motion by indicating the direction of the motion. Each eigenvector describing the motion has a corresponding eigenvalue that represents the energetic contribution of that particular component to the motion. The eigenvalue is the average square displacement of the structure in the direction of the eigenvector. Projection of the trajectory on a particular eigenvector highlights the time dependent motions that the component performs in the particular vibrational mode. The time average of the projection shows the contribution of components of the atomic vibrations to this mode of concerted motion. The eigenvalues are placed in a descending order where the first eigenvector and eigenvalue describe the largest internal motion of the structure. The eigenvalues decline sharply, showing the possibility of separating the dynamics into a small essential space and a relatively large space, containing only small atomic fluctuations. In simpler terms, on average only about 5% of eigenvectors are necessary to describe 90% of the total dynamics.

1. Conventional MD is carried out on the structure obtained from NDB or PDB, using protocols and parameters described above using the *sander* program in the AMBER package.

2. MD trajectories are extracted using the *ptraj* program in the AMBER package.
3. Principal components analysis on the trajectory is then carried out using the PCAZIP (60) software on the last 5 ns employing 500 frames.

3.10. Quantum Mechanical Calculations on G-quadruplexes

Quantum mechanical calculations (*ab initio*) are more accurate and physically complete than molecular mechanics force field calculations. These calculations, however, do not take into account any forces that arise from long-range electrostatics or salvation effects. QM calculations of multiple quartets can be problematic even while estimating single point energy. The conventional density function theory (DFT) method is much superior to molecular mechanics force fields and can accurately calculate hydrogen bonding patterns within a G-quartet and guanine-cation interactions. DFT however, does not account for base stacking and therefore cannot describe interactions between guanines in different quartets. In order to accurately calculate stacking interactions, one must employ the MP2 method with a large basis set of atomic orbitals or by expanding the basis set limit. This is followed by a cluster correction which scales computer requirements with ~6th power of the number of atoms included, thus making it highly computationally expensive. Gradient optimization of a two G-quartet structure results in a mathematical artifact known as the basis set superposition error (BSSE) that originates from the incompleteness of the basis set of atomic orbitals and causes an artefactual stabilization of complexes. This can be corrected for single-point calculations by employing the standard counterpoise method (61).

3.10.1. Hartree-Fock and Density Function Study of Interactions Between Metal Cations and Hoogsteen Hydrogen Bonded G-Quartets

The Hartree-Fock self-consistent field (HF-SCF) method and the DFT (B3LYP approach) in conjunction with the valence triple-zeta basis set (with *d*- and *p*- like polarization functions) are employed to study the hydrogen bonding pattern within a G-quartet (62).

1. The initial structure of G-quartets can be prepared from the coordinates of a G-quadruplex structure downloaded from PDB or NDB.
2. The bases in the quartets are capped with hydrogen atoms.
3. The C_4- and S_4-symmetric G-quartets are studied for comparison with the coplanar complex structure.
4. The metal ions are positioned in the centre of the G-quartet for C_{4h}- and S_4- symmetry, at a distance of 1.6 Å below the centre for C_{4h}-symmetry.
5. The initial structures used for optimizations consist of four G-monomers with a C_{4h}-symmetric complex geometry except for pyramidal amino groups.

6. The amino hydrogens in the C_4-symmetric quartet are all on the same side of the base planes, where as the S_4-symmetric has hydrogen atoms above and below the base plane in an alternating sequence.
7. The G-quartets are optimized using the B3LYP hybrid density function method (62).
8. The basis sets used are 6-31G(d,p), 6-311G(d,p), and 6-311+G(d,p).
9. The individual bases are also investigated using the MP2/6-31(d,p) method (63, 64).
10. The HF method is employed in order to compare results with the DFT method. This is to ensure that the DFT approach does not overestimate the H-bonding interaction between bases resulting in the hydrogen bond lengths being too short.
11. Force field calculations are carried out using the MMFF94 force field (65) as implemented in the Sybyl 7.0 suite of programs (66).
12. A dielectric constant of 1.0 is used throughout.
13. The optimization is terminated when a gradient of 0.0001 kcal/mol is reached.
14. For metal ions, average relativistic potentials with a large orbital basis and a small core are used (67, 68).
15. A 6-31G(d,p) basis was used for base atoms in complex with metal cations.
16. All calculations are carried out using the GAUSSIAN 03 program (69).
17. The energy minimum structures of the cation-G-quartet complexes are located both at the HF and the B3LYP levels by the analytic gradient techniques.
18. The interaction energy and the frequency of the G-quartets are corrected for the basis set superposition error (BSSE) by the standard counterpoise method (61) implemented in GAUSSIAN 03. In general, this method accounts for the exchange, dispersion, and polarization contributions (70).

References

1. Burge S, Parkinson GP, Hazel P, Todd AK, Neidle S (2006) Quadruplex DNA: sequence, topology and structure. Nucleic Acids Res 34:5402–5415
2. Davies JT (2004) G-quartets 40 years later: from 5'-GMP to molecular biology and supramolecular chemistry. Angew Chem Intl Edit 43:668–698
3. Neidle S, Parkinson GN (2008) Quadruplex DNA crystal structures and drug design. Biochimie 90:1184–1196
4. Webba da Silva M (2007) Geometric formalism for DNA quadruplex folding. Chemistry 13:9738–9745
5. Hazel P, Huppert J, Balasubramanian S, Neidle S (2004) Loop-length-dependent folding of

G-quadruplexes. J Amer Chem Soc 126:16405–16415
6. Phan AT, Kuryavyi V, Burge S, Neidle S, Patel DJ (2007) Structure of an unprecedented G-quadruplex scaffold in the human c-kit promoter. J Amer Chem Soc 129:4386–4392
7. Huppert JL, Balasubramanian S (2005) Prevalence of quadruplexes in the human genome. Nucleic Acids Res 33:2908–2916
8. Todd AK, Johnston M, Neidle S (2005) Highly prevalent putative quadruplex sequence motifs in human DNA. Nucleic Acids Res 33:2901–2907
9. Huppert JL, Balasubramanian S (2006) G-quadruplexes in promoters throughout the human genome. Nucleic Acids Res 35:406–413
10. Sun D, Thompson BE, Cathers M, Salazar SM, Kerwin JO, Trent TC, Jenkins SN, Hurley LH (1997) Inhibition of human telomerase by a G-quadruplex-interactive compound. J Med Chem 40:2113–2116
11. Neidle S, Parkinson GN (2002) Telomere maintenance as a target for anticancer drug discovery. Nat Rev Drug Discovery 1:383–393
12. Shay JW, Wright WE (2006) Telomerase therapeutics for cancer: challenges and new directions. Nature Rev Drug Discovery 5:577–584
13. Oganesian L, Bryan TM (2007) Physiological relevance of telomeric G-quadruplex formation: a potential drug target. Bioessays 29:155–165
14. De Cian A, Lacroix L, Douarre C, Temime-Smaali N, Trentesaux C, Riou JF, Mergny JL (2008) Targeting telomeres and telomerase. Biochimie 90:131–155
15. Monchaud D, Teulade-Fichou MP (2008) A hitchhiker's guide to G-quadruplex ligands. Org Biomol Chem 6:627–636
16. Clark GR, Pytel PD, Squire CJ, Neidle S (2003) Structure of the first parallel DNA quadruplex-drug complex. J Amer Chem Soc 125:4066–4067
17. Haider SM, Parkinson GN, Neidle S (2003) Structure of a G-quadruplex-ligand complex. J Mol Biol 326:117–125
18. Parkinson GN, Lee MP, Neidle S (2002) Crystal structure of parallel quadruplexes from human telomeric DNA. Nature 417:876–880
19. Parkinson GN, Ghosh R, Neidle S (2007) Structural basis for binding of porphyrin to human telomeres. Biochemistry 46:2390–2397
20. Campbell NH, Parkinson GN, Reszka AP, Neidle S (2008) Structural basis of DNA quadruplex recognition by an acridine drug. J Amer Chem Soc 130:6722–6724
21. Phan AT, Kuryavyi V, Gaw HY, Patel DJ (2005) Small-molecule interaction with a five-guanine-tract G-quadruplex structure from the human MYC promoter. Nature Chem Biol 1:167–173
22. Ambrus A, Chen D, Dai J, Bialis T, Jones RA, Yang D (2006) Human telomeric sequence forms a hybrid-type intramolecular G-quadruplex structure with mixed parallel/antiparallel strands in potassium solution. Nucleic Acids Res 34:2723–2735
23. Luu KN, Phan AT, Kuryavyi V, Lacroix L, Patel DJ (2006) Structure of the human telomere in K^+ solution: an intramolecular (3 + 1) G-quadruplex scaffold. J Amer Chem Soc 128:9963–9970
24. Phan AT, Luu KN, Patel DJ (2006) Different loop arrangements of intramolecular human telomeric (3+1) G-quadruplexes in K^+ solution. Nucleic Acids Res 34:5715–5719
25. Dai J, Carver M, Punchihewa C, Jones RA, Yang D (2007) Structure of the hybrid-2 type intramolecular human telomeric G-quadruplex in K^+ solution: insights into structure polymorphism of the human telomeric sequence. Nucleic Acids Res 35:4927–4940
26. Šponer J, Špačková N (2007) Molecular dynamics simulations and their application to four-stranded DNA. Methods 43:278–290
27. Yang L, Tan CH, Hsieh MJ, Wang J, Duan Y, Cieplak P, Caldwell J, Kollman PA, Luo R (2006) New-generation amber united-atom force field. J Phys Chem B 110:13166–13176
28. Foloppe N, MacKerell AD (2000) All-atom empirical force field for nucleic acids: I. parameter optimization based on small molecule and condensed phase macromolecular target data. J Comput Chem 21:86–104
29. Goodfellow JM, Levy R (1998) Theory and simulation. Curr Opin Struct Biol 8:209–210
30. Sagui C, Darden TA (1999) Molecular dynamics simulations of biomolecules: long-range electrostatic effects. Annu Rev Biophys Biomol Struct 28:155–179
31. Norberg J, Nilsson L (2000) On the truncation of long-range electrostatic interactions in DNA. Biophys J 79:1537–1553
32. Steinbach PJ, Brooks BR (1994) New spherical-cutoff methods for long-range forces in macromolecular simulation. J Comput Chem 15:667–683
33. Ross WS, Hardin CC (1994) Ion induced stabilization of the G-DNA quadruplex: free energy perturbation studies. J Amer Chem Soc 116:6070–6680
34. York DM, Darden TA, Pedersen LG (1993) The effect of long-range electrostatic interactions

in simulations of macromolecular crystals: A comparison of the Ewald and truncated list methods. J Chem Phys 99:8345–8348
35. Cheatham TE III, Miller JL, Fox T, Darden TA, Kollman PA (1995) Molecular dynamics simulations on solvated biomolecular systems: the particle mesh Ewald method leads to stable trajectories of DNA, RNA and proteins. J Amer Chem Soc 117:4193–4194
36. Hunenberger PH, McCammon JA (1999) Ewald artifacts in computer simulations of ionic salvation and ion-ion interaction: a continuum electrostatics study. J Chem Phys 110:1856–1872
37. MacKerell Jr., A. D., Brooks, B., Brooks III, C. L., Nilsson, L., Roux, B., Won, Y. and Karplus, M. (1998) CHARMM: The energy function and its parameterization with an overview of the program. In: *The Encyclopedia of Computational Chemistry* (J. Wiley and Sons). **1**, 271–277.
38. Soares TA, Hunenberger PH, Kastenholz MA, Krautler V, Lenz T, Lins RD, Oostenbrink C, Van Gunsteren WF (2005) An improved nucleic acid parameter set for the GROMOS force field. J Comput Chem 26:725–737
39. Van Wynsberghe AW, Cui Q (2005) Comparison of mode analyses at different resolutions applied to nucleic acid systems. Biophys J 89:2939–2949
40. Perez A, Marchan I, Svozil D, Sponer J, Cheatham TE III, Laughton CA, Orozco M (2007) Refinement of AMBER force field for nucleic acids: Improving the description of α/γ conformers. Biophys J 92:3817–3829
41. Haider S, Parkinson GN, Neidle S (2002) Crystal structure of the potassium form of an Oxytricha nova G-quadruplex. J Mol Biol 320:189–200
42. Gallagher T, Taylor MJ, Ernst SR, Hackert ML, Poonia NS (1991) Dipotassium and sodium/potassium crystalline picrate complexes with the crown ether. Acta Crystallogr B 47:362–368
43. Phillips K, Dauter Z, Murchie AI, Lilley DM, Luisi B (1997) The crystal structure of a parallel-stranded guanine tetraplex at 0.94 Å resolution. J Mol Biol 273:171–182
44. Schultze P, Smith FW, Feigon J (1994) Refined solution structure of the dimeric quadruplex formed from the Oxytricha telomericoligonucleotided(GGGGTTTTGGGG). Structure 2:221–233
45. Hud NV, Schultze P, Sklenar V, Feigon J (1999) Binding sites and dynamics of ammonium ions in a telomere repeat DNA quadruplex. J Mol Biol 285:233–243
46. Cavallari M, Calzolari A, Garbesi A, Di Felice R (2006) Stability and migration of metal ions in G4-wires by molecular dynamics simulations. J Phys Chem 110:26337–26348
47. Ponomarev SY, Thayer KM, Beveridge DL (2004) Ion motions in molecular dynamics simulations in DNA. Proc Natl Acad Sci USA 101:14771–14775
48. Haider S, Parkinson GN, Neidle S (2008) Molecular dynamics and principal components analysis of human telomeric quadruplex multimers. Biophys J 95:296–311
49. Spackova N, Berger I, Sponer J (2001) Structural dynamics and cation interactions of DNA quadruplex molecules containing mixed guanine/cytosine quartets revealed by large-scale MD simulations. J Amer Chem Soc 123:3295–3307
50. Hazel P, Parkinson GN, Neidle S (2006) Predictive modeling of topology and loop variations in dimeric DNA quadruplex structures. Nucleic Acid Res 34:2117–2127
51. Cornell WD, Cieplak P, Bayly CI, Gould IR, Merz KM Jr, Ferguson DM, Spellmeyer DC, Fox T, Caldwell JW, Kollman PA (1995) A second generation force field for the simulation of proteins, nucleic acids, and organic molecules. J Amer Chem Soc 117:5179–5197
52. Cheatham TE III, Cieplak P, Kollman PA (1999) A modified version of the Cornell et al. force field with improved sugar pucker phases and helical repeat. J Biomol Struct Dyn 16:845–862
53. Humphrey W, Dalke A, Schulten K (1996) VMD - Visual Molecular Dynamics. J Molec Graphics 14:33–38
54. Mazur AK (1998) Accurate DNA dynamics without accurate long-range electrostatics. J Amer Chem Soc 120:10928–10937
55. Bashford D, Case D (2000) Generalised Born models of macromolecular solvation effects. Ann Rev Phys Chem 51:129–152
56. Fadrna E, Spackova N, Stefl R, Koca J, Cheatham TE III, Sponer J (2004) Molecular dynamics simulations of guanine quadruplex loops: advances and force field limitations. Biophys J 87:227–242
57. BIOSYM. San Diego, CA.
58. Bondi A (1964) van der Waals volumes and radii. J Phys Chem 68:441–451
59. Kelly LA, Gardner SP, Sutcliffe MJ (1996) An automated approach for clustering an ensemble of NMR-derived protein structures into conformationally related subfamilies. Protein Eng 9:1063–1065
60. Meyer T, Ferrer-Costa C, Perez A, Rueda M, Bidon-Chanal A, Luque FJ, Laughton CA,

Orozco M (2006) Essential dynamics: A tool for efficient trajectory compression and management. J Chem Theory Comp 2:251–258

61. Boys SF, Bernardi F (1970) Calculations of small molecular interaction by the difference of separate total energies. Some procedures with reduced error. Mol Phys 19:553–566
62. Becke AD (1993) Density function thermochemistry. The role of exact exchange. J Chem Phys 98:5648–5652
63. Gu J, Leszczynski J (1999) Influence of the oxygen at the C8 position on the intramolecular proton transfer in C8-oxidative guanine. J Phys Chem 103:577–584
64. Sponer J, Leszczynski J, Hobza P (1996) Structures and energies of hydrogen-bonded DNA base pairs. A nonempirical study with inclusion of electron correlation. J Phys Chem 100:1965–1974
65. Halgren TA, Damm W (2001) Polarizable force fields. Curr Opin Struct Biol 11:236–242
66. At www.tripos.com
67. Pacios LF, Christiansen PA (1985) Ab initio relativistic effective potentials with spin-orbit operators. I. Li through Ar. J Chem Phys 82:2664–2671
68. Hurley MM, Pacios LF, Christiansen PA (1986) Ab initio relativistic effective potentials with spin-orbit operators. II. K through Kr. J Chem Phys 84:6840–6853
69. Gaussian 03, Revision C.02, Frisch, M. J.; Trucks, G. W.; Schlegel, H. B.; Scuseria, G. E.; Robb, M. A.; Cheeseman, J. R.; Montgomery, Jr., J. A.; Vreven, T.; Kudin, K. N.; Burant, J. C.; Millam, J. M.; Iyengar, S. S.; Tomasi, J.; Barone, V.; Mennucci, B.; Cossi, M.; Scalmani, G.; Rega, N.; Petersson, G. A.; Nakatsuji, H.; Hada, M.; Ehara, M.; Toyota, K.; Fukuda, R.; Hasegawa, J.; Ishida, M.; Nakajima, T.; Honda, Y.; Kitao, O.; Nakai, H.; Klene, M.; Li, X.; Knox, J. E.; Hratchian, H. P.; Cross, J. B.; Bakken, V.; Adamo, C.; Jaramillo, J.; Gomperts, R.; Stratmann, R. E.; Yazyev, O.; Austin, A. J.; Cammi, R.; Pomelli, C.; Ochterski, J. W.; Ayala, P. Y.; Morokuma, K.; Voth, G. A.; Salvador, P.; Dannenberg, J. J.; Zakrzewski, V. G.; Dapprich, S.; Daniels, A. D.; Strain, M. C.; Farkas, O.; Malick, D. K.; Rabuck, A. D.; Raghavachari, K.; Foresman, J. B.; Ortiz, J. V.; Cui, Q.; Baboul, A. G.; Clifford, S.; Cioslowski, J.; Stefanov, B. B.; Liu, G.; Liashenko, A.; Piskorz, P.; Komaromi, I.; Martin, R. L.; Fox, D. J.; Keith, T.; Al-Laham, M. A.; Peng, C. Y.; Nanayakkara, A.; Challacombe, M.; Gill, P. M. W.; Johnson, B.; Chen, W.; Wong, M. W.; Gonzalez, C.; and Pople, J. A.; Gaussian, Inc., Wallingford CT, 2004
70. Chatasinski G, Szesniak M (1994) Origins of structure and energetics of van der Waals clusters from ab initio calculations. Chem Rev 94:1723–1765
71. Fadrna E, Spackova N, Sarzynska J, Koca J, Orozco M, Cheatham TE III, Kulinski T, Sponer J (2009) Single Stranded Loops of Quadruplex DNA As Key Benchmark for Testing Nucleic Acids Force Fields J Chem Theory Comput 5:2514–2530

Chapter 3

Computational Approaches to the Detection and Analysis of Sequences with Intramolecular G-Quadruplex Forming Potential

Paul Ryvkin, Steve G. Hershman, Li-San Wang, and F. Brad Johnson

Abstract

Sequences with the potential to form intramolecular G-quadruplexes (G4-structures) are found in highly nonrandom distributions in the genomes of diverse organisms. These sequences are associated with nucleic acid metabolic processes ranging from transcription and translation to recombination and telomere function. Here we review different computational methods for identifying potential G4-forming sequences and provide protocols for their implementation. We also discuss methods for assessing the significance and specificity of associations between the sequences and different biological functions.

Key words: G-quadruplex, G4-DNA, Bioinformatics, Computational biology

1. Introduction

G4-structures, including G4-DNA and G4-RNA, comprise stacked quartets of Hoogsteen hydrogen bonded guanines stabilized by small monovalent cations. Interest in G4-structures has been sparked by recent findings that suggest that they function in processes ranging from transcription and translation to recombination and telomere maintenance (1, 2). A large number of particular G4-structures are possible, even given the same starting sequences (3, 4). For example, G4-structures can be assembled from guanines within one nucleic acid strand (intramolecular) or from different strands (intermolecular). They can also differ on the basis of the glycosidic bond angles of the guanines, the type of coordinating cations, the number of stacked quartets, the polarity of phosphodiester backbone strands, and the length, nucleotide composition, and arrangement of the sequences that

P. Baumann (ed.), *G-Quadruplex DNA: Methods and Protocols*, Methods in Molecular Biology, vol. 608
DOI 10.1007/978-1-59745-363-9_3, © Humana Press, a part of Springer Science+Business Media, LLC 2010

do not contribute to the quartets. This structural diversity has made it challenging to predict which sequences are most likely to form stable G4-structures, although studies of oligonucleotides in vitro have generated some simple rules, e.g. intramolecular G4-DNA is favored when one or more of the intervening loops are short (5). Furthermore, G-quadruplex formation in vivo is likely to be modulated by chromatin factors, including proteins and helicases that can stabilize or unwind G4-structures (6). Therefore, it is currently not possible to predict with certainty whether a given sequence will form G4-structures within a living cell.

Computational approaches can nonetheless be of value for indicating the potential of a sequence to form G4-structures. There are some sequences that have essentially no potential for forming G4-structures, and computational approaches can be used to separate these from other sequences with higher G4-forming potential. The question can then be asked whether the two sets of sequences are differentially associated with any biological activities that might be of interest, e.g. expression under certain circumstances, chromatin modifications, recombination hotspots, polarity of replication, conservation among or within species, etc. Although such an association will not prove that it is the capacity of the sequences to form G4-structure per se that explains the association, it does provide a starting point for hypothesis testing. Moreover, if the biology being explored involves factors that a priori are expected to influence G4-structure levels or function, an association with sequences having high G4-forming potential provides an argument for the importance of the G4-structures themselves. This interpretation can be strengthened if the associations with sequences having G4-forming potential are stronger than with carefully selected control sequences.

Computational approaches for identifying sequences with G4-forming potential are currently based primarily on three simple rules derived from studies of intramolecular G4-DNA formation in vitro (7–11). (1) At least two and preferably three or more G-quartets must be stacked to form a stable G-quadruplex (3); for intramolecular G4-DNA this translates to runs of two-to-three or more consecutive Gs separated by three loop sequences. (2) Short loops (one to two nucleotides (nt)) favor G4-DNA formation, although longer loops are possible (5). Indeed, at least when one or two of the loops are short, the other(s) can extend beyond 20 nt (12). (3) To a first approximation, the loop sequences are not critical determinants of stability, although recent investigations are beginning to define their contributions (13). Here we describe several of these computational approaches and how to calculate the statistical associations of particular biological variables with the identified potential G4-forming sequences, and discuss controls that can help evaluate the specificity of any observed associations. Because of the exploratory nature of these approaches, the methods presented should not be considered definitive, but it is hoped that they will stimulate collaborative efforts among molecular and computational biologists. In addition, we emphasize that in many cases these approaches should be considered only a starting point for generation of hypotheses that can then be tested using molecular genetic and biochemical approaches.

2. Methods

2.1. Detecting Potential Intramolecular G-quadruplex-Forming Sequences

Three types of computational approaches, based on string pattern matching, have been used in the literature for analyzing the genomic distribution of sequences with the potential to form G-quadruplexes. Although G4-structures could form between different nucleic acid strands, e.g. conceivably between the two strands of a denatured DNA duplex, we limit our discussion to sequences with the potential to form intramolecular G-quadruplexes because all the algorithms published to date are for identifying intramolecular structures.

2.1.1. Regular Expression

This method specifies a regular expression that the G-quadruplex-forming sequence should take. For example, the first analyses of G-quadruplex forming potential of the human genome (8, 9) used the regular expression $G_{3+}N_{1-7}\ G_{3+}N_{1-7}\ G_{3+}N_{1-7}\ G_{3+}$, which requires a matching sequence to satisfy two properties: (1) each of the four guanine runs has a length of at least three nucleotides, and (2) lengths of the three loops are all between 1 and 7 nucleotides (N means any nucleotide). Many programming languages provide regular expression matching; the following example of Perl code:

```
# search this sequence
$s = "AATACGGGACATGGGGATAGAGGGCGCGGGGTT";
if($s=~ m/G{3,}.{1,7}?G{3,}.{1,7}?G{3,}.
  {1,7}?G{3,}/){print "Match";
}
```

Detecting G4-forming candidate sequences on the complementary strand can be done easily by replacing Gs with Cs in the regular expression above. The output of a regular expression analysis is discrete, with any particular stretch of DNA either conforming to the pattern or not. However, once individual potential G4-forming sequences have been identified, their density within any region (e.g. a promoter or intron) can be assessed to provide a more continuous estimate of the overall G4-forming potential of the region. For example, the fraction of the region contained within sequences having G4-forming potential can be calculated. Alternatively, the number of such sequences in a region can be divided by the length of the region; with this approach an arbitrary decision needs to be made for whether overlapping sequences will be counted as individual or distinct occurrences.

2.1.2. G4P

Eddy and Maizels (10) used a sliding window approach to assess the G-quadruplex-forming potential of genomic regions. The algorithm generates a continuous estimate of G4-forming potential that depends on three parameters, *k* (length of the guanine runs; default is 3 nt), *w* (window size; default is 100 nt), and *s* (step; default is 20 nt). Starting from the beginning of the input sequence, the algorithm checks windows of length *w* starting every *s* nucleotides; the G4P is the fraction of these windows containing four guanine runs of length *k* separated by at least one nucleotide. This approach is more flexible than the regular expression, because it only limits the total length of the candidate sequence and not the lengths of individual loops. The authors have made public a program for Microsoft Windows (http://depts.washington.edu/maizels9/G4calc.php); the program calculates G4P on either strand, or the average of the G4P on both strands. The following example shows how G4P (and C4P) can alternatively be computed using the `gregexpr` command in R.

```
ins=paste(sample(c("A","C","G","T"),
    1000000,prob=c(0.3,0.2,0.2,0.3),repl=T),
    collapse="")                    # random sequences
gpat="G{3}.+?G{3}.+?G{3}.+?G{3}";
cpat="C{3}.+?C{3}.+?C{3}.+?C{3}";
n=nchar(ins);
k=n/20
fwdcnt=rep(0,k+1); revcnt=rep(0,k+1); #
        window match results
for  (i in 0:k) {
     ins.k=substring(ins, i*20+1, min(i*20+100,
         n)) # window i
     if (gregexpr(gpat, ins.k, perl=T)[[1]][1]
        !=-1) {fwdcnt[i+1]=1}
     if (gregexpr(cpat, ins.k, perl=T)[[1]][1]
        !=-1) {revcnt[i+1]=1}
}
g4p=sum(fwdcnt)/length(fwdcnt)
c4p=sum(revcnt)/length(revcnt)
```

2.1.3. QFP

Recently we described the distribution of sequences with G-quadruplex forming potential (QFP) within the *Saccharomyces cerevisiae* genome (11). We used the sliding window approach, but instead of returning a continuous estimate of G4-forming potential, we identified the discrete sites that may form G-quadruplexes. In other words, for $k=3$ and $w=100$, the sequence takes the form $G_3\ N_a\ G_3\ N_b\ G_3\ N_c\ G_3$ where *a*, *b*, $c>0$, and $12 + a + b + c \leq 100$. This approach is more sensitive than the regular expression and G4P approaches, which are useful for an organism like yeast that generally has rare sequences with G4-forming potential. An efficient algorithm examines every sliding window that starts from a GGG run, instead of running sliding windows starting at every nucleotide position. Software is available from hershman@fas.harvard.edu.

Comparisons of the three methods are shown in Fig. 3.1 and Table 3.1.

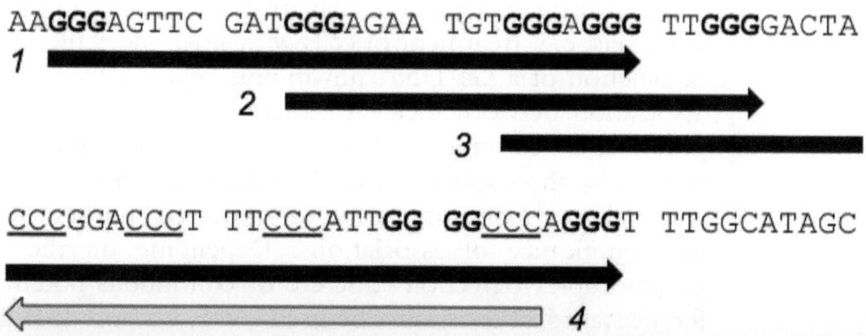

	Condition	1	2	3	4
RegEx	Loop length <=7, G run length>=3	No (first loop length is 8)	Yes	No (third loop length is 23)	Yes (reverse strand)
G4P	Overall length <=100	Will be counted	Will be counted	Will be counted	Will be counted (reverse strand)
QFP	Overall length <=100	Yes	Yes	Yes	Yes (reverse strand)

Fig. 3.1. Example using three pattern matching criteria for intramolecular G-quadruplex candidate sequence detection.

Table 3.1
Comparison of three intramolecular G-quadruplex detection criteria

Criterion	Output format	Example criterion	Note
Regular expression	Matched sequence motif	G-run length ≥3, loop length ≤7	Most stringent of the three; imposes direct constraint on loop length
G4P	Percentage (between 0 and 100%)	Number of windows (100 bp size, 20 bp step) containing at least one G-quadruplex (without loop constraint)	Returns potential rather than putative locations; imposes constraint on total length
QFP	Matched sequence motif	Overall length of G-quadruplex ≤100	Less stringent than regular expression criterion; imposes constraint on total length

2.2. Analysis of the Genomic Distribution of G-quadruplex-Forming Sequences

Once G4-forming candidate sequences are identified or G4 forming potentials (G4Ps) are determined, it may be of interest to compare their distributions with other genomic regions, e.g. promoters, and, moreover, with the behavior of these genomic regions in particular biologic settings. For example, if genes that are differentially expressed under certain experimental conditions have a statistically significant tendency to have G-quadruplex forming sequences in their promoter regions, this raises the possibility that the G4-forming sequences play some role in the regulation of expression. Furthermore, if the experimental conditions being investigated are those that are expected a priori to influence G4 levels, e.g. treatment of cells with small molecule G4 ligands or deletion of a G4-DNA unwinding helicase, then a statistical association between loci with altered expression and loci with potential G4-forming sequences indicates that G-quadruplex formation by the sequences may be part of the mechanism of differential gene expression. Care should be taken when computing the significance of associations. Depending on the types of G-quadruplex detection (discrete or continuous potential), different statistical tests must be used. Also, we note that although we are using transcription as the biological variable in these examples, similar approaches can be applied to other processes (e.g. polarity of replication, recombination rates, etc).

2.2.1. Continuous Variables

If the association to be tested involves continuous *t*-test statistics for gene expression and a continuous measure of G4-forming potential (i.e. G4P or a continuous derivative of output from a regular expression), one can run a test of correlation using Pearson's correlation (assuming normally distributed error) or Spearman's correlation (nonparametric test without assumption of normality). The following is an example in R code:

```
# vector tstat stores the t-statistics for
    all genes in a
# control/treatment microarray experiment
# vector g4p stores the G4P scores for all
     genes
p.cor  =cor(tstat,g4p)          #Pearson
     correlation
p.pval = cor.test(tstat,g4p)$p.value
                                #Significance
s.cor  = cor(tstat,g4p,method="spearman")
                                #Spearman
                                correlation
s.pval = cor.test(tstat,g4p,
     method="spearman")$p.value
                                #Significance
```

2.2.2. Discrete Variables

At the other end of the spectrum, associations can be tested between discrete categories for altered gene expression and for G4-forming potential, e.g. QFP or discrete output from a regular expression. In such cases each gene will fall into one of four categories based on whether the gene is differentially expressed and whether the gene has a G-quadruplex-forming sequence. A 2-by-2 contingency table can be formed for the number of genes falling into the four combinations, and then either a one-sided Fisher's test (if there is an expectation for the direction of the association, e.g., G-quadruplexes activate gene expression) or a two-sided Fisher's test (if no expectation regarding the direction of association; in which case the odds ratio returned from Fisher's test shows the direction of association) can be applied. Many computer programs, such as R (the `fisher.test` function), provide this capability:

```
# vector diff_expressed is a 0/1 vector for
    the status of expression
# change (1 if differentially expressed) for
    all genes in a
# control/treatment microarray experiment
# vector g4p stores the G4P scores for all
    genes
tbl=data.frame(diff_expressed, g4p)
                # 2-column table, one row per gene
r=glm(diff_expressed~g4p, data=tbl,family=
                "binomial") # logistic regression
summary(r) # summary of regression; check the
           p-value
           # for the coefficient of g4p for
           # association significance
```

2.2.3. Mixed Continuous and Discrete Variables

A middle ground also exists, involving mixtures of continuous and discrete categories for G4-forming potential and altered expression. For example, the association of G4P with whether each gene is differentially expressed or not can be tested. One might be tempted to divide the G4Ps into two groups and run a *t*-test, but statistically this is not correct as genes in the two groups are not independent observations from two populations. A better approach is to run a logistic regression using the G4P as the independent variable and the differential expression status as the dependent variable. We can test similarly if we test the association using a discrete G4 criterion (QFP or regular expression) and a continuous score for differential gene expression such as the *t*-statistic:

```
# vector has_gquad is a 0/1 vector on whether
    each gene has a
# G-quadruplex motif (1 if the gene has a
```

```
    G-quadruplex motif)
tbl=data.frame(diff_expressed, has_gquad) #
    two-column table,# one row per gene

pval=fisher.test(tbl)$p.value # Significance
pval2=fisher.test(tbl, alt="greater")$p.value
                        # Significance, one-sided
```

Usually 0.01 or 0.05 is used as a threshold for significance. When performing multiple tests, the threshold of significance can be set using the Bonferroni correction, which involves simply dividing the threshold of individual tests, e.g., 0.05, by the number of tests to give a lower threshold. Alternatively, more sophisticated multiple testing correction procedures such as false discovery rate (FDR) can be used (14); the `fdrtool` package for R (15) includes an implementation.

Note that each statistical test returns the test statistic in addition to the statistical significance. While the significance shows whether the association arises by chance, the interpretation of the statistic is equally important. For example, the Pearson's correlation coefficient test returns the *p*-value indicating that the actual correlation coefficient is different from zero, and a very weak correlation may be statistically significant when many observations are made, which is often the case in genome-scale analyses such as microarray experiments. Thus biological and statistical significance are separate entities that must be interpreted carefully.

2.3. Controls to Assess the Specificity of Associations with G4-Forming Sequences

In addition to the association significance introduced in the last section, a common approach to assess the specificity of associations between potential G4-forming sequences and particular genomic features or biological functions is to make similar comparisons with control sequence patterns. The premise of this approach is that if G4-forming sequences are the sole (or primary) mediators of the biological function then there should be no (or less significant) association with control sequences. This premise will not always be true, e.g. the binding of transcription factors as well as G4-DNA formation might regulate transcription from a given promoter, and thus these control approaches can suffer from being too stringent. However, they can still be very informative, and we describe several commonly used controls below. We note that it is important that the frequency of control patterns is on the same order as the frequency of the G4 pattern; otherwise, the more frequent pattern will have the potential to show greater statistical significance only on the basis of a larger number of observations. We will describe several commonly used controls as follows.

2.3.1. A4/T4 Control

One seemingly simple control is to examine the distribution of sequences having the form of the G4-forming potential strings, but with the Gs replaced by As or Ts, e.g. in the case of the regular expression the sequence $A_{3+}N_{1-7}$ $A_{3+}N_{1-7}$ $A_{3+}N_{1-7}$ A_{3+} (8). Assuming there is no biological significance for such "A4" or "T4" sequences, these control sequences should not be associated with the biological variables under investigation. In practice this approach may be problematic for two reasons: (a) in A/T-rich genomes the number of control sequences will be very high, e.g. A4/T4 sequences are nearly an order of magnitude more frequent in the human genome than G4/C4 sequences and this may inflate the significance of even very weak associations, and (b) it is possible that A4 or T4 sequences will have some functional connection to the biology being investigated.

2.3.2. Randomized Models

A second control approach is to generate random sequences that share characteristics of real genomes (8, 11). This approach is most useful for assessing whether the observed distribution of potential G4-forming sequences is nonrandom, but may also have some applications in association studies. For example, G4-forming sequences within randomized model genomes should be less associated with the biology under investigation than similar sequences in the real genome if the G4-forming potential and not some other related feature of the sequences (e.g. GC-richness) is responsible for the association. There are several possible approaches to generating randomized genomes. For example, a higher-order Markov chain based on real genomic sequence can be used to generate random sequences that share the same base composition, and diad, triad, etc., frequencies as the real genome. There are many software programs that can accomplish this (16), and one can implement first or second order Markov chains easily using statistical software such as R. This control is limited by the level of realism achieved by the Markov model and is more complex to implement than the others we describe. We therefore suggest researchers with limited experience in programming to seek collaboration with a statistician or a computer scientist for this control.

2.3.3. Partial G4 Controls

A third approach is to make comparisons to a sequence motif that is very close to a potential G4-forming sequence but which should not be able to form G4-structures. We suggest two methods. One is to examine strings with three G_{3+} runs instead of four. For example, in the case of the regular expression $G_{3+}N_{1-7}$ $G_{3+}N_{1-7}$ $G_{3+}N_{1-7}$ G_{3+}, the pattern $G_{3+}N_{1-7}$ $G_{3+}N_{1-7}$ G_{3+} could be examined with the added requirement that there be no G_{3+} run within eight nucleotides of the pattern. A second approach is to examine strings that have G-runs interrupted by non-G bases. The idea is to find sequences that have an edit distance of n (point mutations) from putative quadruplex-forming sequences, where $n = 1$ or 2 is ideal. The following Perl code can perform such a search:

```
$s="AATACGGGACATGAGGATAGAGGGCGCGGGGTT";
                                # search string
                                # interrupted G-runs
if ($s=~ /(G.G).{1,7}(G.G).{1,7}(G.G).{1,7}
   (G.G)/) {
   $runs="$1$2$3$4";            # concatenate runs
   $dist=()=$runs=~ /[^G]/g;    # count non-G's in
                                  runs
   if ($dist == 1) { print "Match"; }
}
```

For association studies it is important to consider that loci can have both potential G4-forming sequences and the control sequences, and therefore loci having the control sequence but not the G4-sequence must be identified to provide the control set.

2.3.4. Relaxed G4-Potential

A fourth approach is to relax the stringency for detection of potential G4-forming sequences, to see if the statistical significance dissipates, as the stability of intramolecular G4-structures should depend on how close the GGG runs are to one another. For example the loop lengths can be increased beyond a regular expression or the size of the sliding window can be increased in G4P or QFP detection. We used this method in our paper using QFP, and observed that the significance increases when the window size increases from 35 nt to 100 nt, but the significance drops afterwards (11). The initial positive correlation may be due to the low frequency of QFP sequences in the *S. cerevisiae* genome, and the loss of significance at windows larger than 100 nt due to more random targets (i.e. those not actually capable of forming G4 structures) being included.

2.3.5. Transcription Factor Comparisons

In cases involving associations between potential G4-forming sequences and gene expression an alternative explanation for the associations is that the G4 patterns simply reflect clusters of binding sites for transcription factors that bind to sequences containing G-runs. Similarly, the Gs in CpG islands can flank G runs and thus contribute to G4-forming potential sequences. Recently, Eddy and Maizels tested this idea by subtracting loci containing such patterns from their analyses and showed that this virtually eliminated the apparent associations with potential G4-forming patterns (17). However, it is an expected result that removal of sequences having the same pattern as those with G4-forming potential will diminish the significance of G4 associations. Furthermore, the correspondence between the presence of a potential binding site for a transcription factor and actual binding of the transcription factor is imperfect. Thus we argue that such approaches do not distinguish between the possibilities that the binding of transcription factors or G4-structure formation explains the associations. Data based on actual transcription factor occupancy (i.e. using chromatin immunoprecipitation), which are available on a large scale for model organisms like yeast (18), could help resolve this issue. Genetic approaches that manipulate the activity of such transcription factors could also provide valuable insights. More work is needed before the proper application of this control can be realized.

Determining which of the control approaches described above are most appropriate will require additional research. We suggest that the readers try several controls to better gauge the significance and specificity of their findings.

3. Resources on the Web

The G4P criterion was first defined in (10), and the program (Microsoft Windows only) can be downloaded from (http://depts.washington.edu/maizels9/G4calc.php). A regular expression program by Huppert et al. (8), *Quadparser*, can be obtained here: http://www.quadruplex.org/?view=quadparser. There is also a flexible web tool called *QGRS Mapper* (19) available at http://bioinformatics.ramapo.edu/QGRS/analyze.php. G4P and other methods in this chapter can be entirely implemented using either Perl (http://www.perl.org) or R (http://www.r-project.org). Example codes in this chapter can be downloaded from the companion website at http://people.pcbi.upenn.edu/~lswang/gquad/chapter_ryvkin_etal/.

Acknowledgments

We thank Jay Johnson, Kajia Cao, Marina Kozak, Alex Chavez, Jasmine Smith, and Qijun Chen for advice and discussions. This work was supported by NIH grants R01-AG021521, P01-AG031862, and a U. Penn Institute on Aging Pilot Grant.

References

1. Maizels N (2006) Dynamic roles for G4 DNA in the biology of eukaryotic cells. Nat Struct Mol Biol 13:1055–1059
2. Johnson JE, Smith JS, Kozak ML, Johnson FB (2008) In vivo veritas: using yeast to probe the biological functions of G-quadruplexes. Biochimie 90:1250–1263
3. Lane AN, Chaires JB, Gray RD, Trent JO (2008) Stability and kinetics of G-quadruplex structures. Nucleic Acids Res 36:5482–5515
4. Webba da Silva M (2007) Geometric formalism for DNA quadruplex folding. Chemistry 13:9738–9745
5. Bugaut A, Balasubramanian S (2008) A sequence-independent study of the influence of short loop lengths on the stability and topology of intramolecular DNA G-quadruplexes. Biochemistry 47:689–697
6. Fry M (2007) Tetraplex DNA and its interacting proteins. Front Biosci 12:4336–4351

7. Rawal P, Kummarasetti VB, Ravindran J, Kumar N, Halder K, Sharma R et al (2006) Genome-wide prediction of G4 DNA as regulatory motifs: role in *Escherichia coli* global regulation. Genome Res 16:644–655
8. Huppert JL, Balasubramanian S (2005) Prevalence of quadruplexes in the human genome. Nucleic Acids Res 33:2908–2916
9. Todd AK, Johnston M, Neidle S (2005) Highly prevalent putative quadruplex sequence motifs in human DNA. Nucleic Acids Res 33:2901–2907
10. Eddy J, Maizels N (2006) Gene function correlates with potential for G4 DNA formation in the human genome. Nucleic Acids Res 34:3887–3896
11. Hershman SG, Chen Q, Lee JY, Kozak ML, Yue P, Wang L-S et al (2008) Genomic distribution and functional analyses of potential G-quadruplex-forming sequences in *Saccharomyces cerevisiae*. Nucleic Acids Res 36:144–156
12. Bates P, Mergny JL, Yang D (2007) Quartets in G-major. The First International Meeting on Quadruplex DNA. EMBO Rep 8:1003–1010
13. Guedin A, De Cian A, Gros J, Lacroix L, Mergny JL (2008) Sequence effects in single-base loops for quadruplexes. Biochimie 90:686–696
14. Storey JD, Tibshirani R (2003) Statistical significance for genomewide studies. Proc Natl Acad Sci U S A 100:9440–9445
15. Strimmer K (2008) A unified approach to false discovery rate estimation. BMC Bioinformatics 9:303
16. Ponty Y, Termier M, Denise A (2006) GenRGenS: software for generating random genomic sequences and structures. Bioinformatics 22:1534–1535
17. Eddy J, Maizels N (2008) Conserved elements with potential to form polymorphic G-quadruplex structures in the first intron of human genes. Nucleic Acids Res 36:1321–1333
18. Harbison CT, Gordon DB, Lee TI, Rinaldi NJ, Macisaac KD, Danford TW et al (2004) Transcriptional regulatory code of a eukaryotic genome. Nature 431:91–104
19. Kikin O, D'Antonio L (2006) QGRS Mapper: a web-based server for predicting G-quadruplexes in nucleotide sequences. Nucleic Acids Res 34:W676–W682

Chapter 4

Preparation of G-Quartet Structures and Detection by Native Gel Electrophoresis

Ian K. Moon and Michael B. Jarstfer

Abstract

Mounting evidence supporting the existence of DNA structures containing G-quartets in vivo makes these unique and diverse nucleic acid structures an important research subject, and future investigations aimed at elucidating their biological significance are expected. The purification and characterization of G-quartet structures can be challenging because their inherent structural diversity, complexity, and stability are sensitive to an array of variables. The stability of G-quartet structures depends on many factors including number of DNA strands involved in G-quartet formation, the identity of the stabilizing cation(s), the number and sequence context of the guanosines involved in stacking, the presence of single-stranded overhangs, the intervening loop size, and the identity of nucleosides in the loop. Here we detail current methods used in G-quartet preparation and their purification and characterization by native gel electrophoresis.

Key words: G-quartet, G-quadruplex, Electrophoresis, Purification, Folding, Detection

1. Introduction

Interest in G-quadruplexes lies in the functionality associated with the various unique structures that these guanosine-rich oligonucleotides can form. While initial interest expressed by biologists in G-quadruplexes stemmed from studies of telomere biology, new potential functions and applications for G-quadruplexes are emerging in the areas of gene regulation, therapeutics, biotechnology, and nanotechnology.

Early research directed at investigating the structures formed by G-rich DNA was dominated by the development and characterization of chemical and biological agents that preferentially interact with G-quadruplexes. These have been sought to probe

P. Baumann (ed.), *G-Quadruplex DNA: Methods and Protocols*, Methods in Molecular Biology, vol. 608
DOI 10.1007/978-1-59745-363-9_4, © Humana Press, a part of Springer Science+Business Media, LLC 2010

telomere biology in vitro and to act as anticancer therapeutics by inhibiting telomerase-mediated telomere maintenance in cancer cells. A wide variety of small molecules, most commonly aromatic cations, have been explored as telomerase inhibitors by virtue of their G-quadruplex stabilizing properties (1). More recently, these have been found to directly affect telomere structure and function by preventing telomere binding proteins like Pot1 from associating with the telomere (2). Evidence supporting a role for G-quadruplexes as *cis*-acting regulatory elements in gene regulation has been computationally examined, and greater than 40% of human gene promoters were predicted to contain one or more G-quadruplex motifs (3). The ability of promoter regions to fold into G-quadruplex structures has been substantiated for promoters of several oncogenes including c-*Myc* (4), k-*RAS* (5), and *VEGF* (6). A functional role for these G-quadruplex structures has been tested biochemically using G-quadruplex-stabilizing ligands to bind the c-*Myc* and k-*RAS* promoters in cell-based assays, resulting in suppression of transcriptional activation (4, 5).

Therapeutic applications for G-quadruplexes rely on defined tertiary structures that recognize specific epitopes on the therapeutic target. Generation of G-quadruplex based ligands has been accomplished by modeling on the basis of inherent gene regulatory elements and by using selective evolution of ligands by exponential enrichment (SELEX). Connor et al. utilized a G-quadruplex DNA, containing a two repeat sequence of the insulin-linked polymorphic region of the human insulin gene promoter region, to capture insulin (7), while Tasset et al. performed SELEX to isolate a potent thrombin binding DNA aptamer, containing a highly conserved G-quadruplex structure that inhibits clot formation (8). In the areas of biotechnology and nanotechnology, G-quadruplexes have been designed as sensors and mechanical devices which incorporate fluorescence resonance energy transfer as the reporter. He et al. used a cationic, conjugated polymer in combination with a dual fluorescein chromophore labeled G-quadruplex DNA as a platform to detect potassium ions in solution (9). Alberti et al. constructed a DNA-fuelled nanodevice by utilizing the quadruplex-duplex cycling of the human telomeric sequence 5′-$G_3(T_2AG_3)_3$-3′ (10).

2. Materials

2.1. G-quartet Folding

1. Folding buffer components include but are not limited to stock buffers of 10× Tris-EDTA (100 mM Tris-HCl pH 7.4, 10 mM EDTA pH 8.0); 1 M $MgCl_2$; 1 M DTT; 500 mM TrisOAc pH 7.5; 2 M potassium glutamate; and 2 M sodium

glutamate. These should be autoclaved or sterile filtered and stored at room temperature.

2. Oligonucleotide(s): G4 DNA and marker DNA containing the desired sequences and lengths can be purchased (Integrated DNA Technologies) and purified by denaturing polyacrylamide gel electrophoresis (PAGE) followed by extraction and precipitation procedures as detailed in Subheading 4.3.1.

2.2. Denaturing Polyacrylamide Gel Electrophoresis (PAGE)

1. Denaturing acrylamide solution: 20% acrylamide (acrylamide: bisacrylamide, 19:1), 7 M urea, 1× Tris-borate-EDTA. Store at 4 °C.
2. Dilution buffer: 7 M urea, 1× Tris-borate-EDTA. Store at 4 °C.
3. Ammonium persulfate: 10% (w/v) in water (*see* Note 1). Store at 4 °C.
4. N, N, N′, N′-tetramethylethylenediamine (TEMED). Store at 4 °C.
5. Running buffer: 10× Tris-borate-EDTA: Dilute to 1× in water and store at room temperature.
6. Formamide-loading buffer: 80% (w/v) deionized formamide, 10 mM EDTA pH 8.0, 1 mg/mL xylene cyanol FF, and 1 mg/mL bromophenol blue. Store at room temperature.
7. Extraction buffer: Prepare a stock of 10× TEN buffer (100 mM Tris-HCl pH 7.4, 10 mM EDTA pH 8.0, 1 M NaCl) and store at room temperature. Alternatively, if potassium is used to form G-quartets, substitute NaCl with KCl. Dilute to 1× in water before use.
8. Handee™ spin cup columns with cellulose acetate filter (Pierce) for removal of solid acrylamide from preps. Store at room temperature.

2.3. Native Polyacrylamide Gel Electrophoresis (PAGE)

1. 40% Acrylamide solution (acrylamide: bisacrylamide, 19:1). Store at 4 °C.
2. Ammonium persulfate (APS): Prepare a 10% (w/v) in water and store at 4 °C.
3. N, N, N′, N′-tetramethylethylenediamine (TEMED). Store at 4 °C.
4. Running buffer: 10× Tris-borate-EDTA (Fisher): Dilute to 1× in water and store at room temperature. Components that are in the folding buffer (i.e. NaCl or KCl) should be added to the running buffer at the same concentrations.
5. DNA markers: Poly T DNA (*see* Note 2) or other DNA that is not expected to fold in the presence of cations may be used to compare changes in migration patterns of G4 DNA.
6. Loading buffer: 50% (v/v) glycerol, 2× Tris-borate-EDTA, or other buffer appropriate for folding conditions. Store at room temperature.

2.4. 5′-End Radiolabelling of Oligonucleotides

1. T4 Polynucleotide kinase (Promega). Store at –20 °C.
2. Adenosine 5′-triphosphate, [γ-^{32}P] (10 mCi/ml, 6,000 Ci/mmol, Perkin Elmer). Store at –20 °C in a radiation designated freezer.
3. MicroSpin™ G-25 column (GE Healthcare). Store at room temperature.

2.5. Oligonucleotide Staining

1. SYBR® Green I (10,000× concentration in DMSO, Invitrogen): Aliquot into 50 μl volumes and store at –20 °C.
2. Shallow plastic tray (13 cm × 11 cm × 4 cm).

3. Methods

During the assembly (Subheading 4.3.2), isolation, and characterization (Subheadings 4.3.3 and 4.3.4, respectively) of G-quadruplex structures, much care must be taken to ensure that the integrity of the structures is not compromised if a homogenous G-quadruplex sample is desired. After the initial denaturing/folding of the G4 DNA, the conditions in the workup must allow for maintaining maximum stability of the G-quadruplex. Factors that contribute to G-quadruplex instability are changing stabilizing cation concentrations by dilution, warming of native gels during electrophoresis, using unchilled buffers, and leaving G-quadruplexes at room temperature for extended time periods. To avoid these issues, the native gels and buffers used in electrophoresis, extractions, and reconstitution of G-quadruplexes should contain the same components as the initial folding buffer. These processes should also be performed at 4 °C, and samples should be kept on ice. For long term storage between experiments, samples should be stored at –20 °C.

3.1. Purification of G4 DNA and DNA Markers

1. Upon receiving DNA, dilute DNA pellet in manufacturer's tube to 1 mM using 1× Tris-EDTA (pH 7.4) buffer. Vortex well and briefly centrifuge.
2. Add ~20 μL of each DNA sample to an equal volume of formamide-loading buffer in new microcentrifuge tubes. Vortex well, briefly centrifuge, and heat at 95 °C for 3 min using a heating block or boiling water bath.
3. Remove denatured sample(s), briefly centrifuge, and place on ice.
4. Prepare a 20 cm × 20 cm, 1.5 mm-thick 10% denaturing gel by mixing 35 mL denaturing acrylamide solution, 35 mL dilution solution, 700 μL 10% APS, and 25 μL TEMED. Use a gel comb with wide wells (*see* Note 3) to minimize overloading effects. The gel should polymerize within 45 min.

5. Once the gel has set, carefully remove the comb and use a 3 ml syringe fitted with a 22-gauge needle to flush the wells with 1× running buffer.
6. Place the gel into the electrophoresis apparatus, and add 1× running buffer to the upper and lower reservoirs of the apparatus.
7. Complete the assembly of the electrophoresis unit and connect to a power supply. Equilibrate the gel by running for 10 min at 800 V (*see* Note 4).
8. Remove the cover of the electrophoresis unit and reflush the wells (as above) using running buffer from the top reservoir.
9. Load each DNA sample into individual wells.
10. Replace the electrophoresis cover and run for 30 min at 800 V. The amount of time needed to run all of the DNA samples on a single gel can be optimized by comparing the known sizes of your DNA sequences with the migration of xylene cyanol (~55 nucleotide DNA sequence) and bromophenol blue (~12 nucleotide DNA sequence) in a 10% polyacrylamide gel.
11. Upon completion, the electrophoresis unit is disconnected from the power source and disassembled.
12. The gel is removed from its casting, and the upper corner of the gel is cut to track its orientation. The gel is then placed onto a clear plastic wrap covered fluor-coated TLC plate or on top of an intensifying screen.
13. Using a clean razor blade and a low intensity shortwave UV lamp in the dark, DNA bands that appear as shadows in the gel can be excised (*see* Note 5).
14. Each slice of gel containing a unique DNA fragment is passed through a 3 mL syringe into a 2 mL round-bottom snap-cap eppendorf tube to efficiently crush the gel. All tubes are labeled with names and dates.
15. 1× TEN buffer is added to each eppendorf tube at a ratio of 2 mL per mL of gel.
16. The DNA extractions are incubated overnight at 4 °C on an orbital shaker.
17. The DNA extractions are briefly centrifuged, and the extraction buffer is removed and placed into a Handee™ spin cup column that rests in a collection microcentrifuge tube. Retain the crushed gel for a second extraction if necessary (*see* Note 6).
18. Spin columns are then centrifuged at room temperature for 2 min at maximum speed (≥10,000×*g*).

19. Each flow-through is transferred in ~500 μL aliquots to labeled 1.5 mL pelleting microcentrifuge tubes.
20. 2.5 volumes of absolute ethanol, prechilled at −20 °C, are added to each flow-through and the mixtures are vortexed well.
21. The DNA mixtures are then incubated on dry ice for 6 min or until the viscosity of the mixture resembles syrup.
22. The precipitated DNA samples are then centrifuged, with hinges pointing upward (*see* Note 7), for 25 min at maximum speed (≥10,000×*g* at 4 °C).
23. Each supernatant is then carefully removed, making sure not to touch the area of the tube where the DNA pellet is expected to reside (*see* Note 7).
24. The microcentrifuge tubes are placed on the bench top for 5 min to air-dry.
25. A small volume (≤ 100 μL) of 1× TE buffer is added to each tube to reconstitute the DNA pellet. Vortex well and briefly centrifuge samples.
26. Concentrations for each sample are determined by measuring the absorbance at 260 nm using a UV spectrophotometer and the calculated extinction coefficient.
27. Rerun 100 ng of each sample on a 10 cm × 10 cm, 0.75 mm-thick 15% denaturing polyacrylamide gel to validate a single band of DNA. Image using SYBR® Green I staining or radiolabelling as detailed in Subheading 4.3.5.

3.2. Assembly of G-quartet Structures

1. Purified G4 DNA should be in 1× TE buffer at a high micromolar concentration.
2. Determination of folding mixture: The folding mixture can contain any component at any concentration that the user desires. The concentration of mono and or dications and the concentration of the G4 DNA are defined by the experimentalist. (*see* Note 8). The following are examples of the formation of two different G-quadruplex structures in the same folding buffer (Fig. 4.1) as previously described (11).

3a. A final concentration of purified 1 μM Oxy3.5 DNA (5′-$(G_4T_4)_3G_4$-3′) is added to a folding buffer containing 20 mM TrisOAc pH 7.5, 50 mM potassium glutamate, 10 mM $MgCl_2$, and 1 mM DTT.

3b. A final concentration of purified 200 μM Oxy1.5 DNA (5′-$G_4T_4G_4$-3′) is added to a folding buffer containing 20 mM TrisOAc pH 7.5, 50 mM potassium glutamate, 10 mM $MgCl_2$, and 1 mM DTT.

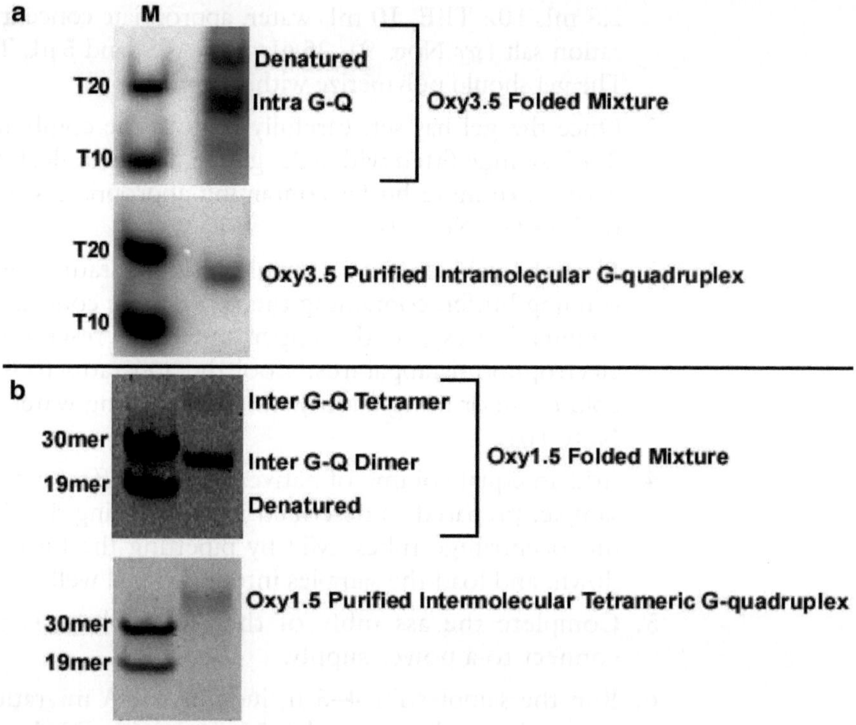

Fig. 4.1. Assembly and purification of G-quadruplex structures as described in (11). Oligonucleotides were visualized by phosphorimaging of 5′-[^{32}P] end labeled DNA (**a**) or SYBR® Green I stained unlabelled DNA (**b**). Lane designated "M" denotes DNA length markers: T_{20}, T_{10}, nonfolding 30-mer, and nonfolding 19-mer sequences. An intramolecular Oxy3.5 G-quadruplex (**a**) and an intermolecular tetrameric Oxy1.5 G-quadruplex (**b**) were purified from a mixture of structures.

4. Each mixture of G4 DNA in folding buffer is placed onto a heating block at 95 °C for 5 min.
5. Samples are cooled to 25 °C by removing the heating block from the heat source and placing it on the lab bench, allowing a cooling rate of ~2 °C/min.
6. Each sample is removed from the heating block, briefly centrifuged, and is placed on ice to preserve G-quadruplex structure.
7. Concentrations for each sample are estimated by observing the absorbance at 260 nm using a UV spectrophotometer.
8. Run 100 ng of each sample on a 10 cm × 10 cm, 0.75 mm-thick 20% native polyacrylamide gel to validate successful G-quartet formation. Image using SYBR® Green I staining or radiolabelling as detailed in Subheading 4.3.5.

3.3. Native Polyacrylamide Gel Electrophoresis

1. Prepare a 10 cm × 10 cm, 1.5 mm-thick (for quadruplex purification) or 0.75 mm-thick (for quadruplex verification and assays) 20% native gel by mixing 12.5 mL 40% acrylamide solution,

2.5 mL 10× TBE, 10 mL water, appropriate concentration of cation salt (*see* Note 9), 25 μL 10% APS, and 5 μL TEMED. The gel should polymerize within 45 min.

2. Once the gel has set, carefully remove the comb and use a 3 mL syringe fitted with a 22-gauge needle to flush the wells with 1× running buffer containing appropriate salt concentrations (*see* Note 9).
3. Place the gel into the electrophoresis apparatus, and add 1× running buffer, containing the appropriate concentration of counter ion (s), to the upper and lower reservoirs of the electrophoresis apparatus. Cool the apparatus to 4 °C in a cold room or using a refrigerated circulating water bath (*see* Note 10).
4. Add an equal volume of native loading buffer to each DNA sample, prepared as described in Subheading 4.3.2, in new microcentrifuge tubes. Mix by pipetting the buffer up and down, and load the samples into individual wells.
5. Complete the assembly of the electrophoresis unit and connect to a power supply.
6. Run the samples for 4–5 h, judging DNA migration using the markers xylene cyanol (~55 nucleotide DNA sequence) and bromophenol blue (~12 nucleotide DNA sequence), at 100 V at 4 °C.
7. Upon completion, the electrophoresis unit is disconnected from the power source and disassembled.
8. The gel is removed from its casting, and one of the upper corners of the gel is cut to track its orientation.
9. At this point the detection of G4 DNA and DNA markers can be achieved using a scanning phosphorimager (*see* Note 11). If nonradiolabelled DNA was used, the SYBR Green I staining protocol in Subheading 4.3.5.1 can be used.
10. The image of the gel and its DNA products should be scaled to actual size.
11. The gel is then placed on top of the image of the gel, with a piece of clear plastic wrap between.
12. Excise areas of the gel that overlay with shifted G4 DNA bands on the gel image.
13. Extraction of the G-quartet DNA is detailed in Subheading 4.3.4.

3.4. Extraction of G-quartet Structures

3.4.1. Crush and Soak

1. Each slice of gel containing a unique G-quartet DNA structure is passed through a 3 mL syringe into a 2 mL round-bottom snap-cap eppendorf tube to efficiently crush the gel. All tubes are labeled with names and dates.

2. Extraction buffer (1 mL) (*see* Note 12), prechilled to 4 °C, is added to each eppendorf tube.
3. The DNA extractions are incubated overnight at 4 °C on an orbital shaker.
4. The DNA extractions are briefly centrifuged, and the extraction buffer is removed and placed into a Handee™ spin cup column that rests in a collection microcentrifuge tube (*see* Note 13).
5. Spin columns are then centrifuged for 2 min at maximum speed (≥10,000×*g*) at 4 °C.
6. Each flow-through is transferred in ~500 μL aliquots to labeled 1.5 mL microcentrifuge tubes on ice.
7. Absolute ethanol (2.5 volumes), prechilled at 4 °C, is added to each flow-through and the mixtures are gently vortexed.
8. The DNA mixtures are then incubated on dry ice for 6 min or until the viscosity of the mixture resembles syrup.
9. The precipitated DNA samples are then centrifuged, with hinges pointing upward (*see* Note 7), for 25 min at maximum speed (≥10,000×*g*) at 4 °C.
10. Each supernatant is then removed carefully (*see* Note 14), making sure not to touch the area of the tube where the DNA pellet is expected to reside (*see* Note 7).
11. The microcentrifuge tubes are placed on the bench top for 5 min to air-dry.
12. A small volume (≤ 50 μL) of folding buffer is added to each tube to reconstitute the DNA pellet. Mix by pipetting the buffer up and down and briefly centrifuge samples.
13. Concentrations for each sample are determined by measuring the absorbance at 260 nm using a UV spectrophotometer.
14. Rerun 100 ng of each sample on a 10 cm × 10 cm, 0.75 mm-thick 20% native polyacrylamide gel to validate successful G-quartet isolation. Image using SYBR® Green I staining or radiolabelling as detailed in Subheading 4.3.5.

3.4.2. Electroelution

Electroelution can be a useful method for isolating G-quadruplex structures when the extraction and precipitation conditions used in the "crush and soak" method do not produce sufficient yields. One example of this is when using short lengths of G4 DNA to form tetrameric structures. Tetrameric structures formed by short oligonucleotides, such as 5′-$G_4T_4G_4$-3′, prove to be more difficult to precipitate than structures formed by longer G4 DNA, such as 5′-$(G_4T_4)_3G_4$-3′. Electroelution allows smaller G-quadruplexes to be directly eluted from a piece of polyacrylamide gel into a concentrating chamber containing a small volume (200 μL – 3.6 mL) of the folding buffer, provided the complex is larger than the membrane molecular weight cut-off. It is important to perform this procedure at 4 °C.

3.5. Detection of G-quartets Utilizing Electrophoresis

3.5.1. SYBR® Green I Staining

1. To a shallow plastic tray 50 mL of 1× TBE is added.
2. An aliquot of SYBR® Green I is thawed and 5 µL is added to the 1× TBE. The remaining SYBR® Green I aliquot is placed back in the freezer.
3. The gel is removed from its casting, and one of the upper corners of the gel is cut to track its orientation.
4. The gel is then placed into the tray containing the 1× SYBR® Green I solution, and is incubated on an orbital shaker at room temperature for 20 min while being protected from exposure to light.
5. The gel is then placed onto a Molecular Dynamics chemiluminescence/blue fluorescence scanner and is imaged.

3.5.2. Radiolabelling

1. Purified DNA (4 µL of 25 µM) (*see* Note 15), 4 µL of 10× kinase buffer, 10 µL of [γ-^{32}P] ATP, 20 µL of water, and 2 µL of T4 polynucleotide kinase are added to a new microcentrifuge tube.
2. The reaction mixture is incubated at 37 °C for 45 min.
3. Before the reaction is complete, a MicroSpin™ G-25 column (GE Healthcare) is prepared by resuspending the resin by vortexing gently. The cap to the column is then loosened one-fourth turn and the bottom is snapped off. The column is placed into a 1.5 mL microcentrifuge tube and is centrifuged for 1 min at 735×*g* at room temperature to remove resin storage buffer.
4. The column is then placed into a new 1.5 mL microcentrifuge tube and the reaction mixture is loaded drop-wise onto the column, taking care not to disturb the resin.
5. The reaction mixture is centrifuged for 2 min at 735×*g* at room temperature.
6. The radiolabelled oligonucleotide present in the flow-through is quantified using a scintillation counter.
7. After running PAGE analysis on radiolabelled oligonucleotides, gels can be imaged by scanning phosphorimaging.

4. Notes

1. Water for buffers is deionized water (18.2 Mohm) purified through a Barnstead NANOpure DIamond™ purifier.

2. Poly T DNA is best used as 5′-[^{32}P] labeled markers, as SYBR® Green I staining does not efficiently stain A and T rich sequences.
3. Wide wells are needed in the purification of oligonucleotides because of the large amount of DNA being loaded onto the gel. Overloading effects, such as streaking, can make purifying a single DNA difficult. Plastic tape can be applied to gel combs to create the desired well width if appropriate combs are not available.
4. Prewarming a gel before running samples allows for the gel to equilibrate, minimizing artifacts that may arise.
5. UV light damages DNA, so it is important to use the lamp sparingly when UV shadowing. Take care in excising full length DNA from lower migrating products.
6. A second extraction can improve an initially poor yield; however it need not go overnight. The crushed gel can be resuspended in 1× TEN buffer, placed on dry ice for 5 min and then heated at 95 °C for 5 min. Alternating these extreme temperatures helps to maximize the second extraction yield. The extraction then continues at step 7.
7. During ethanol precipitations, the hinges of the microcentrifuge tubes are positioned upward. This allows the user to predict that the oligonucleotide pellet will form on the same side as the hinge.
8. G4 DNA concentration will depend on the type of G-quadruplex that the user wishes to form. Higher concentrations of G4 DNA favor intermolecular structures, while lower concentrations favor intramolecular interactions if the sequence permits formation of intramolecular structures.
9. Salts at the same concentrations as those used during G-quartet formation are normally present in the buffer used during polymerization of the native gel and added to the running buffer to keep the experimental conditions constant throughout. The type and amount of salt to be used depend on the given application. Sodium or potassium, at 20–150 mM concentrations, is generally used. Mixtures of monovalent and/or divalent cations such as Mg^{2+} can also be added if desired; however, it should be noted that during electrophoresis use of $MgCl_2$ can be associated with excessive heat and emission of chlorine vapors.
10. By maintaining the native gel and running buffer at 4 °C the denaturing effects of gel heating, which could arise during electrophoresis, are minimized.
11. If imaging the gel for radiolabelled G-quartet DNA isolation, the gel can be placed between two pieces of clear plastic

wrap and imaged for a short time in a phosphorimaging cassette. The actual size image can then be placed under the gel and the regions containing G4 DNA can be excised for isolation as detailed in Subheading 4.3.4. If imaging the gel for a quantitative assay, initially dry the 20% native gel on a gel drier under vacuum and at high heat for ≥4 h (prematurely removing the gel before it completely dries can result in the gel cracking). Remove the gel and place into a phosphorimaging cassette and expose overnight.

12. The extraction buffer is usually modified 1× TE buffer, containing the same salt concentrations as those in the folding and electrophoresis protocol. Use of 10× TE buffer, stocks of sterile monovalent and divalent salts, and water, provides a 1× extraction buffer.
13. A secondary extraction for obtaining G-quadruplex DNA is usually impractical, as a second extraction would need to incubate overnight at 4 °C (increasing the amount of time for potential unfolding) and because less DNA is used in the folding process than in the initial oligonucleotide purification (yielding a more dilute secondary extraction). It is better to reassemble G-quadruplexes in a second attempt and pool the like structures after characterization.
14. If the G-quadruplex being isolated is radiolabelled, precipitation efficiency can be gauged using a Geiger counter to follow counts/min in the pellet and the supernatant, respectively.
15. When [^{32}P] 5′-end radiolabelling oligonucleotides for the purpose of forming intermolecular G-quadruplexes, the user must consider the destabilizing effects resulting from the positioning of multiple phosphate groups at guanosine termini on the same end of the quadruplex (12). For example, if the user wishes to form an intermolecular, four-stranded, parallel G-quadruplex using the sequence 5′-G_5-3′, gel staining would be preferred as the 5′-terminal phosphates would compromise quadruplex stability. Use of SYBR® Green I to detect this subset of G-quadruplexes bypasses this limitation of radiolabeled oligonucleotides.

Acknowledgments

The authors thank Laura Bonifacio for critical reading of the manuscript and Dr. Tracy Bryan for critical discussions. This work was funded by a grant from the National Science Foundation (MCB-0446019).

References

1. Neidle S, Read MA (2001) G-quadruplexes as therapeutic targets. Biopolymers 56:195–208
2. Gomez D, O'Donohue M-F, Wenner T, Douarre C, Macadre J, Koebel P, Giraud-Panis M-J, Kaplan H, Kolkes A, Shin-ya K, Riou J-F (2006) The G-quadruplex ligand telomestatin inhibits POT1 binding to telomeric sequences in vitro and induces GFP-POT1 dissociation from telomeres in human cells. Cancer Res 66:6908–6912
3. Huppert JL, Balasubramanian S (2007) G-quadruplexes in promoters throughout the human genome. Nucleic Acids Res 35: 406–413
4. Siddiqui-Jain A, Grand CL, Bearss DJ, Hurley LH (2002) Direct evidence for a G-quadruplex in a promoter region and its targeting with a small molecule to repress c-MYC transcription. Proc Natl Acad Sci U S A 99:11593–11598
5. Cogoi S, Xodo LE (2006) G-quadruplex formation within the promoter of the KRAS proto-oncogene and its effect on transcription. Nucleic Acids Res 34:2536–2549
6. Sun D, Guo K, Rusche JJ, Hurley LH (2005) Facilitation of a structural transition in the polypurine/polypyrimidine tract within the proximal promoter region of the human VEGF gene by the presence of potassium and G-quadruplex-interactive agents. Nucleic Acids Res 18: 6070–6080
7. Connor AC, Frederick KA, Morgan EJ, McGown LB (2006) Insulin capture by an insulin-linked polymorphic region G-quadruplex DNA oligonucleotide. J Am Chem Soc 128: 4986–4991
8. Tasset DM, Kubik MF, Steiner W (1997) Oligonucleotide inhibitors of human thrombin that bind distinct epitopes. J Mol Biol 272:688–698
9. He F, Tang Y, Wang S, Li Y, Zhu D (2005) Fluorescent amplifying recognition for DNA G-quadruplex folding with a cationic conjugated polymer: a platform for homogeneous potassium detection. J Am Chem Soc 127:12343–12346
10. Alberti P, Bourdoncle A, Sacca B, Lacroix L, Mergny J (2006) DNA nanomachines and nanostructures involving quadruplexes. Org Biomol Chem 4:3383–3391
11. Oganesian L, Moon IK, Bryan TM, Jarstfer MB (2006) Extension of G-quadruplex DNA by ciliate telomerase. EMBO J 25: 1148–1159
12. Uddin MK, Kato Y, Takagi Y, Mikuma T, Taira K (2004) Phosphorylation at 5′ end of guanosine stretches inhibits dimerization of G-quadruplexes and formation of a G-quadruplex interferes with the enzymatic activities of DNA enzymes. Nucleic Acids Res 32:4618–4629

Chapter 5

Biochemical Techniques for the Characterization of G-Quadruplex Structures: EMSA, DMS Footprinting, and DNA Polymerase Stop Assay

Daekyu Sun and Laurence H. Hurley

Abstract

The proximal promoter region of many human growth-related genes contains a polypurine/polypyrimidine tract that serves as multiple binding sites for Sp1 or other transcription factors. These tracts often contain a guanine-rich sequence consisting of four runs of three or more contiguous guanines separated by one or more bases, corresponding to a general motif known for the formation of an intramolecular G-quadruplex. Recent results provide strong evidence that specific G-quadruplex structures form naturally within these polypurine/polypyrimidine tracts in many human promoter regions, raising the possibility that the transcriptional control of these genes can be modulated by G-quadruplex-interactive agents. In this chapter, we describe three general biochemical methodologies, electrophoretic mobility shift assay (EMSA), dimethylsulfate (DMS) footprinting, and the DNA polymerase stop assay, which can be useful for initial characterization of G-quadruplex structures formed by G-rich sequences.

Key words: G-quadruplex, Transcriptional regulation, DMS footprinting, EMSA, DNA polymerase stop assay

1. Introduction

G-rich sequences have been reported to form noncanonical four-stranded secondary structures called G-quadruplexes, which consist of two or more G-tetrads in the presence of monovalent cations such as Na^+ and K^+, as shown in Fig. 5.1 (1). The G-rich sequences capable of forming G-quadruplexes were initially found in telomeric sequences, the insulin gene, the control region of the retinoblastoma susceptibility gene, fragile X syndrome triplet repeats, and HIV-1 RNA (2–6) and were later also found in the proximal promoter region of many TATA-less mammalian genes, including c-Myc, Hmga2, EGF-R, VEGF, BCL-2, PDGF-A,

P. Baumann (ed.), *G-Quadruplex DNA: Methods and Protocols,* Methods in Molecular Biology, vol. 608
DOI 10.1007/978-1-59745-363-9_5, © Humana Press, a part of Springer Science+Business Media, LLC 2010

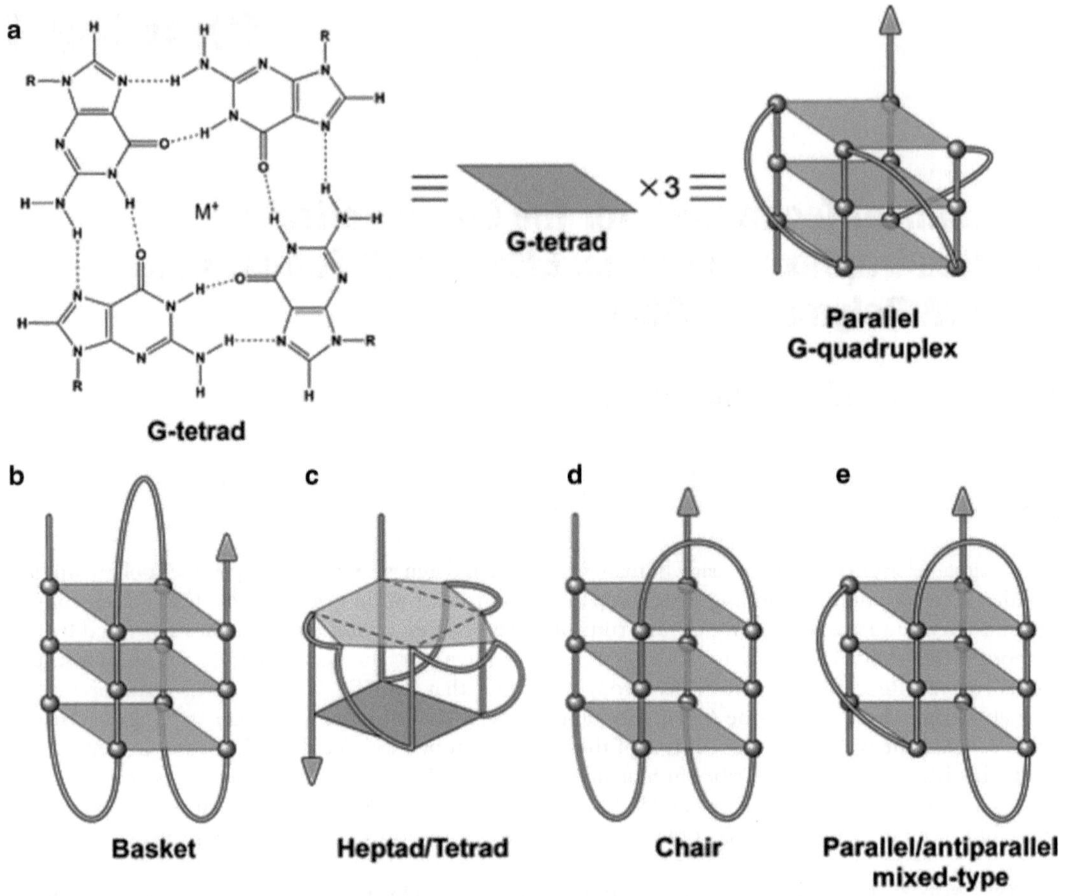

Fig. 5.1. G-tetrad and G-quadruplexes. (**a**) Four guanine residues form a planar structure G-tetrad through Hoogsteen hydrogen bonding to form an intramolecular parallel G-quadruplex. Models are shown for an intramolecular antiparallel basket (**b**), an intramolecular parallel heptad-tetrad (**c**), an intramolecular antiparallel chair (**d**), and a mixed-type intramolecular quadruplex (**e**). Each parallelogram in (**b–f**) represents a G-tetrad.

c-Myb, malic enzyme, I-R, AR, c-Src, c-Ki-Ras, TGFβ, and PDGF A-chain (reviewed in (7)). In particular, the G-rich sequences from the promoter region of these genes have been proposed to be very dynamic in their conformation, easily adopting nonB-DNA conformations, such as melted DNA, hairpin structures, slipped helices, or others, under physiological conditions, provided that there is conformational or torsional stress (8, 9). Direct evidence for the existence of G-quadruplexes in vivo is beginning to emerge, and the ability of these important sequences to form very stable G-quadruplex structures in vitro suggests that G-quadruplex DNA may play an important role in several biological events including telomere maintenance, DNA replication, and transcription. For instance, a recent study provided compelling evidence that a specific G-quadruplex structure formed in the c-Myc promoter functions as a transcriptional repressor element (10), establishing the principle that c-Myc transcription could be

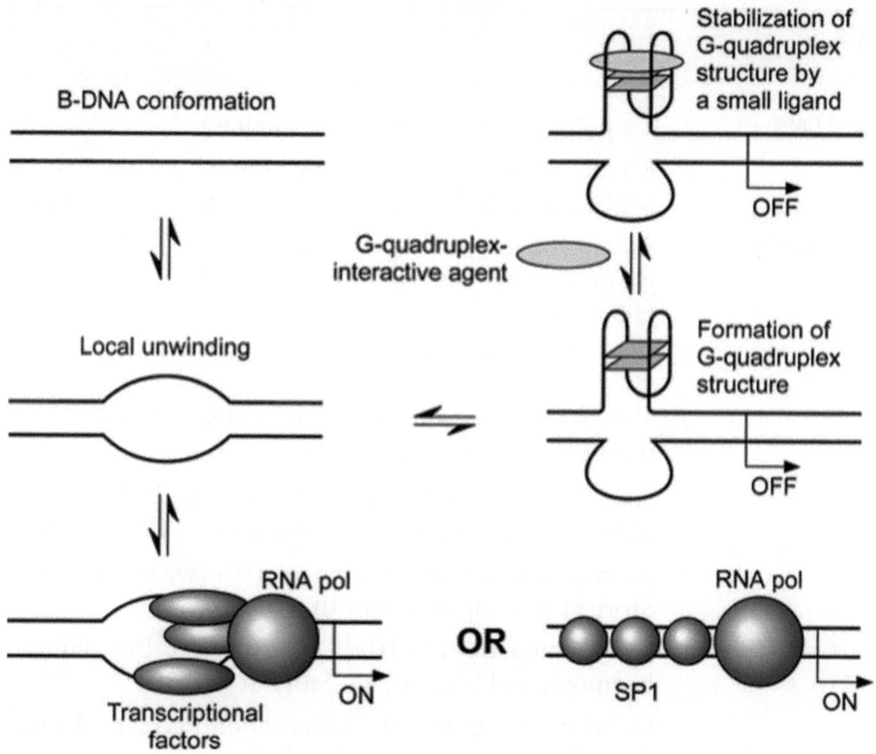

Fig. 5.2. Model for the transition of a duplex strand into atypical secondary structures and repression of gene transcription by the stabilization of a G-quadruplex structure with a small ligand.

controlled by ligand-mediated G-quadruplex stabilization (Fig. 5.2). Recent studies of both crystal and solution structures of various G-quadruplexes revealed that their structures are very stable under physiological conditions and very diverse in their folding patterns (11–13). Therefore, there are high expectations that specific interactions can be achieved between different types of G-quadruplexes and small molecular weight ligands.

With improved understanding of the structures and potential biological functions of G-quadruplexes, there is increased demand for simple but reproducible and reliable biochemical tools best suited for studying G-quadruplexes. Our previous studies on the structures and functions of G-quadruplex structures suggested that the combined use of EMSA, DMS footprinting, and the DNA polymerase stop assay is very useful for the initial characterization of G-quadruplex structures from any origin (14–18). Therefore, in this chapter, we will discuss the application of these biochemical techniques in studying the formation of G-quadruplex structures from various G-rich sequences and their potential application in studying the effects of novel classes of small molecular weight compounds, on the basis of their ability to bind to and stabilize G-quadruplexes.

2. Materials

2.1. EMSA and DMS footprinting

2.1.1. Labeling 5′-Termini of Nucleic Acids with [^{32}P]

1. T4 polynucleotide kinase (Fermentas)
2. Kinase buffer (10×): 500 mM Tris-HCl (pH 7.6), 100 mM $MgCl_2$, 50 mM DTT, 1 mM spermidine, and 1 mM EDTA.
3. Adenosine 5′-gamma ^{32}P triphosphate (γ-^{32}P ATP), triethylammonium salt (6,000 Ci/mmole, 10 mCi/mL, GE, Healthcare).
4. Micro Bio-Spin™ 30 Columns (Bio-Rad)

2.1.2. Native Polyacrylamide Gel Electrophoresis (PAGE)

1. TBE electrophoresis buffer (10×): 0.89 M Tris-HCl (pH 8.0), 0.89 M boric acid, 20 mM EDTA. Store at room temperature.
2. Sixteen percent acrylamide/bisacrylamide (29:1 with 3.3% C) and N,N,N′,N′-tetramethylethylenediamine (TEMED) (Bio-Rad).
3. Ammonium persulfate: prepare 10% (w/v) solution in water. Store at 4 °C up to 1 month.
4. Gel loading buffer (10×): 50% glycerol by volume, 0.005% bromophenol blue (w/v). Store at –20 °C.
5. Gel elution buffer: 0.4 M ammonium acetate, 1 mM $MgCl_2$, 0.2% SDS. Store at room temperature.
6. 100% and 75% ethanol

2.1.3. Chemical DNA Sequencing and DMS Footprinting

1. Formic acid (Sigma-Aldrich); hydrazine (Sigma-Aldrich).
2. DNA sequencing stop solution: 0.5 M sodium acetate (pH 6.0) and 50 µg/mL calf thymus DNA. Store at 4 °C.
3. 1 M piperidine solution in water (freshly prepared).
4. 10% dimethylsulfate solution in 50% ethanol.

2.1.4. Denaturing PAGE

1. TBE electrophoresis buffer (10×): 0.89 M Tris-HCl (pH 8.0), 0.89 M boric acid, 20 mM EDTA. Store at room temperature.
2. Sixteen percent acrylamide/bisacrylamide (29:1 with 3.3% C) with 8 M urea and N,N,N′,N′-TEMED, Bio-Rad, Hercules, CA.
3. Ammonium persulfate: prepare 10% solution in water and store at 4 °C up to one month.
4. Alkaline gel loading dye (1×): 80% formamide by volume, 10 mM NaOH, 0.005% bromophenol blue (w/v). Store at –20 °C.

2.2. DNA Polymerase Stop Assay

2.2.1. Labeling 5′-Termini of Primer with [^{32}P] (see Subheading 2.1.1)

1. T4 polynucleotide kinase (Fermentas).
2. Kinase buffer (10×): 500 mM Tris-HCl (pH 7.6), 100 mM $MgCl_2$, 50 mM DTT, 1 mM spermidine, 1 mM EDTA.
3. Adenosine 5′-gamma ^{32}P triphosphate, triethylammonium salt (6,000 Ci/mmole, 10 mCi/mL, GE, Healthcare).
4. Micro Bio-Spin™ 30 Columns (Bio-Rad).

2.2.2. Native PAGE (see items 1–4 of Subheading 2.1.2)

1. TBE electrophoresis buffer (10×): 0.89 M Tris-HCl (pH 8.0), 0.89 M boric acid, 20 mM EDTA. Store at room temperature.
2. Eight percent acrylamide/bisacrylamide (29:1 with 3.3% C) and N,N,N′,N′-TEMED (Bio-Rad).
3. Ammonium persulfate: prepare 10% solution in water and store at 4 °C up to one month.
4. Gel loading buffer (10×): 50% glycerol by volume, 0.005% bromophenol blue (w/v). Store at –20 °C.

2.2.3. DNA Polymerase Reaction

1. Taq DNA Polymerase (Fermentas)
2. DNA polymerase buffer (10×): 500 mM Tris-HCl (pH 7.6), 10 mM $MgCl_2$, 50 mM DTT.
3. dNTP solution: 2 mM of dATP, dGTP, dTTP, and dCTP.

2.2.4. Dideoxy Sequencing Reaction

1. A termination mix: 2 μM dATP, 250 μM ddATP, 800 μM dGTP, 800 μM dTTP, and 800 μM dCTP.
2. G termination mix: 2 μM dGTP, 250 μM ddGTP, 800 μM dATP, 800 μM dTTP, and 800 μM dCTP.
3. T termination mix: 2 μM dTTP, 250 μM ddTTP, 800 μM dATP, 800 μM dGTP, and 800 μM dCTP.
4. C termination mix: 2 μM dCTP, 250 μM ddCTP, 800 μM dATP, 800 μM dGTP, and 800 μM dTTP.

2.2.5. Denaturing PAGE (see Subheading 2.1.4)

1. TBE electrophoresis buffer (10×): 0.89 M Tris-HCl (pH 8.0), 0.89 M boric acid, 20 mM EDTA. Store at room temperature.
2. Sixteen percent acrylamide/bisacrylamide (29:1 with 3.3% C) with 8 M urea and N,N,N′,N′-TEMED (Bio-Rad).
3. Ammonium persulfate: prepare 10% solution in water, store at 4 °C up to one month.
4. Alkaline gel loading dye (1×): 80% formamide by volume, 10 mM NaOH, 0.005% bromophenolblue (w/v). Store at –20 °C.

3. Methods

The G-rich strand of the promoter of many mammalian oncogenes is characterized by the presence of more than four runs of at least three adjacent guanines (7). To determine which guanine repeats are required for folding into intramolecular G-quadruplex structures, we prepared a series of oligonucleotide DNAs spanning various portions of the G-rich sequence. Each 5′-end-radiolabeled oligonucleotide was subjected to annealing by heating and slowly cooled to room temperature in the presence of KCl, allowing the tandem repeats of guanines to fold into G-quadruplexes. The resulting structures were treated with DMS to methylate the guanine residues in the oligonucleotides. The methylated oligonucleotides were then subjected to native PAGE to separate intramolecular forms of G-quadruplexes from intermolecular forms or unfolded structures on the basis of differences in the electrophoretic mobility (19). A native PAGE is routinely used for separation of nucleic acids on the basis of the difference in their shape and size, resulting in a difference in the electrophoretic mobility. In general, the mobility of G-quadruplex DNA is determined by the type of G-quadruplex structures as well as the number of DNA strands involved in folding into G-quadruplexes. Often, intramolecular forms of G-quadruplexes showed faster mobility on native PAGE, as is evident in G-quadruplex structures formed from the G-rich sequence of the BCL-2 and PDGF-A genes (16, 17). In some cases, intramolecular G-quadruplexes are indistinguishable from unfolded forms in their electrophoretic mobility, although the slowly migrating bands are believed to be an intermolecular G-quadruplex (14, 18). To determine the guanine bases involved in the formation of G-quadruplex structures, each DNA band was excised from the gel and treated with piperidine to produce specific DNA strand breakage at methylated guanine residues, and the cleavage products were resolved on a denaturing PAGE gel. The guanine bases involved in the formation of G-quadruplex structures can be deduced by DMS footprinting, as the N7 position of the guanines involved in Hoogsteen bonding to form the G-tetrad are inaccessible to methylation (19).

We also used a DNA polymerase stop assay to confirm that the G-rich sequence consisting of multiple G-tracts could form intramolecular G-quadruplex structures (20, 21). The DNA polymerase stop assay provides a simple and rapid way to identify DNA secondary structures in vitro, on the basis of the principle that DNA polymerase is incapable of traversing these structures. DNA polymerase, traversing toward the 5′-end of the template and unable to efficiently resolve quadruplex DNA, pauses or stops 3′ to the first guanine involved in a stable G-quadruplex. For the DNA polymerase stop assay, the template DNAs containing various G-quadruplex-forming regions are annealed with radiolabeled primers, and the primer-annealed template DNAs are used in a primer extension assay by Taq DNA polymerase, as described below. This assay has also proven useful in identifying potential G-quadruplex-interactive compounds.

An overall strategy to characterize G-quadruplexes formed by G-rich sequences using EMSA, DMS footprinting, and the DNA polymerase stop assay is shown schematically in Fig. 5.3.

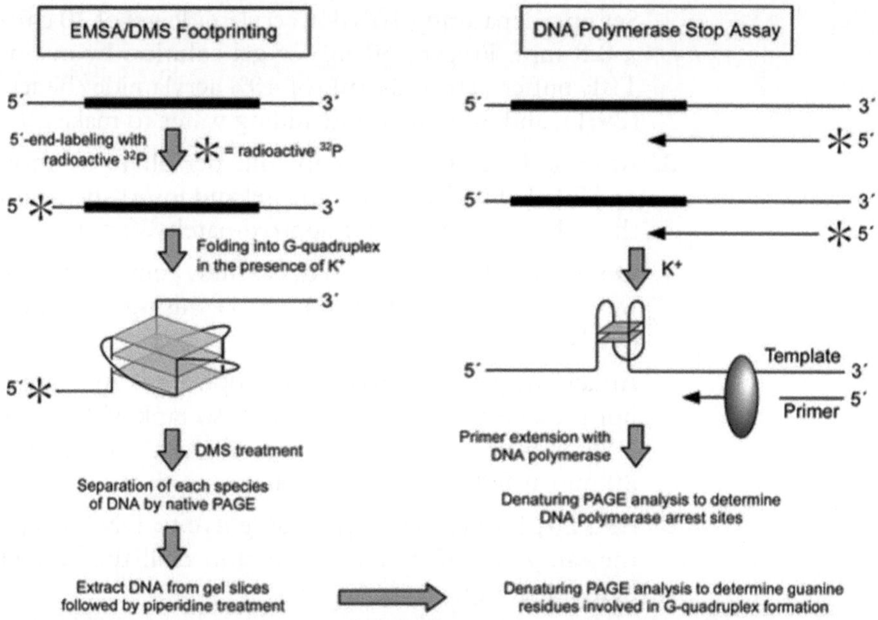

Fig. 5.3. Schematic diagram showing an overall strategy to characterize G-quadruplexes formed by G-rich sequences using EMSA, DMS footprinting, and the DNA polymerase stop assay.

3.1. EMSA and DMS Footprinting

3.1.1. Labeling 5′-Termini of Oligonucleotides with [^{32}P]

1. Preparing a reaction mixture (25 µL), containing oligonucleotide (4 µM), 3 µL γ-^{32}P ATP (6,000 Ci/mmole, 10 mCi/mL), T4 polynucleotide kinase (10 U), 2.5 µL 10× kinase buffer, and water.
2. Incubate the reaction mixture in a water bath at 37 °C for 1 h for labeling 5′-termini of oligonucleotides with [^{32}P].
3. After completion of the reaction, use Micro Bio-Spin™ 30 Columns (Bio-Rad) to remove unincorporated radioactive γ-^{32}P ATP from labeled DNA. The instructions for use of Bio-Spin™ 30 Columns are on the basis of recommendations from the manufacturer. In brief, the reaction mixture (25 µL) is loaded at the top of the column after centrifuging the column at 1,000 × *g* for 4 min in a swinging bucket rotor and removing the packing buffer. The column is then centrifuged for 4 min at 1,000× *g* to collect the purified 5′-end-labeled oligonucleotide in water (*see* Note 1).

3.1.2. Purification of a Desired Full-length Oligonucleotide Using a Denaturing 16% Polyacrylamide Gel

1. The 5′-labeled oligonucleotides should be purified prior to use in footprinting experiments as oligonucleotides made by automated DNA synthesizers in the laboratory or obtained commercially are often contaminated with products of incomplete synthesis or other unknown impurities. Routinely, a denaturing polyacrylamide-urea gel electrophoresis is used to separate a desired full-length oligonucleotide from other contaminants.

2. Set up a denaturing 16% polyacrylamide gel of 20 cm × 16 cm × 0.8 mm. Prepare 60 mL of gel solution by mixing 6 mL TBE buffer (10×), 24 mL of 40% acrylamide/bisacrylamide (29:1), and 30 g urea, and adding water to make 60 mL.
3. After adding 100 μL ammonium persulfate solution (10%) and 20 μL TEMED, pour the gel and insert the comb. Allow the gel to polymerize for approximately 30 min.
4. Once the gel is polymerized, carefully remove the comb and wash the well with TBE buffer (1×) using a pasture pipette (*see* Note 2).
5. Attach the gel plates to the electrophoresis apparatus, and fill both reservoirs of the electrophoresis tank with 1× TBE. Use a DC power supply to prerun and warm the gel for at least 30 min at 500 V (constant voltage).
6. Add 20 μL of alkaline gel loading dye to DNA samples, heat the sample at 95°C for 3 min, and chill the sample on ice before loading.
7. Run the gel at about 500 V until the desired resolution has been obtained as determined empirically (*see* Note 3).
8. After the completion of electrophoresis, turn off the power supply, detach the gel plates from electrophoresis apparatus, and carefully separate both plates while keeping the gel attached to one plate.
9. Wrap the gel and plate with plastic wrap. Autoradiography is often used to visualize the location of DNA bands within the gel.
10. If the amount of DNA is 1 μg or greater, visualize the DNA bands by UV shadowing after wrapping the gel and plate with plastic wrap, inverting, and placing the gel onto a TLC plate containing fluorophores. The DNA fragment of interest can be located with a portable shortwave UV illuminator (*see* Note 4).
11. Cut out the desired DNA band with a razor blade, place the gel fragment inside of a 1.5-mL eppendorf tube, and crush the gel fragment into small pieces by gently touching it with a metal spatula with a narrow blade. Recover the DNA from the gel by adding 400 μL of water and incubating the tube with rotation or in a shaking air incubator at room temperature for 1 h.

3.1.3. Annealing of the ^{32}P-Labeled Oligomer DNAs into G-quadruplex Structures, DMS Methylation, and EMSA

1. Anneal the ^{32}P-labeled oligomer DNAs by heating at 90 °C for 5 min and then cooling slowly to room temperature in 20 μL of 20 mM Tris-HCl (pH 7.4) buffer with or without 100 mM KCl.
2. While annealing reaction is in progress, set up a native 16% polyacrylamide gel of 20 cm × 16 cm × 0.8 mm. Prepare gel solution by mixing 6 mL TBE buffer (10×), 24 mL of 40%

acrylamide/bisacrylamide (29:1), and adding water to 60 mL. Add 100 μL ammonium persulfate solution and 20 μL TEMED, pour the gel, and insert the comb.

3. Once the gel is polymerized after approximately 30 min, carefully remove the comb, and wash the well with TBE buffer (1×) using a pasteur pipette. Attach the gel plates to the electrophoresis apparatus and fill both reservoirs of the electrophoresis tank with 1× TBE. Use a DC power supply to prerun and warm the gel for at least 30 min at 150 V (constant voltage).
4. After the annealing reaction is completed, treat each annealed DNA with DMS (0.5%) for 2 min to methylate the DNA.
5. Stop the DMS modification reaction by adding a tenth volume of a gel loading buffer containing 1 μg calf thymus DNA and immediately load the reactions on a 16% native polyacrylamide gel.
6. Run the gel until the desired resolution is obtained, detach the gel plates from electrophoresis apparatus, and separate both plates while the gel is still attached to one plate.
7. Visualize the location of DNA bands within the gel via autoradiography. Figure 5.4a is an example of the results from EMSA analysis of G-quadruplex structures formed by the G-rich sequence (HIFX) from the polypurine/polypyrimidine tract of the promoter region of the HIF-1α gene (14).
8. Cut out the desired DNA band from the gel with a razor blade and insert in a 1.5-mL eppendorf tube containing 250 μL of a gel elution buffer. DNA can be eluted from the gel without crushing it by incubating the tube overnight at 37 °C in a water bath.
9. Recover the supernatant carefully without touching the gel fragment and transfer to a new tube containing 750 μL of 100% ethanol.
10. Mix the tube well using the vortex and store the samples at –20 °C overnight (or 3 h at –80 °C).
11. Centrifuge the tubes for 30 min at 12,000×*g* at 4 °C to collect the DNA pellet, and wash the recovered DNA pellet once with 250 μL of ice-cold 75% ethanol.
12. Air-dry DNA pellets, resuspend in 100 μL 1 M piperidine solution, and heat at 95 °C for 30 min.
13. Dry the samples in a speed vac, resuspend dried DNA pellets in 100 μL water, and dry the samples again in a speed vac.
14. Resuspend dried DNA pellets in 20 μL alkaline sequencing dye, and resolve cleaved DNA products on a 16% denaturing polyacrylamide gel.

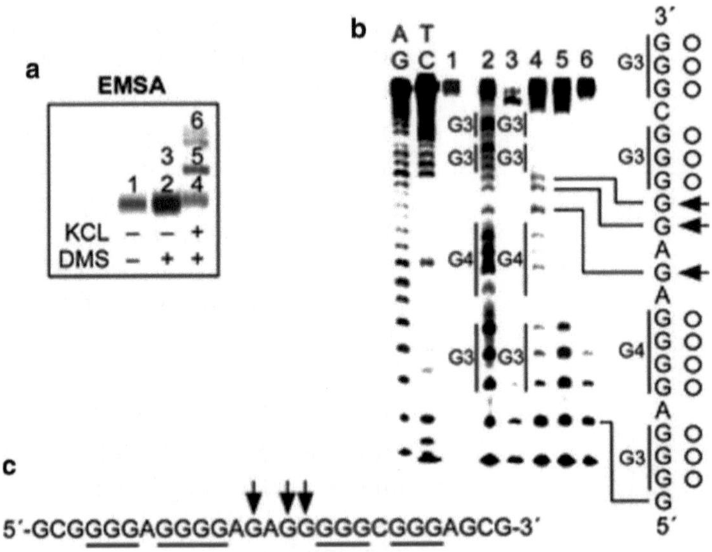

Fig. 5.4. EMSA and DMS footprinting of oligonucleotide HIFX derived from the G-rich sequence of the HIF-1α promoter region. (**a**) EMSA of HIFX preincubated under the conditions specified in the figure. 5′-End-radiolabeled oligonucleotide HIFX was subjected to annealing by heating and was slowly cooled to room temperature in the presence or absence of KCl, allowing the guanine repeats to fold into G-quadruplexes. The resulting structures were treated with 0.5% dimethylsulfate for 2 min to methylate the guanine residues in the oligonucleotides. The methylated oligonucleotides were then subjected to a 16% native PAGE to separate intramolecular forms of G-quadruplexes from intermolecular forms or unfolded structures by differences in the electrophoretic mobility. The numbers indicate the bands that were excised from the gel and treated with piperidine to induce strand breaks at methylated guanine residues. (**b**) Pattern of N7 guanine methylation produced by each band (lanes 1–6) isolated from EMSA described in Fig. 5.4(**a**). AG and TC represent chemical cleavage reaction specific to purine and pyrimidine bases, respectively. The vertical bars to the left of lane 4 correspond to DMS-protected guanine repeats. The protected guanines from DMS are indicated by open circles, and arrows indicate the guanine residues hypermethylated by DMS. (**c**) Summary of DMS footprinting of HIFX in the presence 100 mM KCl. The protected guanines from DMS are *underlined*, and *arrows* indicate the guanine residues hypermethylated by DMS.

3.1.4. Chemical DNA Sequencing Reactions

Sequence ladders are always required for footprinting experiments, allowing clear assignments of cleaved residues. These ladders can be produced by chemical sequencing of the same DNA fragment used for footprinting experiments. The chemical cleavage of DNA by formic acid or hydrazine at a purine or pyrimidine residue, respectively, is typically used in DMS footprinting experiments to generate a cleavage ladder.

1. Add an aliquot of DNA (approximately 100,000 cpm) in water into 1.5 mL eppendorf tubes labeled Pu and Py, in which the final volume is adjusted to 20 μL with water.
2. Add 0.4 μg calf thymus DNA to each tube as a carrier to prevent excessive modification of the bases by chemical reagents.

3. Add 20 µL of formic acid and hydrazine to purine- and pyrimidine-specific reactions, respectively. Mix the reactions well, and incubate at room temperature for 20 min (*see* Note 5).
4. Terminate the reactions by adding 60 µL DNA sequencing stop solution. Mix the stopped reaction well with 400 µL of 100% ethanol, and store the samples at –20 °C overnight.
5. Centrifuge the tubes for 30 min at 12,000×*g* at 4 °C to collect the DNA pellet, and wash the recovered DNA pellet once with 75% ethanol.
6. Air dry DNA, resuspend the pellets in 100 µL 1 M piperidine solution, and heat the solutions at 95°C for 30 min. After piperidine treatment, dry the samples in speed vac, resuspend the dried pellets in 1,000 µL water, and dry again in speed vac.
7. Resuspend cleavage DNA pellets in 20 µL alkaline sequencing dye.

3.1.5. Separation of Cleavage Products on Denaturing PAGE

1. Set up a denaturing 16% polyacrylamide gel of 30 cm× 30 cm × 0.4 mm and prepare 60 mL of gel solution by mixing 6 mL TBE buffer (10×), 24 mL of 40% acrylamide/bisacrylamide (29:1), 30 g urea and adding water to make 60 mL. After adding 100 µL ammonium persulfate solution and 20 µL TEMED, pour the gel and insert the comb.
2. Once the gel is polymerized, carefully remove the comb, and wash the well with TBE buffer (1×) using a pasteur pipette.
3. Attach the gel plates to the electrophoresis apparatus, and fill both reservoirs of the electrophoresis tank with 1× TBE. Prerun and warm the gel for at least 30 min at 1,600 V (constant voltage) using a DC power supply.
4. Heat the samples and sequencing ladders at 95 °C for 3 min, and chill the sample on ice before loading. Run the gel at about 1,600 V.
5. After the desired resolution is obtained, detach the gel plates from the electrophoresis apparatus, and carefully separate both plates, leaving the gel attached to one plate.
6. Place a piece of a thin chromatography paper (DE81) on top of the gel, and slowly pull back on the paper to transfer gels to the paper.
7. Place a piece of Whatman paper (3MM) underneath, and cover the wet gel with plastic wrap on top.
8. Put the gel sandwich in a dryer between a plastic fiber mat and clear plastic sheet, and dry the gel at 80 °C for at least 1 h with a vacuum.
9. Place the dried gel in an x-ray film cassette. Obtain an autoradiogram by exposing x-ray film to the dried gel.

Alternatively, the image can be obtained by exposing the dried gel to a phosphor-imager screen for an appropriate time and scanning the screen. Figure 5.4b is an example of an autoradiogram of a 16% polyacrylamide sequencing gel, showing the results of DMS footprinting experiments carried out with the G-rich sequence (HIFX) from the polypurine/polypyrimidine tract of the promoter region of the HIF-1α gene (14).

3.2. DNA Polymerase Stop Assay

3.2.1. Labeling 5′-Termini of Primer with [^{32}P]

1. Label 5′-termini of primer with [^{32}P] by preparing a reaction mixture (25 μL) containing water, kinase buffer (1×), primer (4 μM), 3 μL γ-^{32}P ATP (6,000 Ci/mmole, 10 mCi/mL), and T4 polynucleotide kinase (10 U) in a single tube and incubating the reaction mixture at 37 °C for 1 h in a water bath.
2. Use a Micro Bio-Spin™ 30 Column (Bio-Rad) to remove unincorporated radioactive γ-^{32}P ATP from labeled DNA, as described in Subheading 5.3.1.1.

3.2.2. Annealing of the ^{32}P-Labeled Primer DNAs into the Template DNA

1. Mix equimolar amounts of the ^{32}P-labeled primer DNA and the template DNA containing G-quadruplex-forming regions together in a single tube in 25 μL of an annealing buffer.
2. Anneal the ^{32}P-labeled primer DNA to the template DNA by heating at 90 °C for 5 min and then cooling slowly to room temperature.
3. Set up a native 8% polyacrylamide gel of 20 cm × 16 cm × 0.8 mm using 60 mL of gel solution (6 mL TBE buffer (10×), 12 mL of 40% acrylamide/bisacrylamide (29:1), and 42 mL water) as described in Subheading 5.3.1.3 to separate the primer-annealed template DNA from excess labeled primer or remaining template DNA.
4. Prerun and warm the gel for at least 30 min at 150 V (constant voltage).
5. Add a tenth of a gel loading buffer to the annealing reaction mixture, mix well, and load the samples onto a native 8% polyacrylamide gel.
6. After running the gel to the desired resolution, detach the gel plates from the electrophoresis apparatus, and carefully separate both plates, leaving the gel attached to one plate.
7. Visualize the location of DNA bands within the gel by autoradiography. Figure 5.5b is an example of an autoradiogram obtained after exposure of x-ray film to the gel.
8. Cut out the desired DNA band with a razor blade, and crush the gel fragment using a spatula with a thin blade inside of a 1.5 mL eppendorf tube.
9. Elute DNA from the gel by incubating the gel fragments overnight in 400 μL annealing buffer at room temperature.

3.2.3. DNA Polymerase Reaction

1. Prepare reaction mixtures (20 μL) containing water, DNA polymerase buffer (1×), DNA template plus primer (5–10 nM), 200 μM dNTP, and Taq DNA polymerase (1 U), and incubate at 37 °C for 30 min in a water bath.
2. Stop the reactions by adding 20 μL of alkaline dye, and dry the samples down to 20 μL in a speed vac.
3. Dideoxy sequencing reactions with the same DNA template are used for the DNA polymerase stop assay to provide a sequencing ladder for clear assignment of DNA polymerase arrest sites in the DNA polymerase stop assay.
4. Introduce an aliquot of 10 μL of A, C, G, and T termination mixes into appropriately labeled tubes, and add 10 μL of the remaining reaction mixture, consisting of Taq polymerase (1 U) and DNA template (10 nM) in 2× polymerase reaction buffer, to each termination tube.
5. Mix tubes well, and place in a 37 °C water bath for 30 min.
6. Terminate the reaction by adding 20 μL of alkaline gel-loading dye to each tube, and heat to 95 °C for 5 min prior to loading onto a denaturing PAGE gel.
7. Resolve the reaction products and sequencing ladders on a 16% denaturing polyacrylamide gel of 30 cm × 30 cm × 0.4 mm, as described in Subheading 5.3.1.5. An example result is shown in Fig. 5.5c.

4. Notes

1. Columns containing radioactive material should be properly disposed off.
2. Be sure to wear safety glasses while pouring the gel as unpolymerized acrylamide is known to be neurotoxic.
3. Excessive heating should be avoided during electrophoresis to prevent the breakage of the glass plates.
4. Avoid unnecessarily long UV exposure with a shortwave UV light, which will damage the nucleic acids.
5. Longer incubation is required for DNA fragments shorter than 20 base pairs.

Acknowledgments

This research was supported by grants from the National Institutes of Health (CA109069 and CA94166). We are grateful to David Bishop for preparing, proofreading, and editing the final version of the manuscript and figures.

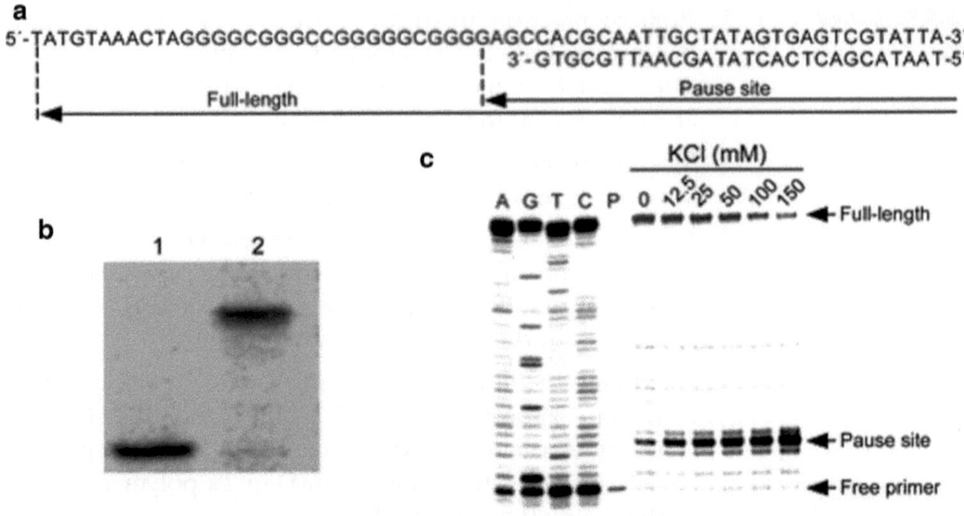

Fig. 5.5. DNA polymerase stop assay to determine the ability of the VEGF promoter to form G-quadruplex structures in the presence of KCl. (**a**) Sequence of the primer-annealed template DNA. The template DNA was designed to contain the G-quadruplex-forming region from the G-rich sequence of the VEGF promoter region. (**b**) Autoradiogram showing the separation of the primer-annealed template DNAs from excess labeled primer or remaining template DNA on an 8% native PAGE. Lanes 1 and 2 represent labeled primer and primer-annealed template DNA, respectively. (**c**) DNA polymerase stop assay showing the effect of KCl on the formation of G-quadruplex structures in the presence of KCl. DNA polymerase reactions were performed with labeled primer-annealed template DNA at increasing concentrations of K^+ (0–150 mM). *Arrows* indicate the positions of the full-length product of DNA synthesis, the G-quadruplex pause sites, and the free primer. Lanes A, G, T, and C represent dideoxysequencing reactions with the same template as a size marker for the precise arrest sites, and P represents primer without enzyme.

References

1. Jin RZ, Breslauer KJ, Jones RA, Gaffney BL (1990) Tetraplex formation of a guanine-containing nonameric DNA fragment. Science 250:543–546
2. Wang Y, Patel DJ (1994) Solution structure of the Tetrahymena telomeric repeat $d(T_2G_4)_4$ G-tetraplex. Structure 2:1141–1156
3. Hammond-Kosack MC, Kilpatrick MW, Docherty K (1993) The human insulin gene-linked polymorphic region adopts a G-quartet structure in chromatin assembled in vitro. J Mol Endocrinol 10:121–126
4. Murchie AI, Lilley DM (1992) Retinoblastoma susceptibility genes contain 5′ sequences with a high propensity to form guanine-tetrad structures. Nucleic Acids Res 20:49–53
5. Fry M, Loeb LA (1994) The fragile X syndrome $d(CGG)_n$ nucleotide repeats form a stable tetrahelical structure. Proc Natl Acad Sci U S A 91:4950–4954
6. Majumdar A, Gosser Y, Patel DJ (2001) ^{1}H–^{1}H correlations across N–H···N hydrogen bonds in nucleic acids. J Biomol NMR 21:289–306
7. Huppert JL, Balasubramanian S (2007) G-quadruplexes in promoters throughout the human genome. Nucleic Acids Res 35: 406–413
8. Michelotti GA, Michelotti EF, Pullner A, Duncan RC, Eick D, Levens D (1996) Multiple single-stranded cis elements are associated with activated chromatin of the human c-myc gene in vivo. Mol Cell Biol 16:2656–2669
9. Rustighi A, Tessari MA, Vascotto F, Sgarra R, Giancotti V, Manfioletti G (2002) Apolypyrimidine/polypurine tract within the *Hmga2* minimal promoter: a common feature of many growth-related genes. Biochemistry 41:1229–1240
10. Siddiqui-Jain A, Grand CL, Bearss DJ, Hurley LH (2002) Direct evidence for a G-quadruplex in a promoter region and its targeting with a small molecule to repress c-MYC transcription. Proc Natl Acad Sci U S A 99:11593–11598
11. Parkinson GN, Lee MP, Neidle S (2002) Crystal structure of parallel quadruplexes from human telomeric DNA. Nature 417:876–880

12. Phan AT, Modi YS, Patel DJ (2004) Propeller-type parallel-stranded G-quadruplexes in the human c-myc promoter. J Am Chem Soc 126:8710–8716
13. Dai J, Chen D, Jones RA, Hurley LH, Yang D (2006) NMR solution structure of the major G-quadruplex structure formed in the human BCL2 promoter region. Nucleic Acids Res 34:5133–5144
14. De Armond R, Wood S, Sun D, Hurley LH, Ebbinghaus SW (2005) Evidence for the presence of a guanine quadruplex forming region within a polypurine tract of the hypoxia inducible factor 1α promoter. Biochemistry 44:16341–16350
15. Sun D, Guo K, Rusche JJ, Hurley LH (2005) Facilitation of a structural transition in the polypurine/polypyrimidine tract within the proximal promoter region of the human VEGF gene by the presence of potassium and G-quadruplex-interactive agents. Nucleic Acids Res 33:6070–6080
16. Dexheimer TS, Sun D, Hurley LH (2006) Deconvoluting the structural and drug-recognition complexity of the G-quadruplex-forming region upstream of the bcl-2 P1 promoter. J Am Chem Soc 128:5404–5415
17. Qin Y, Rezler EM, Gokhale V, Sun D, Hurley LH (2007) Characterization of the G-quadruplexes in the duplex nuclease hypersensitive element of the PDGF-A promoter and modulation of PDGF-A promoter activity by TMPyP4. Nucleic Acids Res 35:7698–7713
18. Guo K, Pourpak A, Beetz-Rogers K, Gokhale V, Sun D, Hurley LH (2007) Formation of pseudo-symmetrical G-quadruplex and i-motif structures in the proximal promoter region of the RET oncogene. J Am Chem Soc 129:10220–10228
19. Akman SA, Lingeman RG, Doroshow JH, Smith SS (1991) Quadruplex DNA formation in a region of the tRNA gene supF associated with hydrogen peroxide mediated mutations. Biochemistry 30:8648–8653
20. Woodford KJ, Howell RM, Usdin K (1994) A novel K(+)-dependent DNA synthesis arrest site in a commonly occurring sequence motif in eukaryotes. J Biol Chem 269:27029–27035
21. Han H, Hurley LH, Salazar M (1999) A DNA polymerase stop assay for G-quadruplex-interactive compounds. Nucleic Acids Res 27:537–542

Chapter 6

Real-Time Observation of G-Quadruplex Dynamics Using Single-Molecule FRET Microscopy

Burak Okumus and Taekjip Ha

Abstract

The potential importance of G-quadruplex structures was implied by the recent findings that the human POT1 disrupts G-quadruplex and stimulates the telomerase activity. A solid understanding of the range of conformations that can be adopted by guanine-rich sequences can potentially shed much light on the molecular mechanisms underlying certain human diseases related to telomeres. Furthermore, structure-based design of chemotherapeutic drugs for cancer might be realized by addressing different types of G-quadruplex structures. Using the unique capabilities of single-molecule spectroscopy, we have recently reported on the intricate dynamic structural properties of a minimal form of human telomeric DNA. Here, we present the detailed step-by-step methods for the real-time observation of G-rich DNA sequences by means of single-molecule FRET microscopy and provide the protocols for vesicle encapsulation and surface immobilization assays. Such assays provide a firm basis for future studies aimed at elucidating the interaction between telomeric DNA and telomere-associated proteins as well as the synthetic therapeutic agents that specifically stabilize certain G-quadruplex topologies.

Key words: Single-molecule, FRET, G-quadruplex, G4, G-tetrad, Telomere, hPOT1, Vesicle, Encapsulation

Abbreviations

FRET	Fluorescence resonance energy transfer
smFRET	Single-molecule FRET
SLB	Supported lipid bilayer
SUV	Small unilamellar vesicle
MLV	Multilamellar vesicle
GQ	G-Quadruplex
TDP	Transition density plot
HaMMy	Hidden Markov model analysis

P. Baumann (ed.), *G-Quadruplex DNA: Methods and Protocols*, Methods in Molecular Biology, vol. 608
DOI 10.1007/978-1-59745-363-9_6, © Humana Press, a part of Springer Science+Business Media, LLC 2010

1. Introduction

In vitro studies have shown that oligonucleotides with tandem repeats of guanines can spontaneously form noncanonical structures called G-quadruplexes (GQs) under physiological conditions (1). Potential quadruplex-forming sequences were then identified within the genome of various organisms (2), e.g., in promoter regions of some human proto-oncogenes (3) and in some bacterial promoters (4). Further in vivo data suggested the presence of quadruplex structures in intracellular contexts including telomeres (5) and the nontemplate strand during transcription (6). Obviously, the quadruplexes forming within the cell must be dynamic and need to be modified by proteins (2). Indeed, both in vivo (7) and in vitro (8) studies suggest specific interaction between miscellaneous proteins (e.g., human POT1, WRN, and BLM helicases) and DNA quadruplexes.

Because accumulating data is highlighting the potential importance of GQ structures in various biologically relevant contexts, a better understanding of the quadruplex behavior under a range of conditions seems essential. Therefore, we have previously used single-molecule spectroscopy to probe the dynamics of a DNA with tandem repeats of human telomeric sequence $(GGGTTA)_n$. One unfolded and two folded structures were observable within a range of potassium or sodium concentrations and temperatures. Each conformation could further be classified as long-lived and short-lived species, based on their characteristic lifetimes of minutes vs. seconds, respectively. Telomeric DNA encapsulated inside vesicles exhibited all of the six states, suggesting that the probed intricate dynamics was intrinsic to the molecule. As a means of further control, replacing a single guanine severely hindered folding and made only the short-lived species detectable. Ensemble measurements also revealed a biphasic dynamics that reflected the long- and the short-lived states. Our earlier work has thus exposed the conformations of GQ in its naked form (9).

In this chapter, we describe the detailed protocols for the single-molecule FRET assay to study GQ DNA. The methodologies discussed herein include direct immobilization of molecules on the surface as well as vesicle encapsulation. We anticipate that our assay will provide a basis for future studies of the interaction between enzymes/ligands and G-rich oligonucleotides. It might, for instance, be possible to dissect the biology at the human telomeres by enabling step-by-step in vitro reconstitution of the system from its individual components. A thorough understanding of the system might provide insights for designing drugs that recognize and interact with G-rich sequences in the desired fashion in vivo.

2. Materials

2.1. Human Telomeric DNA

1. DNA sequences from Integrated DNA Technologies, Coralville, IA (see Note 1).
2. G-quadruplex strand sequence (GQ+B): 5′-Cy5-(**GGGTTA**)$_3$GGG AGA GGT AAA AGG ATA ATG GCC ACG GTG CG-3′-biotin. The human telomeric repeat is highlighted by bold-face type. Biotin moiety makes the surface tethering possible.
3. Complementary stem strand sequence (STEM): 5′-CGC ACC GTG GCC ATT ATC CTT (amino-C6 dT)TA CCT CT-3′. The complementary stem strand is labeled with tetramethylrhodamine internally via amino-modified C6 dT.
4. Mutated GQ studies (GQ MUT): (**GGGTTA**)$_2$ GTGTTAGGG AGA GGT AAA AGG ATA ATG GCC ACG GTG CG-3′-biotin.
5. For encapsulation, a separate DNA strand without biotin (GQ – B): 5′-Cy5-(**GGGTTA**)$_3$GGG AGA GGT AAA AGG ATA ATG GCC ACG GTG CG-3′.
6. Annealing buffer: 10 mM Tris-HCl pH 8.0, 50 mM NaCl (see Note 2).
7. Heating block (a.k.a. Dry-bath incubator).

2.2. Vesicle Prep aration

1. Phospholipids from Avanti polar Lipids, Alabaster, AL.
2. Lipid for encapsulation: DMPC (10 mg/mL in chloroform; Cat # 850345C).
3. Biotinylated lipid for immobilization: DPPE-biotin (N-Biotinyl Cap-PE 16:0, 1 mg/mL in chloroform; Cat # 870277C).
4. Lipid for the supported bilayer: EggPC (10 mg/mL in chloroform; Cat # 840051C).
5. Mini Extruder, two syringes (250 μL), 100 poly carbonate membranes (200 nm diameter), and 100 filter supports (Avanti polar Lipids, Alabaster, AL; Cat #. 610023) (see Note 3).
6. Glass disposable scintillation vials (20 mL) (see Note 4).
7. Centrifuge tube (15 mL) (Corning).

2.3. Sample Chamber

1. Rectangular cover slips (24×40 mm No.1½).
2. Quartz microscope slides, 1″×3″, 1 mm thick (G. Finkenbeiner Inc., Waltham, MA) (see Note 5).
3. Diamond drill bits, 3/4 mm diameter (Kingsley North Inc.; Cat # 1-0500-100).

4. Sonicator, Bransonic B1510 tabletop ultrasonic cleaner.
5. Glass staining dishes (Fisher Scientific).
6. Alconox from Alconox Inc.
7. Acetone.
8. KOH (potassium hydroxide).
9. Basic propane torch (Bernzomatic).
10. Methanol.
11. Epoxy, 5 min epoxy (Devcon).
12. Double-sided tape.
13. Biotinylated bovine serum albumin (BSA) (Sigma-Aldrich; Cat # A8549).
14. Neutravidin, ImmunoPure NeutrAvidin protein (Pierce; Cat # 31000).
15. Streptavidin (Invitrogen; Cat # S-888).
16. DNaseI kit, Amplification-grade DNaseI (Sigma-Aldrich; Cat # AMP-D1).

2.4. Microscopy

2.4.1. Imaging Buffer Base

The base components of the imaging buffer are 10 mM Tris–HCl pH 7.4, 0.4% (w/v) β-D (+) glucose (Sigma-Aldrich) or 0.8% (w/v) D-glucose/dextrose monohydrate (Sigma-Aldrich) (*see* Notes 6 and 7). This combination will be referred to as the "base buffer" throughout the text. For the preparation of oxygen scavenging system, 20 μL of catalase, purified from bovine liver (Roche Applied Science) and 10 mg glucose oxidase, purified from *Aspergillus niger* (Sigma-Aldrich), are gently mixed in 100 μL T50 (we refer to this mix as "gloxy"). Vortexing is not recommended since it might denature the proteins. The gloxy is then centrifuged at for 1 min (13,000 *g*). The supernatant (a gold colored solution) must be used with minimal contamination from the pellet. The gloxy can be used for 2–3 weeks if stored at 4°C. 2-Mercaptoethanol (βME) was from Acros Organics (see Note 8).

2.4.2. Temperature Regulation

1. Water-circulating bath, NESLAB RTE-7 Digital One refrigerated bath from Thermo Scientific.
2. Thermocouple, digital thermometers from Omega, Stamford, CT (part # HH12A).

2.4.3. Data Acquisition

An extensive description of the materials, methodologies, details of data acquisition, and analysis for single-molecule total internal reflection microscopy (TIRM) can be found elsewhere (10).

3. Methods

For our studies, we used a construct similar to that designed by Balasubramanian and coworkers (11), except for an added biotin for specific tethering to a quartz surface via biotin–streptavidin linker to allow prolonged observation periods. The telomeric DNA typically consists of 100–150 tandem repeats of the (GGGTTA) sequence, but the minimal form that is capable of forming a stable GQ structure is chosen for simplicity. Folding of the DNA into the compact GQ structure is expected to yield a smaller average distance between the donor (tetramethylrhodamine) and the acceptor (Cy5), and hence to display higher FRET than the unfolded form. In single-molecule FRET studies, it is essential to ensure that the fluorophore attachment does not interfere with the native behavior. Previous UV melting studies proposed that the dye labeling did not induce substantial alteration to the stability of the construct.

The vesicle encapsulation served here as an alternative immobilization scheme to minimize the potential surface alteration of molecules by the immediate glass surface. The pores on the vesicles enable buffer exchange and make it possible to monitor encapsulated molecules under various salt conditions as a means of comparison with the surface data. Moreover, the pores ensure that the buffer conditions inside the vesicles stay in equilibrium with the bulk and that the individual DNA molecules inside different vesicles reside within identical environments. Thus, using porous vesicles rule out the possibility that the diverse behavior exhibited by the DNA might arise from the variation between the intravesicular salt concentrations (see Note 9). Because the vesicle encapsulation measurements prove the surface immobilization valid, we recommend using the more straightforward scheme of direct surface attachment unless one wants to utilize other aspects of the vesicle encapsulation technique (12).

Finally, it is noteworthy that our surface immobilization assay establishes the platform for single-molecule studies of the interactions between telomeric DNA and chemical agents together with various proteins (see Note 10). As a proof of concept, we studied the interaction of the DNA with a synthetic stabilizing agent. Moreover, we looked at the effect of yeast Replication Protein A (RPA) on the GQ folds, inspired by the ensemble studies that suggested that the human RPA actively disrupts the formation of GQ (13). The data from such studies are yet preliminary, and the systems are currently under investigation.

3.1. DNA Hybridization

1. DNA is received from the vendor in a lyophilized form. The DNA oligos are hydrated with T50 to a final concentration of

100–200 μM. Because it is known that the shelf life of fluorescence dyes is lengthened at low temperatures, the stock solutions are kept at −20°C (or preferably at −80°C).

2. The wells of the heating block are filled with water, and the temperature is adjusted to 95°C. The temperature is monitored by a thermometer in one of the other water-filled wells of the heating block.
3. The GQ strands are hybridized with the stem strand. The stem and the GQ strands are mixed in T50 buffer with a 2:1 molar ratio to final concentrations of 20 and 10 μM. The final volume of the DNA preparation is typically 10 μL. The final solution is vortexed briefly to ensure complete mixing and is then centrifuged to collect all the sample at the bottom of the tube.
4. The tube containing the DNA mixture is placed in water in one of the wells and kept at 95°C for 3 min (see Note 11). This step is needed for breaking apart all the previously mishybridized strands, and making the strands available for proper annealing. At the end of 3 min, the block is taken off the heater and placed in the dark (or covered with aluminum foil) at room temperature. Annealing occurs as the sample is left to slowly cool to room temperature.
5. The partial duplex DNA is then kept in a freezer (−20°C) until use (see Note 12).

3.2. Vesicle Encapsulation

1. Two types of v.esicles are needed for the encapsulation measurements. DMPC (dimyristoylphosphatidylcholine) vesicles are used for the encapsulation because of their spontaneous porosity at room temperature. EggPC vesicles are prepared for the formation of the supported lipid bilayer (SLB) on the surface which acts as an immobilization surface and a cushion for DNA encapsulating vesicles. The presence of a surface passivation for the immobilization of vesicles is vital to keep the vesicles and the encapsulated vesicles undisturbed. Note that the conventional method of BSA–biotin/streptavidin is not suitable for vesicle encapsulation measurements (see Note 13).
2. One milliliter of the 10 mg/mL DMPC (see Note 14) is mixed with 70 μL of 1 mg/mL DPPE-biotin in chloroform to obtain 1 mole percent of the biotinylated lipid (see Note 15). The mixture is then briefly sonicated to ensure proper mixing. The lipid mix is divided into four glass vials (approx. 250 μL, or 2.5 mg per vial). The chloroform is evaporated under stream of nitrogen. The nitrogen blow should be kept at a low pressure to prevent the splashing of the chloroform

until all the macroscopic chloroform is removed. The nitrogen pressure can later be increased to dry more of the solvent. Finally, the glass vials are placed under vacuum for 2 h to remove the residual chloroform. The lipid films (2.5 mg/vial) thus formed can be used directly or be kept at 4°C for up to 1 week.

3. For encapsulation, 250 μL (final volume) of T50 buffer including a final nonbiotinylated DNA concentration of 50 nM (see Note 16) is added to the 2.5 mg DMPC lipid film. The components are mixed until the film is completely hydrated and dissolved (see Note 17). After hydration, the mixture constitutes a 10 mg/mL suspension of multilamellar vesicles (MLVs) and looks turbid (see Note 18).
4. The MLV suspension is transferred to a 15-mL centrifuge tube in order to carry out freeze/thaw (F/T) cycles (typically 7 times), which increases the encapsulation efficiency. The freezing is done in liquid nitrogen, and subsequent thawing is realized in a bath of room temperature water (see Note 19).
5. The suspension looks more transparent after the F/T due to breaking of MLVs into smaller structures.
6. The MLV is transformed into small unilamellar vesicles (SUVs) by means of extrusion which is achieved with a mini extruder. Typically, 200- and 50-nm-diameter polycarbonate membranes are used for making SUV for encapsulation and SLB formation, respectively (see Note 20). A detailed protocol for the extrusion step is available on the vendor's website.
7. The SUV appears transparent due to lower light scattering from the smaller vesicles. The SUV can be used within a week as long as it is stored in the fridge. Ten milligrams per milliliter of lipid yields 520 and 32.5 nM of 50 and 200 nm diameter SUV, respectively (see Note 21). SUV must not be frozen because freezing destroys the vesicles.
8. It is recommended that the vesicles are diluted to 1 mg/mL (in T50) for long-term storage, as high concentrations (>1 mg/mL) of vesicle solutions lead to sample instability due to aggregation and fusion between vesicles (14).

3.3. Slide Preparation and Sample Immobilization

1. The details of the slide cleaning, sample chamber assembly, and DNA immobilization are discussed at great length in a previous paper (10).
2. The surface experiments with quadruplex DNA can be done with a conventional BSA–biotin/streptavidin surface as described earlier (10).

3. For the vesicle encapsulation experiments, first the SLB must be formed on the surface. For the formation of SLB, the 1 mg/mL EggPC (containing 1 molar percent of the biotinylated lipids) SUV solution is injected into the assembled flow chamber, and left to incubate for 10–60 min (see Note 22).
4. The incubation is done in a closed container (e.g., typically in an empty pipette box). The bottom of the container is kept wet in order to keep the interior of the container humid, thus preventing evaporation (see Note 23).
5. The stray vesicles are washed away with T50 buffer (For rinsing, we use 200 μL, i.e., ~5 times the volume of the chamber). Care must be taken not to introduce bubbles into the chamber after this point because air bubbles can destroy the SLB (and the later immobilized vesicles). We therefore recommend gentle pipetting of solutes into the chamber to prevent air bubble formation. Typically, buffer is slowly injected from one of the holes (hole #1) until a drop of buffer comes out of the other hole (hole #2). The pipette tip is then immediately switched to hole #2. The chamber must be monitored during the buffer injection, and the hole switching must be done whenever bubbles appear.
6. At this point, the nonspecific binding between the sample (DNA+SUV) and the membrane can be checked. First the SUV sample is diluted (typically by a factor of 32.5 to yield a final concentration of 1 nM) in T50. This dilution is introduced into the flow chamber, and the surface is monitored. For our particular constructs, we did not see a long-lasting interaction with the membrane and concluded that the nonspecific interaction was insignificant (9). The sample chamber is again rinsed with T50 after the control measurement is complete.
7. Streptavidin or neutravidin (0.2 mg/mL of T50 buffer) is similarly injected to the sample and incubated for 5 min. The unbound streptavidin is rinsed away at the end of the incubation period.
8. For the specific attachment of the vesicles, SUV samples are diluted in T50 and injected into the chamber. Because of the relatively slow diffusion of vesicles, a 10–15 min incubation period is recommended (see Note 24) for the completion of surface tethering. For such large objects (i.e., vesicles), gravity has a considerable effect. For instance, for prism-type TIR experiments the sample chamber must be flipped after SUV injection to ensure that the vesicles sink towards the quartz slide which will be the imaged surface.
9. Note that the spots immobilized on the surface are solely due to encapsulated molecules because, as already observed, the

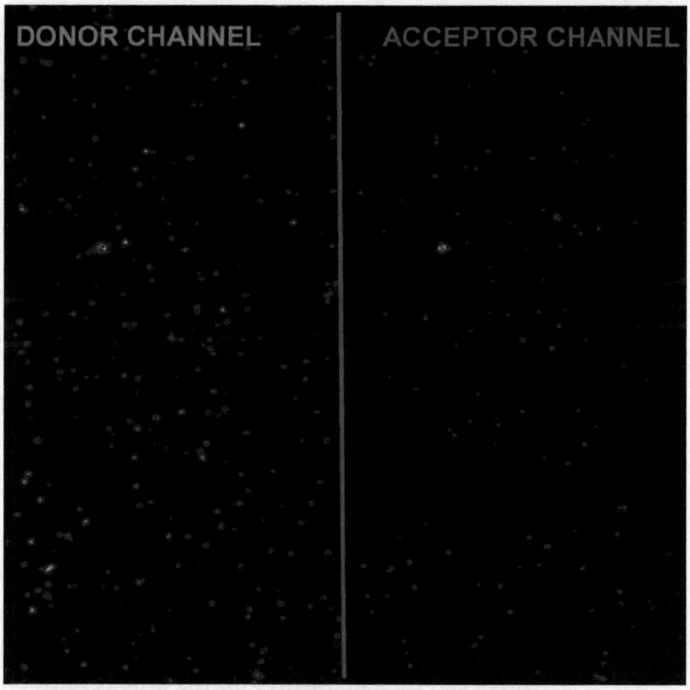

Fig. 6.1. Image of surface tethered vesicles after DNase treatment. DNaseI was injected into the flow chamber, incubated with the sample for 5 min, and rinsed away. Subsequently, the surface was imaged in 2 mM K^+. DNase that would otherwise remove the fluorescent spots from the surface upon digestion (data not shown) did not affect the DNA in this case because the encapsulated DNA was protected by the vesicles.

molecules do not nonspecifically stick to the surface (*see* Item 6 above). Besides, if the DNA was stuck to the SLB, they would diffuse on the surface whereas the vesicles are expected to be totally immobilized due to multiple biotin streptavidin attachments (15).

10. As a further control for the successful encapsulation, DNaseI (0.03 units dissolved in 1× reaction buffer specified by the vendor) is incubated with the sample for 5 min. In the case of proper encapsulation, surface-bound spots should not be removed because the DNA is protected by the vesicles against DNase digestion (*see* Notes 25 and 26). The image of DNaseI-treated surface-immobilized vesicles encapsulating the DNA is shown in Fig. 6.1. DNase treatment is not practiced at all times, and usually the nonspecific binding check provides a good enough control for ensuring proper encapsulation (*see* Item 6 above).

3.4. Imaging

1. To the base buffer, 1% 2-mercaptoethanol (v/v) and 1% gloxy (v/v) together with desired amounts of NaCl or KCl are added (imaging buffer) and injected into the sample chamber prior to imaging (*see* Note 27).

2. Oxygen scavenging also works for vesicle encapsulation. Although neither the glucose oxidase nor the catalase is expected to penetrate into the vesicles, oxygen can rapidly exit the vesicle, as it is constantly removed from the environment by the oxygen scavenging system.
3. The number of molecules per vesicle is expected to exhibit a Poisson distribution (15). For sample preparation, we therefore chose a DNA concentration to yield an average of 0.125 DNA/vesicle such that the probability of having two molecules per vesicle is <1%. However, for further precaution only the time traces that displayed stable signals and/or a single step photobleaching are selected for the analysis.
4. For the titration measurements, it is essential to rinse the sample chamber with the base buffer (see Subheading 2.4.1) between changing the salt concentrations. This intermediate rinsing unfolds all the molecules and enables proper folding for the subsequent salt condition to be tested.
5. Data is taken with 100-ms integration time for building histograms. Thanks to the slow dynamics of the GQ DNA, long time traces can be taken with 900 ms time resolution, which makes it possible to use low laser intensities and monitor the molecules for over 15 min before the fluorophores photobleach.
6. The biggest challenge for taking time traces of over 15 min is spontaneous defocusing during data acquisition. This issue can be resolved as follows: First a region on the slide that displays a good number of spots is identified by moving the microscope stage. The focus is adjusted and the laser is turned off. After 5 min, the focus is checked and readjusted if there was defocusing. This cycle is repeated until there is no substantial defocusing at the end of the 5 min. The focus is then stabilized and data acquisition can start.
7. When measurements are carried out at $T \neq RT$, enough time should be allowed for the stabilization of temperature. Especially, when buffer conditions are changed, we recommend waiting for at least 5 min for the temperature to equilibrate (see Note 28).
8. If long time traces will be taken at elevated temperatures, evaporation of water on the objective and subsequent image loss is likely to occur. Therefore, the maximum amount of water that can be put on the objective must be used under these conditions to prevent signal disappearance in the middle of data recording.
9. Other technical aspects pertaining to imaging are discussed elsewhere (10).

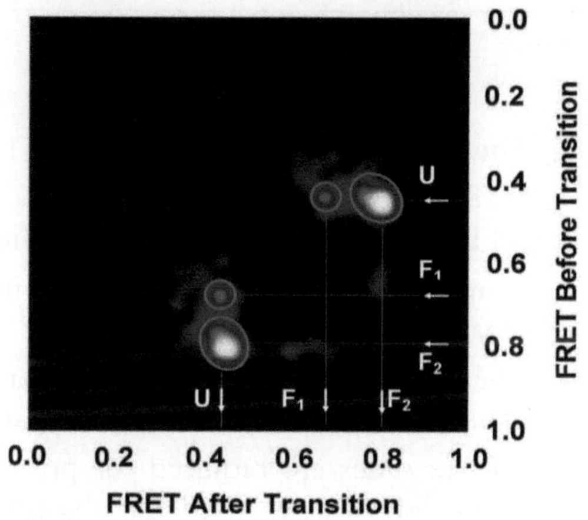

Fig. 6.2. Transition density plot (TDP) for 2 mM K^+. Result of HaMMy analysis carried out for 2 mM K^+ is shown as a TDP. Evidently, the direct transition between high (second folded state: denoted by F_2) and intermediate (first folded state: denoted by F_1) FRET values was insignificant, while the vast majority of the transitions (*white arrows*) took place from high/intermediate FRET (folded GQ) to the low FRET (unfolded GQ: denoted by U), or vice versa.

3.5. Data Analysis

1. The details related to the general smFRET data analysis are described in another paper (10), and the particular approaches taken for handling the GQ data can be found in the supporting material of our previous report on GQ dynamics (9).
2. Recently, a method for analyzing complex smFRET traces based on hidden Markov modeling was established (16) by our group (HaMMy, available for download at http://bio.physics.uiuc.edu/HaMMy.html), which is more reliable than the threshold algorithms that are prone to human bias. Nevertheless, due to the biphasic behavior observed in this system (i.e., the same FRET state shows both short- and long-lived behavior), rate determination via HaMMy is not reliable.
3. HaMMy can still provide useful information about the GQ dynamics. The results of HaMMy analysis can be summarized in a 2D plot (TDP: transition density plot), which displays the frequency of transitions. An example is shown in Fig. 6.2. The density of all transitions was plotted for 2 mM K^+, where the DNA showed three FRET states representing three different structures; i.e., low for unfolded, and middle and high for two distinct folded GQ structures. Strikingly, TDP shows that direct transitions between intermediate and high FRET states were insignificant, which was already anticipated because the transitions between distinct topologies of the folded GQ would require complete unfolding of the DNA.

4. Notes

1. The names of the vendors provided herein are listed as mere suggestions.
2. All buffers are prepared in water (MilliQ 18.5 MΩ).
3. Components needed for lipid extrusion can be purchased separately.
4. There is nothing special about a particular brand, and any glassware can be used for lipid preparation.
5. Quartz slides are required for prism-type TIR microscopy. Glass slides can be used for objective-type TIRM. Because quartz slides are costly, they are recycled and used multiple times (10).
6. The glucose is included in the base buffer for the oxygen scavenger system which is required to remove the oxygen from solution and thus lengthen the lifetimes of the fluorophores (10). The scavenger system consumes only β-D-glucose, and dextrose is a mixture of β- and α-D-glucose. Therefore, compared to glucose, twice as much of dextrose must be used. Because the vendor discontinued the production of glucose, we recently started using dextrose.
7. Because the buffering agent Tris can act as a cation and might induce folding in the absence of salt, we prepared an alternative buffer using HEPES, which is a neutral buffering agent. We did not observe any substantial difference for HEPES buffer and continued our experiments with the Tris buffer.
8. Although we recently reported Trolox to be a superior alternative to BME (17), we do not recommend using Trolox for this system. We observed that the Trolox interacts with the human telomeric DNA used here and stabilizes the GQ formation (unpublished observations).
9. For a 200-nm-diameter vesicle, 2 mM K^+ corresponds to approximately 5,000 ions, and statistical error due to the randomness in encapsulation is ±70, which results in only 1.4% variation in the intravesicular concentration for 2 mM. However, a more significant factor that contributes to the variation turns out to be the size distribution of the vesicles. The size distribution for 200-nm-diameter vesicles displays a variation of ±18 nm around its mean (i.e., 200 nm) (unpublished data), which corresponds to a threefold variation in volume and concentration.
10. For protein studies, it is essential to passivate the surfaces because proteins are more prone to nonspecific alteration due to artificial substrates. Hence, we use PEGylated slides for the RPA studies mentioned here. Despite its ease, using a

supported bilayer is an unlikely possibility. The DNA molecules tethered on an SLB would diffuse when the SLB is fluidic. In contrast, when an SLB in the gel phase is used for hindering the diffusion, discontinuities form within the SLB due to the tighter packing of lipids. Consequently, the proteins stick to the exposed glass surface at the defective regions on the SLB. A relatively easy surface passivation by means of an SLB is yet still achievable and is reported by Graneli et al. (18).

11. We strongly recommend not exceeding 3 min because the high temperatures would have adverse effects on the dyes.
12. Because freeze-thawing the DNA samples gradually degrades the fluorophores, usually aliquots of DNA are prepared right after hybridization.
13. EggPC rather than DMPC is used for the formation of the supported bilayer for its fluidity at room temperature ($T_m = -4\,°C$). A poly ethylene glycol (PEG) coated surface can alternatively serve as a cushion. A detailed protocol for surface PEGylation is described elsewhere (10). pH of the buffer must be monitored when working with the PEG surface, because we found that the DNA stuck to the PEG surface when the pH was lower than pH 7.4. However, it is worth mentioning that the making of the SLB is easier and more straightforward compared to the PEG treatment.
14. EggPC lipid film is prepared similarly but in a separate single vial to a final amount of 10 mg.
15. We observed no significant complications when using plastic pipette tips for measuring out the lipids in chloroform. Nevertheless, care should be taken, and the exposure of the plastic tips to chloroform must not be long.
16. Higher concentration of DNA yields a better encapsulation efficiency (15). However, a very high DNA concentration yields multiple molecules per vesicle and thus must be avoided.
17. Warming up the solution helps dissolving the film, but the temperature of the solution should not exceed the melting temperature of the DNA.
18. EggPC MLV is prepared the same way, except that the T50 does not contain any DNA. One milliliter T50 is used to hydrate 10 mg of EggPC to obtain a 10 mg/mL MLV suspension.
19. F/T cycles must be carefully carried out if a protein is being encapsulated. Although small nucleic acids are not affected by the freezing and thawing, activity of some proteins can be severely disrupted, and F/T cycles must then be skipped (unpublished observations). Vesicle preparation for making the SLB vesicles does not require F/T cycles because the final breakdown is regardless achieved by extrusion.

20. Bigger vesicles are well suited for increasing the encapsulation yield, but smaller vesicles are better for the formation of SLB. Due to their higher surface tension, smaller vesicles are more easily ruptured upon sticking to the surfaces.
21. The molecular weight of both EggPC and DMPC is approximately 760 g/mole. The lipid concentration is converted to vesicle concentration by assuming a 60 Å^2 area of the lipid head group (PC).
22. Vesicles stick to the glass/quartz surface nonspecifically and grow upon fusing with vesicles in suspension until they reach a critical size and rupture. The rupturing vesicles thus form patches, and eventually patches merge and form a continuous planar lipid bilayer on the glass surface (19).
23. The shear at the air–water interface destroys the SLB, and drying or air bubbles must be prevented after the SLB is formed.
24. It is possible to come up with a ballpark estimate for the number of molecules on the surface vs. sample dilution. However, such estimates might not be very reliable because the encapsulation yield depends on several parameters (such as number of F/T cycles, initial encapsulant concentration, type of lipids used, etc.), and this causes sample-to-sample variation. Hence, we instead recommend a trial-and-error approach. We typically start with a 200-fold dilution for the tethering of the vesicles. After 15 min, the surface is monitored for the number of molecules bound to the surface. If the number of spots on the surface is not enough (~500 spots within half of Charge Coupled Device (CCD) image, which corresponds to a real area of ~50 μm × 100 μm, is a good number), we keep adding more until we reach a good number of spots on the surface. It is worth mentioning here that a very dense surface coverage must be avoided in order to be able to distinguish individual molecules.
25. The same amount of DNaseI under the same conditions was shown to have removed the surface-attached DNA almost instantaneously, and therefore 5 min incubation is practiced for the sake of prudence.
26. The pores forming on the DMPC around room temperature are known to be smaller than the smallest oligonucleotide composed of two nucleotides. Therefore, we do not expect the DNaseI to be able to reach the DNA encapsulated inside the vesicle.
27. Gloxy produces an acid at the end of the reaction, so the solution becomes very acidic over a long time. This problem can mostly be remedied by adding the gloxy to the imaging buffer just before measurements are taken. Small holes on

the sample chamber reduce oxygen uptake by the solution, and solution acidification is less of a concern when the imaging buffer is present in the sample chamber compared to when it is kept in the tube.

28. In case of temperature measurements with the DMPC vesicles, one should keep in mind that the vesicles would only stay porous around $23 \pm 1.5°C$ (20).

Acknowledgments

Some experimentation and most of the data analysis was carried out by Jayil Lee. We thank Shankar Balasubramanian for providing us with the GQ stabilizing agent. Yeast RPA was a gift of Tim Lohman. Rahul Roy helped with building the TDP and provided useful discussion about the HaMMy analysis.

References

1. Sen D, Gilbert W (1988) Formation of parallel four-stranded complexes by guanine-rich motifs in DNA and its implications for meiosis. Nature 334:364–366
2. Maizels N (2006) Dynamic roles for G4 DNA in the biology of eukaryotic cells. Nat Struct Mol Biol 13:1055–1059
3. Rankin S, Reszka AP, Huppert J et al (2005) Putative DNA quadruplex formation within the human c-kit oncogene. J Am Chem Soc 127:10584–10589
4. Rawal P, Kummarasetti VB, Ravindran J et al (2006) Genome-wide prediction of G4 DNA as regulatory motifs: role in Escherichia coli global regulation. Genome Res 16:644–655
5. Schaffitzel C, Berger I, Postberg J, Hanes J, Lipps HJ, Pluckthun A (2001) In vitro generated antibodies specific for telomeric guanine-quadruplex DNA react with Stylonychia lemnae macronuclei. Proc Natl Acad Sci U S A 98:8572–8577
6. Duquette ML, Handa P, Vincent JA, Taylor AF, Maizels N (2004) Intracellular transcription of G-rich DNAs induces formation of G-loops, novel structures containing G4 DNA. Genes Dev 18:1618–1629
7. Paeschke K, Simonsson T, Postberg J, Rhodes D, Lipps HJ (2005) Telomere end-binding proteins control the formation of G-quadruplex DNA structures in vivo. Nat Struct Mol Biol 12:847–854
8. Sun H, Karow JK, Hickson ID, Maizels N (1998) The Bloom's syndrome helicase unwinds G4 DNA. J Biol Chem 273:27587–27592
9. Lee JY, Okumus B, Kim DS, Ha T (2005) Extreme conformational diversity in human telomeric DNA. Proc Natl Acad Sci U S A 102:18938–18943
10. Joo C, Ha T (2008) Single molecule FRET with total internal reflection microscopy. In: Selvin P, Ha T (eds) Single molecule techniques: A laboratory manual, Cold Spring Harbor Laboratory Press, Cold Spring Harbor, New York ISBN 978-087969775-4, 507 pp
11. Ying L, Green JJ, Li H, Klenerman D, Balasubramanian S (2003) Studies on the structure and dynamics of the human telomeric G quadruplex by single-molecule fluorescence resonance energy transfer. Proc Natl Acad Sci U S A 100:14629–14634
12. Cisse I, Okumus B, Joo C, Ha T (2007) Fueling protein–DNA interactions inside porous nanocontainers. Proc Natl Acad Sci U S A 104:12646–12650
13. Salas TR, Petruseva I, Lavrik O et al (2006) Human replication protein A unfolds telomeric G-quadruplexes. Nucleic Acids Res 34:4857–4865

14. Zhang L, Hong L, Yu Y, Bae SC, Granick S (2006) Nanoparticle-assisted surface immobilization of phospholipid liposomes. J Am Chem Soc 128:9026–9027
15. Okumus B, Wilson TJ, Lilley DM, Ha T (2004) Vesicle encapsulation studies reveal that single molecule ribozyme heterogeneities are intrinsic. Biophys J 87:2798–2806
16. McKinney SA, Joo C, Ha T (2006) Analysis of single-molecule FRET trajectories using hidden Markov modeling. Biophys J 91: 1941–1951
17. Rasnik I, McKinney SA, Ha T (2006) Nonblinking and long-lasting single-molecule fluorescence imaging. Nat Methods 3:891–893
18. Graneli A, Yeykal CC, Prasad TK, Greene EC (2006) Organized arrays of individual DNA molecules tethered to supported lipid bilayers. Langmuir 22:292–299
19. Johnson JM, Ha T, Chu S, Boxer SG (2002) Early steps of supported bilayer formation probed by single vesicle fluorescence assays. Biophys J 83:3371–3379
20. Bolinger PY, Stamou D, Vogel H (2004) Integrated nanoreactor systems: triggering the release and mixing of compounds inside single vesicles. J Am Chem Soc 126:8594–8595

Chapter 7

Sedimentation Velocity Ultracentrifugation Analysis for Hydrodynamic Characterization of G-Quadruplex Structures

Nichola C. Garbett, Chongkham S. Mekmaysy, and Jonathan B. Chaires

Abstract

Analytical ultracentrifugation (AUC) is a powerful technique for the characterization of hydrodynamic and thermodynamic properties. The intent of this article is to demonstrate the utility of sedimentation velocity (SV) studies to obtain hydrodynamic information for G-quadruplex (GQ) systems and to provide insights into one part of this process, namely, data analysis of existing SV data. An array of data analysis software is available, mostly written and continually developed by established researchers in the AUC field, with particularly rapid advances in the analysis of SV data. Each program has its own learning curve, and this article is intended as a resource in the data analysis process for beginning researchers in the field. We discuss the application of three of the most commonly used data analysis programs, DCDT+, Sedfit, and SedAnal, to the interpretation of SV data obtained in our laboratory on two GQ systems.

Key words: Analytical ultracentrifugation, Sedimentation velocity, Sedimentation coefficient, Frictional ratio, Hydrodynamic, Solution conformation, G-quadruplex (GQ) DNA, Data analysis, DCDT+, Sedfit, SedAnal

Abbreviations

AUC	Analytical ultracentrifugation
CD	Circular dichroism
D	Translational diffusion coefficient
SV	Sedimentation velocity
SE	Sedimentation equilibrium
s	Sedimentation coefficient
$s_{20,w}$	Sedimentation coefficient corrected to a temperature of 20°C and to the density and viscosity of pure water
S	Svedberg unit equal to 10^{-13} s
M	Molecular weight
$\bar{v}$	Partial specific volume
ρ	Density
GQ	G-quadruplex

P. Baumann (ed.), *G-Quadruplex DNA: Methods and Protocols*, Methods in Molecular Biology, vol. 608
DOI 10.1007/978-1-59745-363-9_7, © Humana Press, a part of Springer Science+Business Media, LLC 2010

1. Introduction

The technique of analytical ultracentrifugation (AUC) was developed by Svedberg in 1923 when he equipped a rapidly rotating centrifuge with an optical detection system to observe sedimenting particles and applied his method to determine particle size distributions of gold sols. Use of the technique for the characterization of biochemical systems, especially proteins and their complexes, became widespread from 1950 to 1980 with the ability to determine sedimentation coefficients, molar masses, and shape information (1). However, its use declined substantially in the following decade because of a lack of new instrumentation for digital data acquisition and the adoption of alternative, less demanding, techniques including gel electrophoresis and size exclusion chromatography. The reemergence of modern AUC instrumentation in the early 1990s occurred concomitantly with significant advancements in data analysis methods, making possible a comprehensive characterization of hydrodynamic and thermodynamic properties using AUC studies (1–3).

Sedimentation velocity (SV) is one of two major AUC approaches, the other being sedimentation equilibrium (SE) (1–8). SV experiments are performed at high rotor speeds, up to 290, $304 \times g$, resulting in the establishment of a sample concentration boundary that moves towards the bottom of the centrifuge cell over time. The rate of movement of the boundary is observed through the acquisition of optical scans (absorbance, interference, or fluorescence) as a function of radial position every few minutes over the course of several hours. The rate of movement of the concentration boundary as a function of time yields the sedimentation coefficient s. The rate of spreading of the concentration boundary allows an estimation of the translational diffusion coefficient D, which opposes sedimentation. Molecular weight M can be estimated from the ratio of s and D. SE experiments are performed at much lower rotor speeds and run for much longer periods in order to establish an equilibrium between sedimentation and diffusion. The exponential sample concentration distribution that results can be fit by nonlinear least-squares methods to thermodynamic equations to yield molecular weights, association constants, and stoichiometries of systems. It is essential to perform SV experiments before equilibrium analysis in order to assess sample heterogeneity and polydispersity which could complicate analysis of SE data. With recent advances in the analysis of SV data, these experiments alone are often informative enough to provide the information being sought.

This article focuses on the application of selected data analysis methods to the hydrodynamic characterization of G-quadruplex (GQ) structures using SV ultracentrifugation. Significant literature exists concerning the appropriate design of AUC experiments, protocols for performing the experiments, and methods of data analysis (1–29). A number of comprehensive reviews exist (1–8, 30). This article will not attempt to reproduce information that exists more expansively elsewhere, and the reader is referred to these original papers. The intent is twofold – to illustrate the utility of SV and to provide helpful insights into one small part of this process, namely, data analysis of existing SV data of GQ systems. To this end, we will present SV data obtained in our laboratory on two GQ systems and provide commentary on our path through the data analysis process. A host of data analysis software is currently available for the analysis of sedimentation data, mostly written and continually developed by established researchers in the AUC field, with particularly rapid advances in the analysis of SV data. Examples of available software include Sedfit, Sedphat, SedAnal, DCDT+, Heteroanalysis, Svedberg, and UltraScan (31–37). This article is focused solely on the analysis of SV data with a discussion of analysis using three of these programs, i.e., DCDT+, Sedfit, and SedAnal.

2. Materials

The following programs are needed:

1. DCDT+ (http://www.jphilo.mailway.com/dcdt+.htm) (34)
2. Sedfit (http://www.analyticalultracentrifugation.com/download.htm) (31)
3. SedAnal (http://sedanal.bbri.org/) (33)
4. XLGraph (http://www.jphilo.mailway.com/download.htm) (38)
5. WinMatch (http://www.biotech.uconn.edu/auf/?i=aufftp) (39)
6. Sednterp (http://www.jphilo.mailway.com/download.htm) (40)

3. Methods

3.1. Introduction to the Analysis Approaches

Important differences exist between the analysis methods implemented in these programs, and it is informative to include a brief discussion of the different approaches employed in the analysis of SV data. Early estimation of s values was simply based on the measurement of the movement of the boundary midpoint over time (2). However, the boundary shape is subject to increasing diffusional broadening during the sedimentation experiment, and the midpoint would only be free of diffusion and yield an accurate s value for a single-component sample (1). For multicomponent samples, the midpoint yields only an average *s* value, which is insensitive to multiple solution species. The van Holde–Weischet method addressed this by dividing the sedimenting boundary into equal fractions, calculating an apparent *s* value for each boundary fraction, and extrapolating the apparent *s* value at the same boundary fraction of each scan to infinite time to remove the effects of diffusion and yield the apparent *s* value of that boundary region (1, 3, 4, 14–16). A plot of boundary fraction against extrapolated *s* values yields an integral sedimentation coefficient distribution *G*(*s*), for the sample. For a homogeneous sample, the *G*(*s*) distribution will be vertical since the extrapolated *s* values will be the same at all points in the boundary. For heterogeneous samples, the *G*(*s*) distribution will have a positive slope with the fastest sedimenting species corresponding to the later boundary fractions and hence appearing at the top of the positive slope of the *G*(*s*) plot. The van Holde–Weischet method increases the resolution of components with similar *s* values, which is particularly helpful for complex mixtures, giving a rapid assessment of sample heterogeneity and any interaction behavior. However, the removal of diffusional spreading from the sedimentation profiles means that the method cannot be used to determine diffusion coefficients or molecular weights. The method is available in Sedfit and UltraScan but does not form the basis of the analysis method in these programs.

The differential time derivative dc/dt method, like the van–Holde Weischet method, provides a model-independent assessment of sedimentation behavior. However, the basis of the time-derivative method is very different (4, 17, 18). Closely spaced boundary scans are subtracted in pairs to yield a set of time derivative dc/dt against radius profiles. The time derivative dc/dt is then transformed into dc/ds, which is a differential apparent sedimentation coefficient distribution designated $g(s^*)$, and the radial position is transformed into an apparent sedimentation coefficient (s^*). Subtraction of scans in the time-derivative dc/dt method results in a significant decrease in time-invariant noise. The $g(s^*)$ versus s^* profile is similar in appearance to a chromatogram, although the apparent *s* value is far more informative than a retention time. Visually, the $g(s^*)$ profile is more intuitive than the *G*(*s*) profile. A single-component sample will yield a Gaussian-shaped curve with a sedimentation coefficient corresponding to the peak position and the diffusion coefficient corresponding to the peak width. Determination of both *s* and *D* values allows calculation of *M* (*see* Note 1). Inclusion of the effect of diffusion reduces the resolution compared to the *G*(*s*) method; however, some might argue that this situation is the reality and should be included. The model-independent nature of the method is a significant advantage of the approach and means that it is possible to obtain information from the sedimentation profiles of very complicated mixtures of species which would otherwise be extremely challenging to analyze if it was necessary to determine a suitable model before analysis (4). The time-derivative d*c*/d*t* method is the basis of the program DCDT+ and is implemented in SedAnal.

Around the time of the conception of the time-derivative d*c*/d*t* method, computer programs were developed to fit SV data to approximate solutions of the Lamm equation, which is the underlying transport equation describing the SV process (4, 19, 20, 29). This approach is the basis of fitting $g(s^*)$ data in DCDT+ and Svedberg analysis programs. An alternative approach of using finite-element solutions to the Lamm equation is used in the programs Sedfit, UltraScan, and SedAnal (2, 4, 9, 21–26, 28). (UltraScan was written for Unix systems, whereas all the other programs operate on the Windows platform.) Both approaches allow the estimation of s, D, M, association constants, and stoichiometries; however, a model must be specified that best fits the sedimentation behavior of the sample under analysis, the selection of which can be difficult. DCDT+ allows the possibility of generating a model-independent $g(s^*)$ distribution prior to model-dependent fitting for accurate values of the aforementioned parameters, whereas both Sedfit and SedAnal require the input of a model at the outset of analysis.

The finite-element modeling method used in Sedfit and UltraScan uses a distribution of Lamm equation solutions to directly model the sedimentation boundary and yields a differential sedimentation coefficient distribution $c(s)$ with the deconvolution of diffusion effects (21–26, 28). As previously mentioned, the removal of diffusion information enhances the resolution and quantization of mixtures of solution components. However, unlike the $g(s^*)$ distribution, the peak widths in a $c(s)$ distribution are a function of the signal-to-noise ratio, and the parameters employed by the maximum entropy regularization processes in Sedfit have no physical meaning (4, 41). Also, it is important to be aware of the possibility of the generation of false peaks in the $c(s)$ distribution where a good fit of the raw data cannot be obtained or inappropriate parameters are entered. Sedfit can model a wide range of sedimentation processes including associating systems, nonideal sedimentation, the redistribution of salts, both very low mass (e.g., peptides) and high mass species (e.g., viruses), flotation as well as sedimentation, and can use data near the base of the cell where solutes accumulate (2, 4, 41). SedAnal contains a basic set of models that can be supplemented by an unlimited number of user-defined models incorporated using a separate ModelEditor program, whereas DCDT+ can fit $g(s^*)$ distributions for up to five discrete noninteracting species. Sedfit estimates a weight-average shape factor f/f_0 from the experimental data which forms the basis of the relationship between s and D values. The $c(s)$ distribution in Sedfit can be converted to a $c(M)$ distribution with the caveat of the assumption of f/f_0 which may lead to incorrect M values for some species (2). Whenever a distribution of species is present, the meaning of f/f_0 is complicated.

Having introduced the background to the various approaches used in the analysis of SV data, we will now discuss the application of three data analysis programs, DCDT+, Sedfit, and SedAnal, to the interpretation of SV data obtained in our laboratory for GQs. The user is referred to the program manuals, references, help files, tutorials, and courses applicable to these programs. These provide comprehensive instruction as to the operation of each analysis program. This information would be too expansive to reiterate here; instead, we will attempt to provide useful comments at various stages of the data analysis process to expand upon the information provided by the program authors and assist researchers in the application of these programs to the analysis of their own experimental data. Data presented are absorbance scans obtained on the Beckman XL-A instrument; small differences in the analysis setup are necessary for interference scans but the same overall analysis strategies would apply.

3.2. DCDT+

SV data was obtained for the GQ series $G_nT_4G_n$ where $n = 2–10$. It has been reported that $G_4T_4G_4$ forms a hairpin dimer structure in sodium solution (42, 43) and we hypothesized that with higher n, increased numbers of GQ stacks would form within a dimeric quadruplex structure. To investigate this, we performed circular dichroism (CD) and AUC studies. CD studies indicated that both "parallel" and "anti-parallel" GQ forms were apparent for $n = 2–4$, with only "parallel" forms present for $n = 5–10$; there was also an increase in ellipticity with increase in G-tract length (44) (Fig. 7.1). While CD studies provided useful conformational information, they could not provide an assessment of the molecularity or hydrodynamic properties of the samples and, to this end, AUC studies were undertaken. We will highlight here only some of this work in order to provide a relevant system for the demonstration of the utility of AUC studies for the characterization of GQs. SV studies were performed using a Beckman Coulter XL-A instrument at 290,304 × *g* (the current limit of the instrument) because of the expected low molecular weights of the samples (4.9 kDa for $n = 2$ to 15.5 kDa for $n = 10$) (*see* Note 2). Because of the large number of samples involved in the study, data were initially collected for a single concentration ($A_{260} = 0.8$) of each sample (three samples per rotor = 3 SV runs). Data were collected with no time interval between scans, with a run typically starting towards the end of the day and continued overnight. This resulted in more scans than necessary (with a number of scans collected after the sample had pelleted) but ensured complete sedimentation of all solution contents and allowed the selection of appropriate scans to be made during the data analysis process.

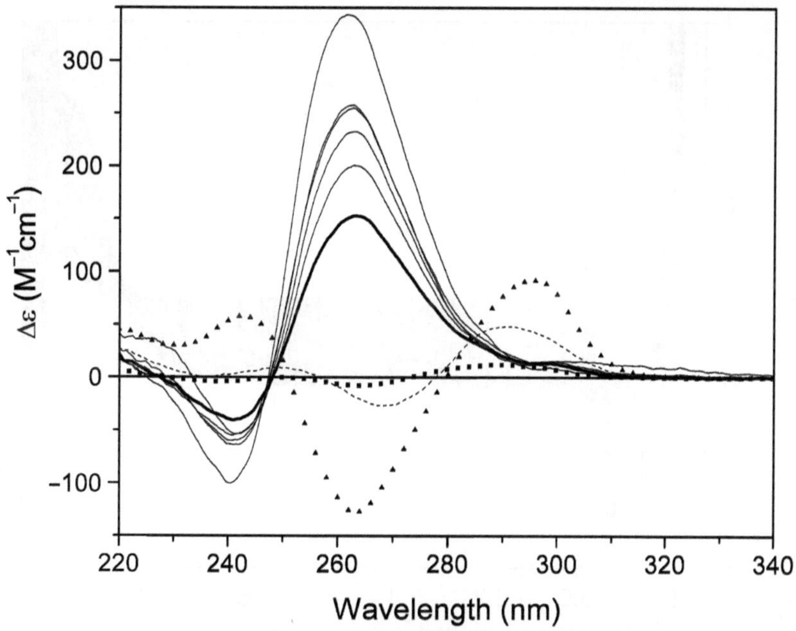

Fig. 7.1. CD spectra of $G_nT_4G_n$ (n = 2–10) in BPES buffer (6 mM Na_2HPO_4, 2 mM NaH_2PO_4, 1 mM Na_2EDTA, 185 mM NaCl, pH 7.0). Both "parallel" and "antiparallel" G-quadruplex forms were apparent for n = 2–4 ($G_2T_4G_2$, *squares*; $G_3T_4G_3$, *dashed black line*; $G_4T_4G_4$, *triangles*). A switch to only "parallel" forms occurs at n = 5 with an increase in ellipticity with increase in G-tract length ($G_5T_4G_5$, *thick black line*; $G_6T_4G_6$ – $G_{10}T_4G_{10}$ *black lines*).

1. Load SV data into DCDT+ according to the software directions. Inspect raw data for initial observations/impressions about the nature of the sample. Look for the rate of depletion of signal at the meniscus, the rate of movement of the boundary, the rate of depletion of the plateau region, and the number of steps or inflection points evident in the boundary (this becomes more difficult to see over time as the effects of diffusion become more apparent).

These general observations provide information about sample heterogeneity, the number of sedimenting species, as well as their relative mass, density, and shape. Inspection of the entire raw dataset for $G_2T_4G_2$ revealed that at 290,304 × g the sample had cleared the meniscus region but had not sedimented to the bottom of the cell, even after overnight data collection (Fig. 7.2). This indicated that the sample either had a low molecular weight, low density, and/or a less compact shape. By $G_4T_4G_4$, the sample sedimented to the bottom of the centrifuge cell indicating that the sample had a higher molecular weight, higher density, and/or a more compact shape. Raw data for $G_5T_4G_5$ through $G_{10}T_4G_{10}$ exhibited larger spacing between scans with obvious "steps" in the earlier scans (Fig. 7.3). This indicated the presence of multiple, faster sedimenting solution components.

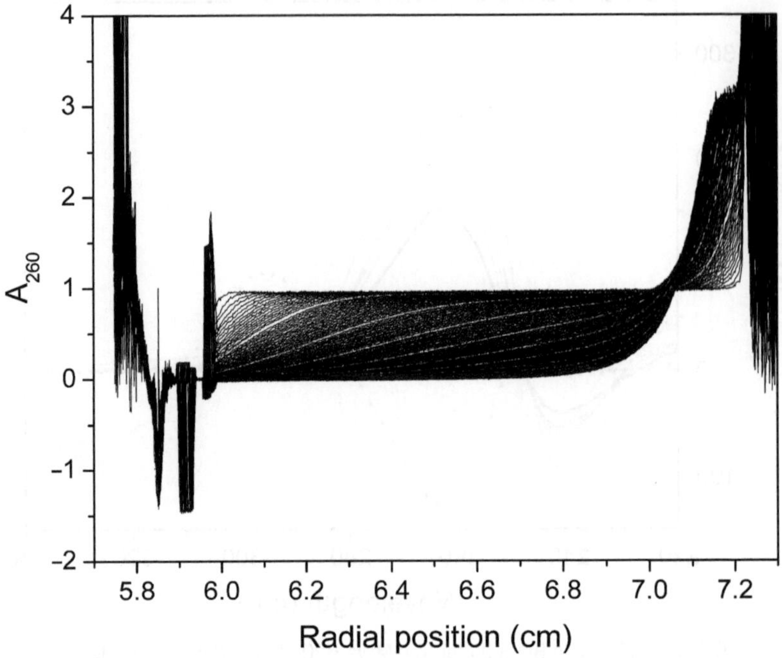

Fig. 7.2. Sedimentation velocity scans for $G_2T_4G_2$ collected on the Beckman XL-A at 290,304 × *g* at A_{260} ~ 0.8 in a 1.2-cm cell. The sample does not sediment to the bottom of the cell even after overnight data collection, indicating the presence of small sedimenting species.

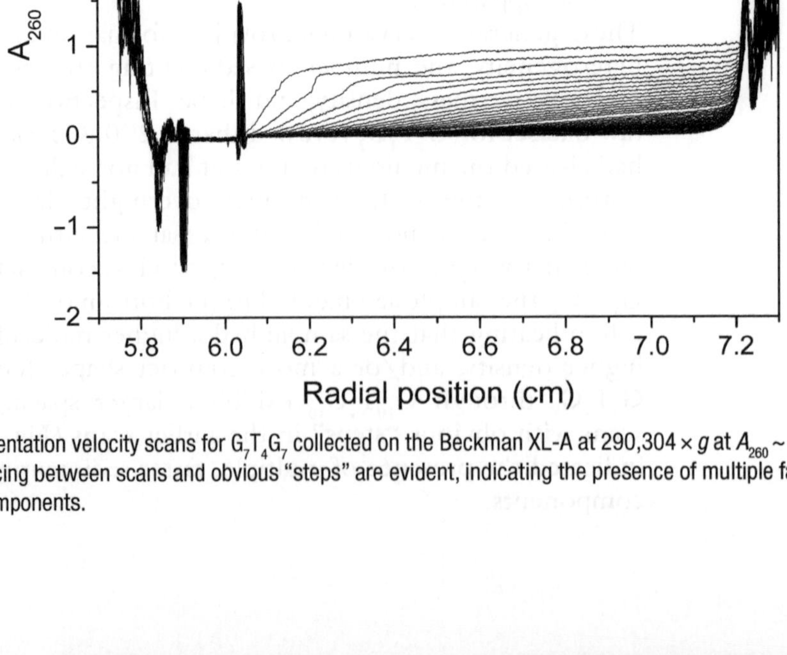

Fig. 7.3. Sedimentation velocity scans for $G_7T_4G_7$ collected on the Beckman XL-A at 290,304 × *g* at A_{260} ~ 0.8 in a 1.2-cm cell. Large spacing between scans and obvious "steps" are evident, indicating the presence of multiple faster sedimenting solution components.

2. Set sample meniscus position indicator and the left and right fitting limits. When prompted, we accept the default of setting the meniscus to the maximum of the meniscus peak. The left fitting limit is set to a position just clear of the meniscus. The right fitting limit is set to a position which will be at the end of the plateau region when scans from the middle of the run are selected. Fine adjustment of all fitting limits is performed in step 3.
3. Use the auto adjust button to provide an initial selection of scans for transformation into the $g(s^*)$ distribution. Adjust the number and position of scans in the run as appropriate. Check that the scan selection is appropriate for the peak broadening limit. Make final adjustments.

This selection is usually quite good, but it is important to use the slider controls to investigate other regions of the data. Move the run position slider to investigate earlier and later scans in the run to look for other sedimenting species (see Note 3). Once an initial range of scans has been selected encompassing the solution species of interest, visually inspect both the dc/dt distribution and the raw data scans, then adjust the scan number slider and the run position slider in turn to optimize the selection of scans. The ultimate goal is to select raw scans that are in the middle of the run with a flat, zero signal after the meniscus peak and a flat plateau region after the sedimentation boundary. Check that the left and right fitting limits are in the flat meniscus and plateau regions and that the meniscus position indicator is still at the maximum of the meniscus peak. The dc/dt distribution should be centered in the window with a fairly symmetrical Gaussian-like shape tailing off to flat, zero regions on either side of the peak. (A nonzero region on the right-hand side of the peak might indicate the presence of aggregates in the sample).

By moving the run position and scan number sliders, the peak broadening limit value changes. This is a numerical assessment of whether too many scans are being averaged to generate the dc/dt distribution. Including scans covering too much boundary movement causes the $\Delta c/\Delta t$ values that are calculated to be a poor approximation of the true derivative dc/dt. The result is excessive peak broadening of the distribution and loss of resolution between species. The trade-off is that decreasing the number of scans to prevent a peak from being broadened too much and ensuring good resolution results in a decrease in the signal-to-noise ratio. (The help section of the DCDT+ program contains useful information on the peak broadening limit.) Having adjusted the scan number slider, check the position of the scans within the run in both the dc/dt and raw data windows. Make final adjustments by checking the meniscus indicator and the left and right fitting limit indicators (see Note 4).

4. Set the range of s values for transformation to the $g(s^*)$ distribution. Trim the data points at the extreme edges of the

distribution in order to have flat, zero baseline in these regions. Depending on the nature of the sample, this may not always be possible (e.g., nonzero baseline at higher *s* values could indicate aggregation) but bad data should certainly be set outside the range for subsequent $g(s^*)$ transformation.

5. Transform the data into a $g(s^*)$ distribution. Normalize. Convert to $s^*_{20,w}$.

Normalization sets the area under the $g(s^*)$ distribution to a value of 1. It is important to run different concentrations of a sample and compare the normalized $g(s^*)$ distributions to provide information about any concentration-dependent properties (e.g., association) of the sample. DCDT+ has a convenient option to prepare a $g(s^*)$ overlay graph for this purpose. Conversion of a solution-dependent *s* value to that observed in water at 20°C ($s^*_{20,w}$) accounts for solvent contributions to the sedimentation coefficient (*see* Note 5). This allows the comparison of data obtained under different experimental conditions. For a concentration series, an extrapolation of $s^*_{20,w}$ values can be made to zero concentration to yield an *s* value which is solely a property of the macromolecule (4). The "Show integrals over distribution" button is a useful option to display number, weight, *z* and *z* + 1 average *s* values at both the time of analysis and extrapolated to $t = 0$.

Figure 7.4 depicts steps 2–5 for the GQ sample $G_4T_4G_4$. The resulting $g(s^*)$ distribution for all nine GQ samples is shown in Fig. 7.5. It is apparent that there are two distinct groups of samples: discrete $g(s^*)$ distributions centered around low *s* values are observed for $G_2T_4G_2$ through $G_4T_4G_4$, with a noticeably smaller *s* value for $G_2T_4G_2$; extremely broad distributions of significantly higher *s* values are apparent for $G_5T_4G_5$ through $G_{10}T_4G_{10}$. These observations are in accordance with those made from an examination of the raw data scans; initial examination of the raw data is useful to provide important clues as to the nature of the sample for subsequent steps in the analysis process. The nature of the $g(s^*)$ distribution for $G_5T_4G_5$ and higher indicates the presence of multiple higher order species and is a significant break in sample behavior from the discrete distributions obtained for $G_4T_4G_4$ and lower.

We typically pause analysis using DCDT+ having generated a model-independent $g(s^*)$ distribution and some initial conclusions about the nature of a sample. We then proceed with a $c(s)$ analysis using Sedfit before entering the final stages of the analysis. It is possible to export from DCDT+ at the $g(s^*)$ stage (or earlier) by right-clicking in the appropriate graph window and selecting either a picture file or a data/text file and appropriate data subsets.

3.3. Sedfit

1. Load data into Sedfit according to the software directions.

Although Sedfit analyzes scans from the entire SV run, it is simply not appropriate to import every single scan collected. Current

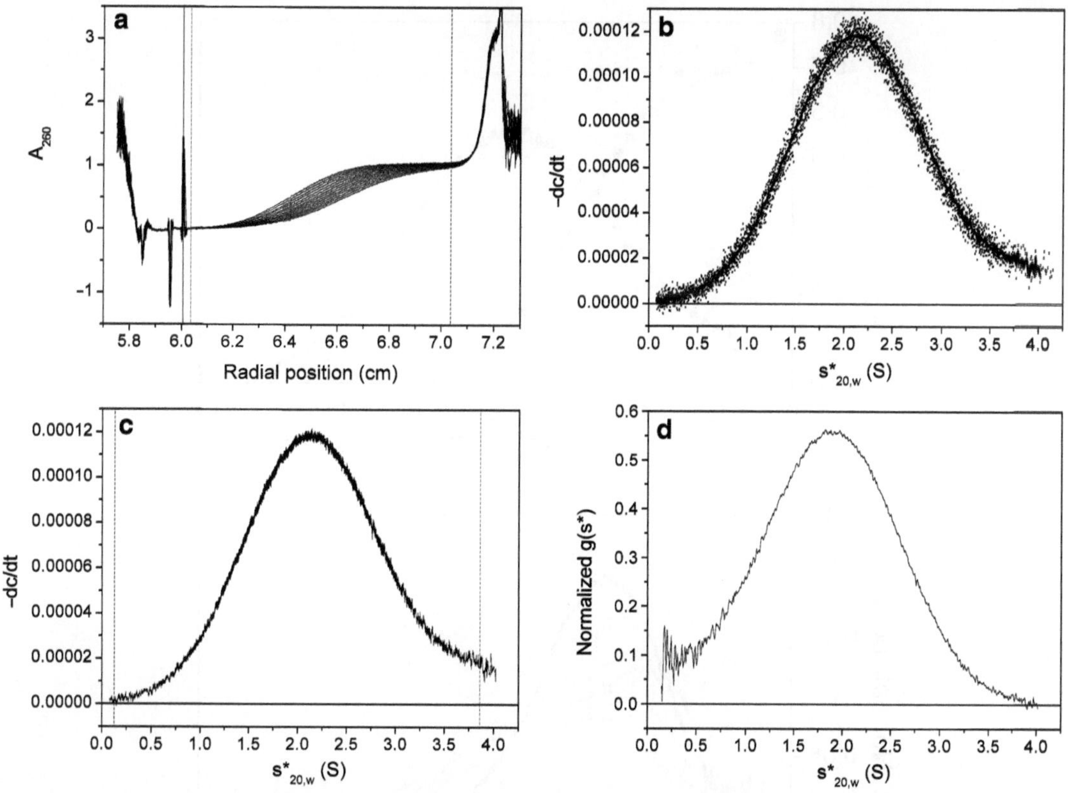

Fig. 7.4. Analysis steps using the DCDT+ software. (**a**) Scans selected for calculation of *g*(s*) distribution (the meniscus position is depicted by the *black vertical line*, and the left and right fitting limits by the *dashed vertical lines*); (**b**) d*c*/d*t* curves from pairwise subtraction of selected scans (the average curve is shown by the *thick black line*); (**c**) selection of *s* range for *g*(s*) transformation (left and right limits are shown by the *vertical dashed lines*); (**d**) normalized and corrected *g*(s*) distribution.

versions of Sedfit have a color-coded gradation from black to red to indicate that a good range of scans has been imported. It is recommended to import scans such that an even color gradation is obtained with one-third of the scans black/blue, one-third green/yellow, and one-third orange/red. For example, if a large number of scans are imported after the sample has completely sedimented, these scans will all overlay at the end of the run and there will be very little orange/red coloration. In this situation, it would be necessary to determine when the sample has completely sedimented and exclude later scans from the analysis. For a slower sedimenting species, it will be possible to increase the interval between imported scans, as there will be smaller differences between subsequent scans than there would be for a faster sedimenting sample. We typically use on the order of 100 scans for analysis, which might mean, for example, importing every third scan from a 300-scan dataset. However, this is not a hard-and-fast rule and it is important to examine different combinations of scans to ensure that appropriate scans are selected that represent

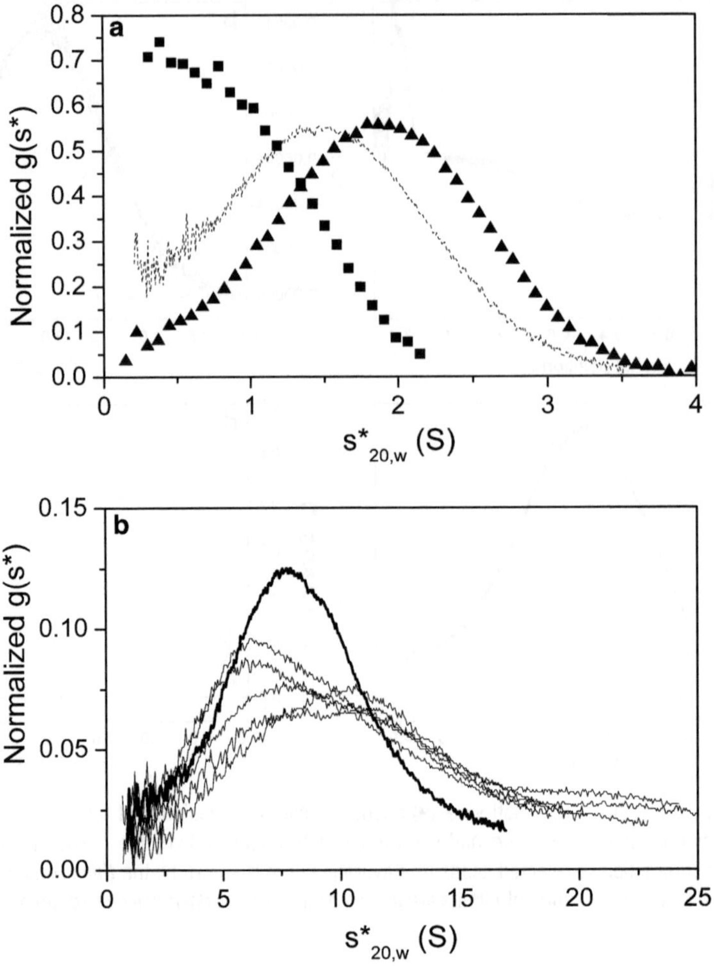

Fig. 7.5. Normalized *g*(*s**) distribution curves for $G_2T_4G_2$ to $G_{10}T_4G_{10}$. (**a**) Discrete *s* distributions are revealed for $G_2T_4G_2$ through $G_4T_4G_4$ with the distribution centered ~1 S for $G_2T_4G_2$ (*squares*) and ~2 1.5-2 S for $G_3T_4G_3$ and $G_4T_4G_4$ (*dashed black line and triangles*, respectively). (**b**) A switch in sedimentation behavior is observed for $G_5T_4G_5$ and higher ($G_5T_4G_5$, *thick black line*; $G_6T_4G_6$ to $G_{10}T_4G_{10}$ *black lines*) with a broad distribution of multiple faster sedimenting species.

the observed sedimentation behavior. To aid in the process of scan selection, we employ XLGraph and WinMatch (38, 39). XLGraph is used to examine all scans and identify bad scans that will not be included in the analysis (this is also useful for DCDT+ but less of an issue given the smaller number of scans selected for analysis). WinMatch is used to assess when the sample has completely sedimented. A group of scans is loaded and a plot made of the differences between each scan and the last scan of the group; in this way the scan number can be identified where there are no further changes in sedimentation.

2. Similarly to DCDT+, positions are marked for the meniscus and the left and right fitting limits. Sedfit also has a cell bottom

position indicator and limits for both the meniscus and cell bottom which need to be adjusted (see Note 6).

The cell bottom position is typically marked close to a radial position of 7.2 cm and is visualized as a distinct break from a high absorbance signal corresponding to material piling up at the bottom of the cell to a lower absorbance, noisy region corresponding to the material of the centerpiece. The meniscus and cell bottom limits are the upper and lower limits within which the meniscus and cell bottom can move when these positions are allowed to float during fitting. It is possible to accept default values for these limits if there is uncertainty about their placement.

3. Subtract all systematic noise.

4. Select continuous *c*(*s*) distribution.

For a first-pass analysis, we select the continuous *c*(*s*) distribution. This model is strictly correct only for mixtures of noninteracting, ideally sedimenting species; other more suitable models can be selected in the later stages of fitting if appropriate.

5. Input appropriate parameter values.

Values for the meniscus and bottom are taken from the positions selected in step 2 and should not be changed. A typical range of *s* values is from 0.5 to 10 S. We typically select a resolution to give a spacing of 0.1 S, so for a 0.5–10 S range the resolution would be 96. For absorbance data, radial-independent and time-independent noise are typically not fitted (unchecked). The baseline is typically floated (checked). The typical value of 1.2 for the frictional ratio is fixed (unchecked). The confidence level is set to 0.68 (1 standard deviation).

6. Execute the "Run" command.

The run command optimizes the linear parameters (baseline, RI noise, TI noise, loading concentrations, and size distributions) to provide better starting guesses for subsequent fitting ("Fit" command – step 7). A simulation of the sedimentation process is made with the entered parameters and compared with the experimental data. At this stage, the residuals and root mean-square deviation (rmsd) will appear, indicating the similarity between the simulated and experimental data, and an initial *c*(*s*) distribution will be displayed. Adjustment of the parameters or fitting limits can be made. For example, large residuals at the base of the cell would indicate that the right fitting limit should be adjusted to account for possible back-diffusion extending further back into the cell than the position of the right fitting limit. Also, partial peaks may appear at the extreme ends of the *c*(*s*) distribution. This could indicate that there might be a sedimenting species outside of the range of *s* values selected for analysis. If this is the case, the *s* range should be extended. If the rmsd is unchanged and the partial peak height increases, this indicates that the partial peak is an artifact and does not correspond to a smaller sedimenting species; the *s* range can be reset to the original range. Alternatively, if the rmsd is reduced and the partial peak height decreases, then there is a smaller sedimenting species and the extended range should be used. If any adjustments are made after the initial "Run" command, the "Run" command should be executed again. It should be noted that the appearance of this initial *c*(*s*) distribution will not necessarily be close to the final *c*(*s*) distribution obtained after the "Fit" function.

7. Execute the "Fit" command.

Having optimized the linear parameters, the meniscus, bottom, and frictional ratio parameters are checked to be fitted and the "Fit" command performed to optimize "these" nonlinear parameters. An initial round of fitting is performed with the simplex algorithm before switching to the Marquardt–Levenberg algorithm followed by a final round of the simplex algorithm. This is continued until there is no further decrease in rmsd.

8. Integrate the final *c*(*s*) distribution.

Integration reveals important information about the peaks in a *c*(*s*) distribution. Unlike DCDT+, peaks in the *c*(*s*) distribution are not adjusted to $s_{20,w}$; however, this information is calculated for each peak using the integration function and can be accessed by clicking on each of the molecular weight boxes that appear on each peak. The molecular weights are calculated using a weight-average best-fit frictional ratio and as such they may not be accurate for all solution species (*see* Note 7). These molecular weights can be graphically represented by transformation into a *c*(*M*) distribution. Similar to DCDT+, the *c*(*s*) distribution should be normalized for comparison of a concentration series. In this version of Sedfit, there is no function to do this and so we export the data to a convenient graphing program for normalization.

At this stage, a comparison of the *g*(*s**) and *c*(*s*) distributions for all GQs can be made (Figs. 7.5 and 7.6). The *g*(*s**) and *c*(*s*) distributions show similar trends. Broad distributions are apparent for $G_5T_4G_5$ through $G_{10}T_4G_{10}$, suggesting multiple higher order sedimenting species. Discrete *s* distributions are revealed for $G_2T_4G_2$ through $G_4T_4G_4$, with Sedfit revealing a second minor peak for $G_3T_4G_3$ and $G_4T_4G_4$. Estimation of *M* values using Sedfit suggested the minor peak to correspond to monomer strand molecular weight and the major peak to dimer molecular weight. Peaks with relatively close *s* values are not resolved in DCDT+ because of the inclusion of diffusion, but could be revealed by fitting to a two species model (see Subheading 3.4).

It is informative to calculate a theoretical *s* value for a sphere using the Svedberg equation and compare this to the experimentally observed values. The *s* value for a sphere will represent the maximum *s* value for a sample of a given molecular weight because

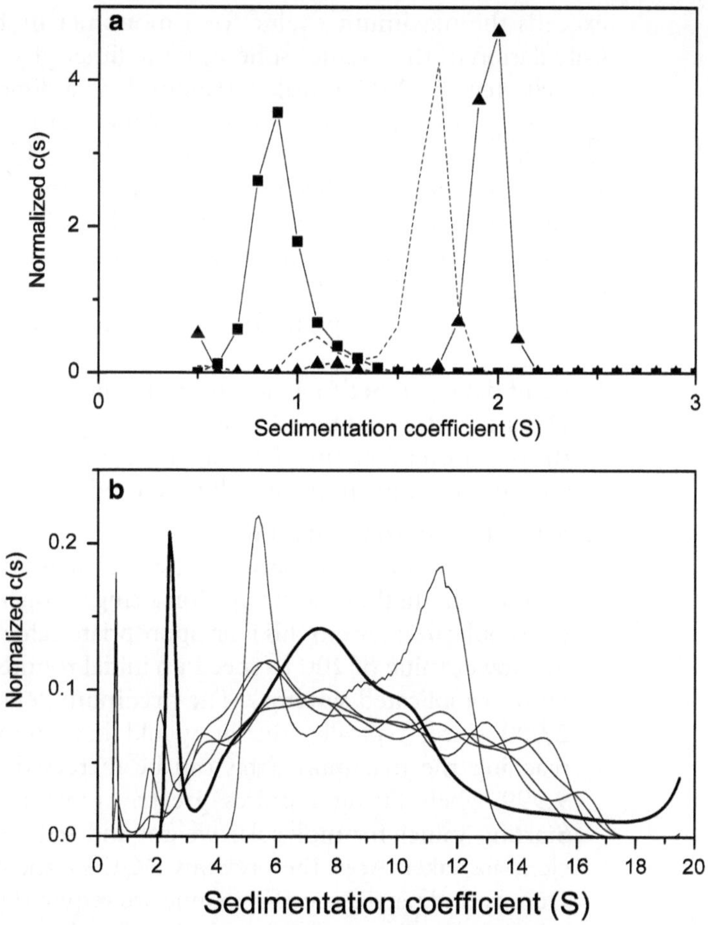

Fig. 7.6. Normalized $c(s)$ distribution curves for $G_2T_4G_2$ to $G_{10}T_4G_{10}$. (**a**) A single peak ~1 S is apparent for $G_2T_4G_2$ (*squares*). For $G_3T_4G_3$ and $G_4T_4G_4$ (*dashed black line and triangles*, respectively) a major peak ~1.5–2 S and a minor peak ~1 S are observed, indicating the presence of two sedimenting species. (**b**) Broad distributions of multiple faster sedimenting species (similar to the $g(s^*)$ distribution) are observed for $G_5T_4G_5$ to $G_{10}T_4G_{10}$ ($G_5T_4G_5$, *thick black line*; $G_6T_4G_6$ to $G_{10}T_4G_{10}$ *black lines*).

of the fact that a sphere has a minimum frictional coefficient and thus represents the shape of the fastest possible sedimenting species for that molecular weight. The comparison of a calculated s value for a sphere and the s value observed experimentally will provide information about the molecularity of a given sample. This process has been outlined by Lebowitz *et al.* (2). As an example, $G_4T_4G_4$ in Fig. 7.6 shows a major peak corresponding to $s_{20,w}$ = 2.02 S and M = 7.9 kDa (close to the expected dimer M of 7.6 kDa) and a minor peak corresponding to $s_{20,w}$ = 1.23 S and M = 3.8 kDa (monomer strand M = 3.8 kDa). For $G_4T_4G_4$, assuming $\bar{\nu}$ = 0.55 mL/g and ρ = 1.00712 g/mL, a strand M of 3.8 kDa would yield an s value (sphere) = 1.59 S. This compares with experimental s values from Sedfit of 1.23 S and 2.02 S, indicating that the 2.02 S peak cannot correspond to a monomer as it exceeds the maximum *s* value for a monomer molecular weight. Calculation of the *s* value (sphere) for a dimer gives 2.52 S, and it is likely that the 2.02 S peak corresponds to a dimer. The ratio of $s_{sphere}/s_{20,w}$ gives a weight-averaged shape factor $f/f_{0=1.25}$, which supports the assumption of a dimer with a moderately extended shape. This indicates that the two peaks in the *c*(*s*) distribution correspond to a monomer ~1.23 S and a dimer ~2.02 S with f/f_0 ~1.3 (similar to the minimized frictional ratio from Sedfit). Examination of the *g*(*s**) and *c*(*s*) distributions in this way provides a useful starting point for model-dependent analysis.

3.4. SedAnal

1. Load data into SedAnal according to the software directions. Having performed detailed analysis using DCDT+ and Sedfit, the selection of meniscus, fitting limits and scans from DCDT+ and the cell bottom from sedfit are used.
2. Select the appropriate model.
   The number of points between the meniscus and base is set to 800 for the highest accuracy (for a single concentration of an uncomplicated system this is an appropriate value; this can be set to a lower value of 200 to speed up initial rounds of analysis for more complicated systems). The maximum iterations are set to 2,000 since, typically, the fit would have minimized before reaching the maximum (this can be increased to the limit of 9,999 when fitting requires greater numbers of iterations). Starting values for molecular weight and sedimentation coefficient are taken from the previous DCDT+ and Sedfit analyses. Density and partial specific volume are required (see Note 5) to calculate the density increment, $1 - \bar{v}\rho$. Mass extinction coefficient is the extinction coefficient per gram of sample. For nucleic acids, a molar extinction coefficient is more typical and can be converted to a mass extinction coefficient by dividing by molecular weight. A number of researchers use nearest neighbor extinction coefficients (routinely supplied by oligonucleotide manufacturers) but we prefer to measure these experimentally. The mass extinction coefficient must be multiplied by 1.2 to account for the 1.2-cm path length of the instrument. The loading concentration is calculated from the sample absorbance before loading using a known extinction coefficient.
3. Select the simplex fitting option for initial rounds of analysis and then switch to the Levenberg–Marquardt algorithm.
4. For initial fitting, fix density increment and mass extinction and allow other parameters to float. Assess the minimized values and standard deviation.

A good model should have a low standard deviation (typically values in the range of 0.003 can be obtained with good-quality optics) and rational minimized parameters. The model can be tested further by fixing and floating other parameters. If the model is correct, similar parameter values should be obtained. Two examples will be presented for $G_2T_4G_2$ and $G_4T_4G_4$ to illustrate different fitting strategies.

For $G_2T_4G_2$, a single-species model is selected and the results of the fitting are shown in Fig. 7.7. The model describes the data well, with the data points evenly spaced about the fitted lines, and the residuals are randomly dispersed about zero; the minimized values for *M*, *s*, and loading concentration are consistent with $g(s^*)$ and $c(s)$ analyses and close to the expected values for the monomer and the standard deviation is low (0.00393).

For $G_4T_4G_4$, Sedfit indicated the presence of the two sedimenting species with a major peak ~2.02 S, suggesting a dimeric specie, and a minor peak ~1.23 S monomeric specie; however, the minor peak was estimated to be on the order of only 5% of the total solution content. Fitting to three different models was attempted with SedAnal: (1) monomer–dimer equilibrium model; (2) noninteracting two-species model; and (3) single-species model. An example of one of the fitting results is shown in Fig. 7.8 for the noninteracting two-species model. Fitting to a monomer–dimer equilibrium model (setting *K* to a starting value of 1×10^5 M^{-1}) returned an *s* value ~ 2 S and an *M* value close to dimer for one specie and an *s* value ~ 4 S and *M* value close to tetramer for the second specie with a good rmsd (0.00387). Analysis with a

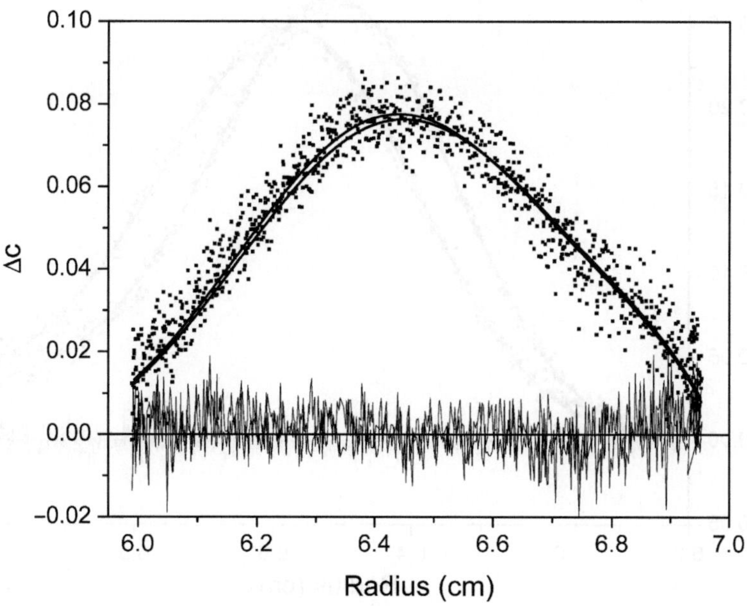

Fig. 7.7. A plot of the minimized fit to a single species model for $G_2T_4G_2$ using SedAnal for. Data are plotted as ΔC versus radius, where ΔC is the difference between pairs of closely spaced boundary scans. The data are shown as black squares and the best fit as *thick black lines*, with the residuals (*thin black lines*) randomly distributed about zero.

noninteracting two-species model (setting the ratio of species at a starting value of 5 %) returned *s* values close to the experimental *c*(*s*) values and *M* values close to monomer and dimer with approximately 5% of monomer and a standard deviation of 0.00385 for this limited dataset (Fig. 7.8). It is impossible to assess the integrity of the two models by fitting to only a single dataset for the sample. The monomer–dimer equilibrium model returns a second species ~4 S, which is obviously not apparent in either the *g*(*s**) or *c*(*s*) distributions but would not be expected to be apparent for a kinetically mediated equilibrium. A comprehensive concentration series is necessary to reveal the behavior of the system. Plots of the normalized *g*(*s**) and *c*(*s*) distributions should superimpose for a noninteracting system but should shift to larger species with increasing loading concentration for self-associating systems. For self-associating systems, it is informative to investigate the kinetics of association through fitting of k_{off}; more information can be found in a recent article by Correia and Stafford (45). A thorough analysis of the solution behavior of this sample is beyond the scope of this introductory tutorial to SV and the reader is referred to the appropriate literature to attempt these studies (11, 45–54). The final model-fitting attempted was to a single-species model which might be appropriate given the low

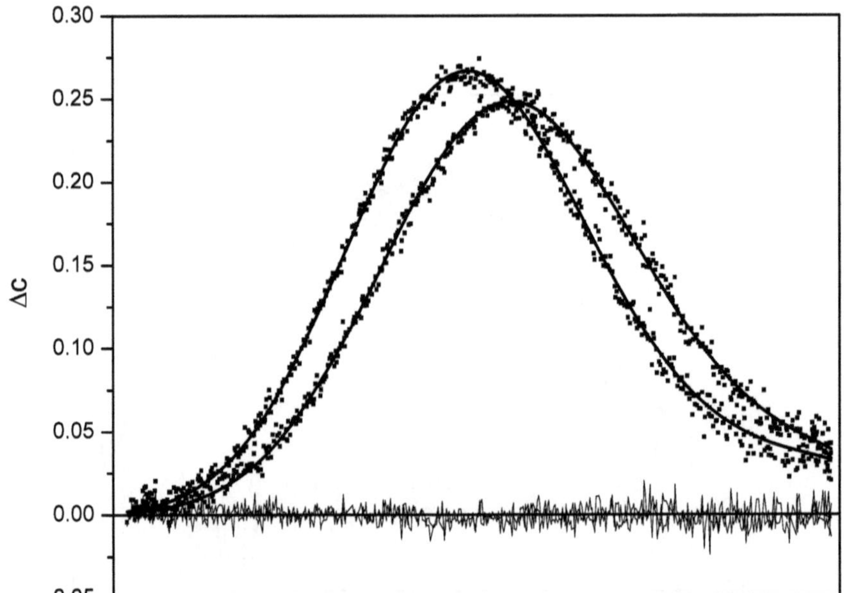

Fig. 7.8. A plot of the minimized fit to a noninteracting two-species model $G_4T_4G_4$ using SedAnal. Data are plotted as ΔC versus radius, where ΔC is the difference between pairs of closely spaced boundary scans. The data are shown as *black squares* and the best fit as *thick black lines*, with the residuals (*thin black lines*) randomly distributed about zero.

percentage of monomer estimated by *c*(*s*) analysis. The final fit from this model was also good, returning an *s* value ~ 2 S and an *M* value close to that of the dimer. The standard deviation was close to that for the other two models (0.00391). For this sample, there is little to choose between the three models, and a more comprehensive analysis is required as outlined above.

Model-fitting can also be performed using DCDT+ and Sedfit. DCDT+ does not offer the range of models incorporated into Sedfit and SedAnal but can be used to fit up to five discrete, noninteracting species; for the three examples highlighted above, DCDT+ would have utility. DCDT+ also has the option of fitting for either molecular weight or diffusion. A discrete-species Lamm equation model would be a useful starting point using Sedfit. Of course, the example data highlighted in this article only represents the starting point for any thorough analysis of sedimentation behavior. It is useful to run the same sample at different concentrations and rotor speeds to reveal the presence of, for example, an interacting system, conformational changes, or hydrodynamic nonideality. In the absence of macromolecular association, a decrease in sedimentation coefficient with concentration can provide information about the asymmetry of a molecule (4). However, an increase in *s* value with concentration is indicative of self-association; if association is present, stability of the complex can be determined by the effect of dilution on *s* value. A recent tutorial article by Correia and Stafford illustrates the rigor that is needed to correctly characterize SV data for a relatively simple equilibrium monomer–dimer system (45).

While sedimentation coefficients, frictional ratios, diffusion coefficients, and sample molecularity can provide useful information about the structural nature of a sample, determination of these properties is a long way from providing the absolute structure of a molecule. One final process that can be of utility in the characterization of the hydrodynamic properties of a sample is to compare experimentally determined sedimentation coefficients with those calculated from published structures and determine whether the structures are consistent with the solution conformation of a molecule (4). This approach has been used successfully in our laboratory to assess the solution conformation of the 22nt human telomere sequence in both sodium and potassium solutions (55). Hydrodynamic properties were calculated using “bead models” from the published atomic-level structures for the crystallographic potassium propeller structure and the NMR-determined sodium basket structure (56–59). Comparison of distributions of calculated sedimentation coefficients with experimentally determined *c*(*s*) distributions revealed striking differences between calculated and experimental *s* values for the potassium form but not for the sodium form. This suggested a hydrodynamically more compact conformation in potassium

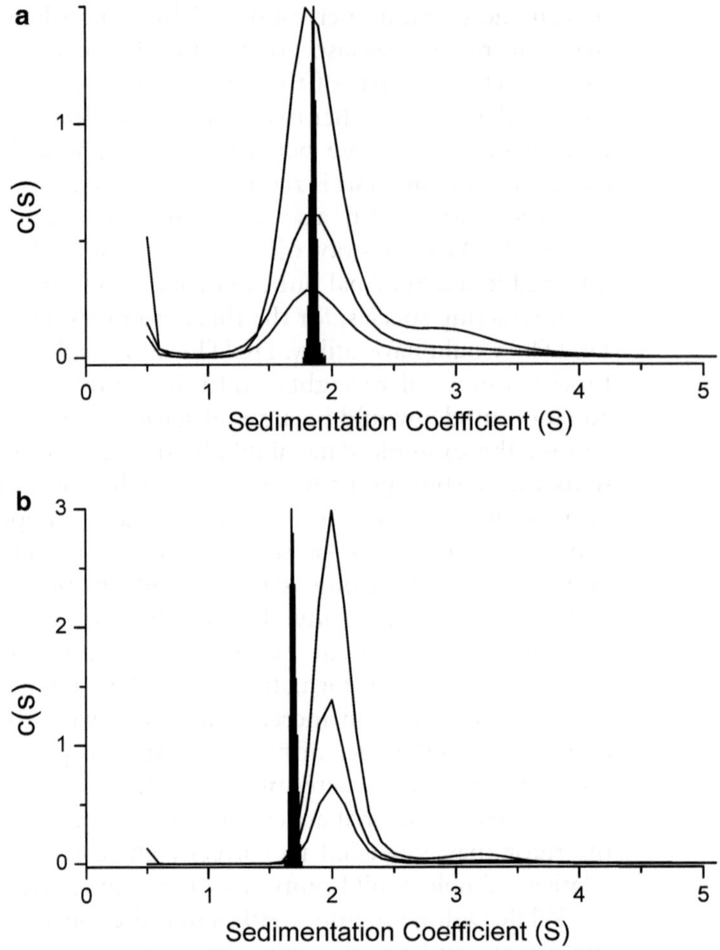

Fig. 7.9. Comparison of distributions of calculated sedimentation coefficients with experimentally determined *c*(*s*) distributions. Data for three different loading concentrations (5.1, 2.55, 1.275 μM strand) are shown by the black lines. The black distributions denote calculated sedimentation coefficients using "bead models" from the published atomic-level structures for the 22nt human telomere in and potassium sodium and potassium solutions. Significant differences are apparent between calculated and experimental *s* values for the potassium form (**b**) but not for the sodium form (**a**) and suggest a hydrodynamically more compact conformation in potassium solution than that observed in the crystal form. Reproduced with permission from Li et al. [55].

solution than that observed in the crystal form (Fig. 7.9). Calculated diffusion coefficients showed the same trend with close similarity to the experimentally determined value for the sodium form but the experimentally determined diffusion coefficient for the potassium form represents a more hydrodynamically compact structure compared to that calculated from the crystal structure. The use of hydrodynamic parameters calculated from atomic structures has therefore proved to be of great utility in discerning the solution structure of GQs.

7.4. Notes

1. Determination of molecular weight is typically more accurate using SE; therefore, in our laboratory we determine *s* values using SV and *M* values from SE experiments and then calculate diffusion coefficients using the Svedberg equation.
2. Runs at 290,304 × *g* were performed with graphite-filled epoxy centerpieces from Spin Analytical, Inc. (SedVel60K). Beckman Coulter, the manufacturer of the modern AUC instruments XL-A or XL-I, supply Epon charcoal-filled centerpieces rated to 142,249 × *g*. We routinely run these at 163,296 × *g* or 201,600 × *g* without problems.
3. It might not be possible to capture multiple species with the selection of one set of data scans. In these cases, it would be necessary to analyze these separately with the selection of different ranges of scans. Judicial scan selection might also be a useful strategy to remove a complicating species (e.g., an aggregate) from the analysis.
4. When running a concentration series for a particular sample, use the same scan selection for all concentrations.
5. Conversion to $s^*_{20,w}$ requires input of solution density, viscosity, and the partial specific volume of the sample. Ideally, these values should be determined experimentally (density and viscosity can vary even between batches of the same buffer) using a density meter and a kinematic viscometer (2); however, in the absence of access to a suitable experimental setup, they can be calculated using the program Sednterp (40). For proteins, the partial specific volume can be calculated with good accuracy from its amino acid composition, which can easily be imported as one or three-letter codes in a text file or from a databank. For DNA, a value of 0.55 mL/g is typically assumed (60). A recent article by Hellman et al. addressed the issue of estimation of partial specific volume for GQ DNA (61).
6. Depending on the nature of the sample, position of the right fitting limit might be a little different between DCDT+ and Sedfit. For Sedfit, where scans are used for the entire run, the limit might be set a little lower to exclude the upward curvature towards the base of the cell. This curvature would extend further into the cell for a sample with a small *s* value that might not completely sediment to the bottom of the cell (similar to that seen for $G_2T_4G_2$). However, for DCDT+ the limit is set to correspond to the end of the plateau region for scans selected in the middle of the run and this might be at a higher radial position than the situation for Sedfit. These differences are typically very small for samples that sediment completely to the bottom of the cell.

7. The frictional coefficient *f* provides a measure of the shape of a molecule (62). For a molecule with the same molecular weight, a compact structure will have a smaller *f* value than a more elongated structure. For a given molecule, reference is made to a smooth, compact sphere which has the minimum surface area in contact with solvent and hence the minimal frictional coefficient, designated f_o. f_o is calculated from the Stokes equation with the particle radius defined as that of a sphere. The frictional ratio f/f_0 is a shape factor providing a measure of the maximal shape asymmetry from that of a sphere. The frictional ratio can be determined by the ratio of the experimental *s* value to the maximal *s* value calculated for a sphere of the same molecular weight under the same solution conditions (*see* Subheading 3.3). The frictional ratio has a theoretical minimum value of 1 for a perfectly spherical molecule; a value of 1.2 is commonly accepted as a reasonable starting value during analysis for moderate molecular asymmetry. It is important to evaluate minimized values of frictional ratios to assess whether they reasonably represent the expected solution structure. Unfortunately, little is known about frictional ratios for DNA; it is possible to generate some information about a molecule's shape from published atomic structures or through molecular modeling; Sednterp and Sedfit also offer convenient shape calculators.

Acknowledgments

We thank Jack Correia for countless helpful discussions. This work was supported by National Cancer Institute grant CA35635 to J.B.C.

References

1. Hansen JC, Kreider JI, Demeler B, Fletcher TM (1997) Analytical ultracentrifugation and agarose gel electrophoresis as tools for studying chromatin folding in solution. Methods 12:62–72
2. Lebowitz J, Lewis MS, Schuck P (2002) Modern analytical ultracentrifugation in protein science: a tutorial review. Protein Sci 11:2067–79
3. Hansen JC, Lebowitz J, Demeler B (1994) Analytical ultracentrifugation of complex macromolecular systems. Biochemistry 33: 13155–63
4. Laue T (2001) Biophysical studies by ultracentrifugation. Curr Opin Struc Biol 11:579–83
5. Howlett GJ, Minton AP, Rivas G (2006) Analytical ultracentrifugation for the study of protein association and assembly. Curr Opin Chem Biol 10:430–36
6. Cole JL, Hansen JC (1999) Analytical ultracentrifugation as a contemporary biomolecular research tool. J Biomol Tech 10:163–76
7. Ralston G (1993) Introduction to analytical ultracentrifugation. Beckman Instruments, Inc., Fullerton, CA
8. Laue TM, Stafford WF III (1999) Modern applications of analytical ultracentrifugation. Annu Rev Biophys Biomol Struct 28:75–100
9. Stafford WF, Sherwood PJ (2004) Analysis of heterologous interacting systems by

sedimentation velocity: curve fitting algorithms for estimation of sedimentation coefficients, equilibrium and kinetic constants. Biophys Chem 108:231–43

10. Schuck P, Perugini MA, Gonzales NR, Howlett GJ, Schubert D (2002) Size-distribution analysis of proteins by analytical ultracentrifugation: strategies and application to model systems. Biophys J 82:1096–111
11. Brown PH, Balbo A, Schuck P (2008) Characterizing protein–protein interactions by sedimentation velocity analytical ultracentifrugation. Curr Protoc Immunol 81; 18.15.1–18.15.39
12. Balbo A, Schuck P (2005) Analytical ultracentrifugation in the study of protein self-association and heterogeneous protein–protein interactions. In: Golemis EA, Adams PD (eds) Protein-Protein Interactions: A Molecular Cloning Manual. Cold Spring Harbor Laboratory, Woodbury, NY, pp 253–77
13. Balbo A, Brown PH, Schuck P (2008) Experimental protocol for sedimentation velocity analytical ultracentrifugation. Available from http://www.analyticalultracentrifugation.com/svprotocols.htm
14. Demeler B, Saber H, Hansen JC (1997) Identification and interpretation of complexity in sedimentation velocity boundaries. Biophys J 72:397–407
15. van Holde KE, Weischet WO (1978) Boundary analysis of sedimentation-velocity experiments with monodisperse and paucidisperse solutes. Biopolymers 17:1387–403
16. Carruthers LM, Schirf VR, Demeler B, Hansen JC (2000) Sedimentation velocity analysis of macromolecular assemblies. Methods Enzymol 321:66–80
17. Stafford WF (1992) Boundary analysis in sedimentation transport experiments: a procedure for obtaining sedimentation coefficient distributions using the time derivative of the concentration profile. Anal Biochem 203:295–301
18. Stafford WF (2000) Analysis of reversibly interacting macromolecular systems by time derivative sedimentation velocity. Methods Enzymol 323:302–25
19. Philo JS (2000) A method for directly fitting the time derivative of sedimentation velocity data and an alternative algorithm for calculating sedimentation coefficient distribution functions. Anal Biochem 279:151–63
20. Philo JS (1997) An improved function for fitting sedimentation velocity data for low-molecular-weight solutes. Biophys J 72:435–44
21. Schuck P, MacPhee CE, Howlett GJ (1998) Determination of sedimentation coefficients for small peptides. Biophys J 74:466–74
22. Schuck P, Millar DB (1998) Rapid determination of molar mass in modified archibald experiments using direct fitting of the Lamm equation. Anal Biochem 259:48–53
23. Schuck P (1998) Sedimentation analysis of noninteracting and self-associating solutes using numerical solutions to the Lamm equation. Biophys. J. 75:1503–12
24. Schuck P (2000) Size-distribution analysis of macromolecules by sedimentation velocity ultracentrifugation and Lamm equation modeling. Biophys J 78:1606–19
25. Schuck P, Rossmanith P (2000) Determination of the sedimentation coefficient distribution by least-squares boundary modeling. Biopolymers 54:328–41
26. Schuck P (1999) Sedimentation equilibrium analysis of interference optical data by systematic noise decomposition. Anal Biochem 272:199–208
27. Laue TM, Shah BD, Ridgeway TM, Pelletier SL (1992) Computer-aided interpretation of analytical sedimentation data for proteins. In: Harding SE, Rowe AJ, Horton JC (eds) Analytical ultracentrifugation in biochemistry and polymer science. Royal Society of Chemistry, Cambridge, U.K, pp 90–125
28. Schuck P, Demeler B (1999) Direct sedimentation analysis of interference optical data in analytical ultracentrifugation. Biophys J 76:2288–96
29. Philo J (2006) Improved methods for fitting sedimentation coefficient distributions derived by time-derivative techniques. Anal Biochem 354:238–46
30. Cole JL, Lary JW, Moody TP, Laue TM (2008) Analytical ultracentrifugation: sedimentation velocity and sedimentation equilibrium. In: Correia JJ, Detrich HW (eds) Methods in cell biology, Vol. 84, biophysical tools for biologists: volume 1, in vitro techniques. Academic, San Diego, CA, pp 143–79
31. Schuck P (2008) SEDFIT (version 11.3b). National Institutes of Health, Bethesda, MD. Available from http://www.analyticalultracentrifugation.com/download.htm.
32. Schuck P (2008) SEDPHAT (version 5.14). National Institutes of Health, Bethesda, MD. Available from http://www.analyticalultracentrifugation.com/sedphat/download.htm.
33. Sherwood P, Stafford W (2008) SedAnal, sedimentation analysis software (version 5.03). Boston Biomedical Research Institute, Watertown, MA. Available from http://sedanal.bbri.org/.

34. Philo JS (2007) DCDT+ (version 2.1.0.28333). John S Philo, Thousand Oaks, CA. Available from http://www.jphilo.mailway.com/dcdt+.htm.
35. Cole JL, Lary JW (2006) HeteroAnalysis (version 1.1.33). Analytical ultracentrifugation facility, Biotechnology Bioservices Center, University of Connecticut, Storrs, CT. Available from http://www.biotech.uconn.edu/auf/?i=aufftp.
36. Philo JS (2001) SVEDBERG (version 6.39). John S Philo, Thousand Oaks, CA. Available from http://www.jphilo.mailway.com/svedberg.htm.
37. Demeler B (2008) UltraScan (version 9.9 (Revision 714)). Borries Demeler and The University of Texas Health Science Center at San Antonio, San Antonio, TX. Available from http://www.ultrascan.uthscsa.edu/.
38. Philo JS (2003) XLGraph (version 3.21). John S Philo, Thousand Oaks, CA. Available from http://www.jphilo.mailway.com/download.htm.
39. Yphantis D, Lary JW (1999) WinMatch V0.99 forWin95+WinNT.ProgramForIntercomparison of Ultracentrifuge Data Files (version 0.99.0039). David A. Yphantis and Jeffrey W Lary, Mansfield Center, CT. Available from http://www.biotech.uconn.edu/auf/?i=aufftp.
40. Hayes DB, Laue T, Philo J (2006) Sedimentation Interpretation Program (version 1.09). University of New Hampshire, Durham, NH. Available from http://www.jphilo.mailway.com/download.htm.
41. Philo, JS DCDT+ Home Page John S Philo, Thousand Oaks, CA. Available from http://www.jphilo.mailway.com/dcdt+.htm.
42. Shafer RH, Smirnov I (2001) Biological Aspects of DNA/RNA Quadruplexes. Biopolymers 56:209–27
43. Scaria PV, Shire SJ, Shafer RH (1992) Quadruplex structure of d(G3T4G3) stabilized by K+ or Na+ is an asymmetric hairpin dimer. Proc Natl Acad Sci USA 89:10336–40
44. Mekmaysy C (2007) The structure and stability of G-quadruplexes. Masters Thesis, University of Louisville, Louisville, KY, 109 p.
45. Correia JJ, Stafford WF (2009) Extracting equilibrium constants from kinetically limited reacting systems. Methods Enzymol 455, in press.
46. Correia JJ (2000) Analysis of weight average sedimentation velocity data. Methods Enzymol 321:81–100
47. Kegeles G, Cann J (1978) Kinetically controlled mass transport of associating-dissociating macromolecules. Methods Enzymol 48:248–70
48. Gelinas AD, Toth J, Bethoney KA, Stafford WF, Harrison CJ (2004) Mutational analysis of the energetics of the GrpE.DnaK binding interface: equilibrium association constants by sedimentation velocity analytical ultracentrifugation. J Mol Biol 339:447–58
49. Stafford WF (2009) Protein-protein and ligand-protein interactions studied by analytical ultracentrifugation. In: Schriver JW (ed) Methods in molecular biology, Vol. 490, Protein structure, stability, and interactions. Humana, Totowa, NJ, pp 83–113
50. Kegeles G (1978) Pressure-jump light-scattering observations of macromolecular interaction kinetics. Methods Enzymol 48:308–20
51. Cox DJ (1978) Calculation of simulated sedimentation velocity profiles for self-associating solutes. Methods Enzymol 48:212–42
52. Cann JR (1978) Measurement of protein interactions mediated by small molecules using sedimentation velocity. Methods Enzymol 48:242–48
53. Cann JR (1978) Ligand binding by associating systems. Methods Enzymol 48:299–307
54. Cann JR, Kegeles G (1974) Theory of sedimentation for kinetically controlled dimerization reactions. Biochemistry 13:1868–74
55. Li J, Correia JJ, Wang L, Trent JO, Chaires JB (2005) Not so crystal clear: the structure of the human telomere G-quadruplex in solution differs from that present in a crystal. Nucleic Acids Res 33:4649–59
56. García de la Torre J (2001) Hydration from hydrodynamics. General considerations and applications of bead modelling to globular proteins. Biophys Chem 93:159–70
57. García de la Torre J, Huertas ML, Carrasco B (2000) Calculation of hydrodynamic properties of globular proteins from their atomic-level structure. Biophys J 78:719–30
58. Fernandes MX, Ortega A, López Martínez MC, García de la Torre J (2002) Calculation of hydrodynamic properties of small nucleic acids from their atomic structure. Nucleic Acids Res 30:1782–88
59. Byron O (2000) Hydrodynamic bead modeling of biological macromolecules. Methods Enzymol 321:278–304
60. Bloomfield VA, Crothers DM, Tinoco I Jr (2000) Nucleic Acids: Structures. Properties and Functions. University Science Books, Sausalito, CA, p 359
61. Hellman LM, Rodgers DW, Fried MG (2009) Phenomenological partial specific volumes for G-Quadruplex DNAs. Eur. Biophys. J, in press.
62. StaffordWF,SchusterTM(1995)Hydrodynamic Methods. In: Glasel JA, Deutscher MP (eds) Introduction to biophysical methods for protein and nucleic acid research. Academic, San Diego, CA, pp 111–45

Chapter 8

2-Aminopurine as a Probe for Quadruplex Loop Structures

Robert D. Gray, Luigi Petraccone, Robert Buscaglia, and Jonathan B. Chaires

Abstract

Fluorescent reporter groups have served for many years as sensitive probes of macromolecular structure. Such probes can be especially useful in comparative studies such as detection of conformational changes and discrimination among structural models. Spectroscopic methods such as fluorescence are attractive because they are rapid, require small amounts of material, are nondestructive, can be carried out with commonly available equipment, and are relatively inexpensive. In addition, there is a rich library of theoretical and practical materials available to aid in data interpretation.

The intrinsic fluorescence of most nucleic acids is too low to be useful in structural studies. Thus, it is necessary to incorporate a suitable reporter group to utilize fluorescence methods involving polynucleotide structure. A highly fluorescent adenine analog, 2-aminopurine, has long served in this capacity. The present article describes our use of 2-aminopurine as a probe of loop structures in quadruplex DNA. In particular, we show how knowledge of the relative intensity of 2-aminopurine emission as well as its sensitivity to exogenous quenching molecules such as acrylamide can aid in comparing crystal and solution structures of an oligonucleotide model of the human telomere and in discrimination among models containing tandem repeats of the telomeric quadruplex.

Key words: Quadruplex DNA, Telomere model, Polynucleotide folding, Loop Structure, Fluorescence, Fluorescence quenching, Quantum yield, Fluorescence lifetime, 2-Aminopurine, Solvent accessibility

1. Introduction

1.1. Quadruplexes

Polynucleotides containing strings of contiguous guanine residues can associate to form quadruple stranded structures referred to as quadruplexes. There are many recent reviews of DNA quadruplexes that one may consult for details (e.g., (2–7)). For convenience, we provide a brief discussion of the structure of DNA quadruplexes below. These macromolecular assemblies consist of helical arrays of stacked planar, cyclic structures of four guanines

P. Baumann (ed.), *G-Quadruplex DNA: Methods and Protocols*, Methods in Molecular Biology, vol. 608
DOI 10.1007/978-1-59745-363-9_8, © Humana Press, a part of Springer Science+Business Media, LLC 2010

linked by eight hydrogen bonds involving the N1, N2, N7, and O6 atoms of the bases within an individual G-quartet. The O6 atoms within a quartet are directed toward the central cavity of the quadruplex and thereby form a coordination site that can accommodate a cation such as Na^+ or K^+ (Fig. 8.1).

Conformational diversity is one of the remarkable characteristics of quadruplexes. This plasticity results from a variety of chemical and physical variables. For example, the number of G-quartets comprising a stack may vary; the individual strands may be parallel, antiparallel, or of mixed polarity; the loops connecting the quartets may be diagonal or lateral; the sequences flanking the stacks may differ; the *N*-glycoside bonds of the G residues may adopt *syn* or *anti* conformations; and the stabilizing cation may vary. The topological diversity of monomolecular quadruplex structures with three G-stacks is illustrated schematically in Fig. 8.2.

Although the structure of G-quartets was first described in 1962 (9), interest in quadruplex forming oligonucleotides has accelerated over the last 20 years. This may be attributed to the observation in 1989 (10) that telomeres (which provide a

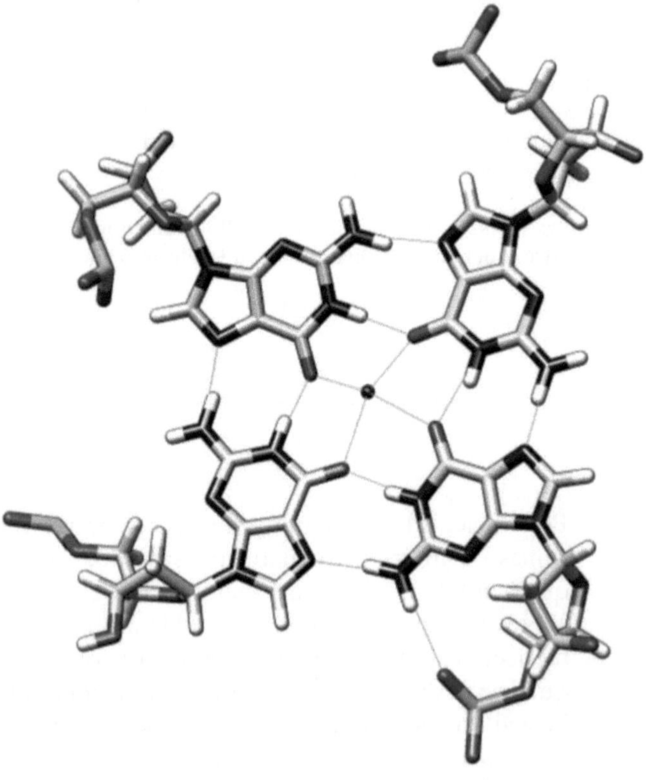

Fig. 8.1. G-quartet showing H-bonding pattern and coordinated sodium ion in center. The figure shows a single G-quartet and is based on the coordinates of HuTel22 in PDB file 143D (1). The hydrogen-bonding pattern is shown by the thin lines connecting the N and O atoms. The D-ribosyl phosphate moieties are shown on the periphery of the quartet.

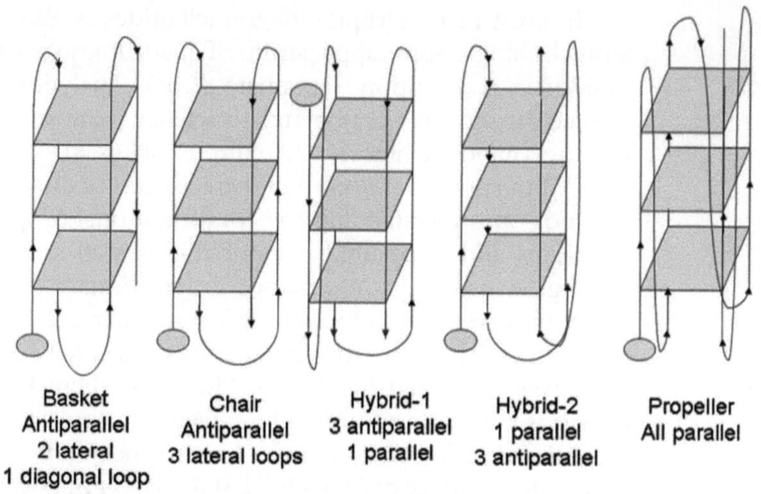

Fig. 8.2. Schematic diagram of potential folding topographies of unimolecular quadruplexes. The Na+ form of HuTel22 has been shown to exist in solution mainly in the form of an antiparallel basket (1). In the presence of K+, HuTel22 in crystals is an all-parallel "propeller" structure (23). In K+ solution, NMR and other physical studies indicate that HuTel22 is present as a mixture of different conformers (4, 8). The circle represents the 5′ end of the oligonucleotide.

mechanism for complete chromosomal replication and protect chromosomal ends from degradation) consist of repeating tracts of G-rich sequences such as d(TTTTGGGG) in the protozoan *Oxytricha nova* and d(TTAGGG) in humans. In human chromosomal DNA, 200–300 bp of these sequences protrude as a single-stranded overhang from the 3′ end, and studies with model oligonucleotides suggest that these sequences form a series of tandem quadruplexes (11). Subsequent investigators have shown that telomeres in vivo contain quadruplexes, and that quadruplex conformational changes may represent a key process in controlling chromosomal replication (12, 13). The relatively facile ability to respond to environmental changes by undergoing conformational transformations (which stands in remarkable contrast to their high degree of thermal stability) suggests that quadruplex motifs may possess regulatory properties in other fundamental biological contexts, for example, in cellular senescence and oncogenesis (14, 15).

Bioinformatic analysis has revealed the presence of G-rich segments in parts of eukaryotic genomes in addition to those described above (16); several of these segments have been shown to be involved in controlling expression of a number of genes, including a variety of oncogenes as well as sequences involved in immunoglobulin switching (17). In addition, based on their putative regulatory roles, quadruplex-containing DNA segments are rational targets for drug development (7, 18, 19). Development of drugs targeted to specific quadruplex sequences requires an appreciation of subtle conformational differences likely to exist among different quadruplexes.

Interest in quadruplex oligonucleotides is also increasing in biotechnology. Some applications of quadruplex structures include molecular recognition (aptamers (20)), analytical applications (molecular sensors (21)), materials science (nanostructures), engineering (nanomachines), (22) and as drugs (19). Full development of the potential of each of these aspects of quadruplex science depends on understanding the conformational properties of quadruplexes and quadruplex assemblies, as well as their ability to undergo controlled conformational transitions.

X-ray crystallography and nuclear magnetic resonance (NMR) methods continue to provide structures of a variety of specific quadruplexes (reviewed in (3, 6)). However, these high-resolution methods have proven not to be universally applicable. For example, NMR yielded an ensemble of conformers of the 22-nt human telomeric sequence d[AGGG(TTAGGG)$_3$] (HuTel22) in the presence of Na^+ or K+ (1). In Na+ structure consists of an antiparallel "basket" structure characterized by two lateral and one diagonal dTTA loops (Fig. 8.2). However, it was not possible to obtain a solution structure for HuTel22 in the presence of K^+ (the more physiologically relevant cation), presumably because of the presence of several conformational isomers. Interestingly, HuTel22 could be crystallized in the presence of K^+ (23). The resulting structure exhibits a parallel-stranded topology with the three dTTA loops packed against the quadruplex core (the "propeller" topology). However, several studies suggest that the all-parallel structure found in the crystal may not be the predominant solution structure ((8) and references therein). In fact, NMR solution studies of related sequences in the presence of K^+ suggest the possibility that so-called hybrid-1 or hybrid-2 (3 parallel strands + 1 antiparallel strand) may be the preferred solution structure in the presence of this cation (24–26).

These examples provide a concrete basis for the importance of applying a variety of lower resolution techniques to test structural hypotheses, to detect conformational changes, and to compare unknown structures with known structures. Of these methods, the use of site-specific spectroscopic probes is particularly useful because of their ability to report specific rather than global structural differences brought about by mutation or changes in solution conditions. In general, fluorescent probes can serve as versatile, site-specific conformational reporters.

1.2. Aminopurine (2AP) as a Site-Specific Oligonucleotide Structural Probe

In contrast to polypeptides, for which the intrinsic fluorescence of the indole and phenolic side chains provide sensitive fluorescent reporters, the fluorescence quantum yield of the purine and pyrimidine building blocks of polynucleotides is too low to be analytically useful (except for the relatively rare Y base found in t-RNA^{Phe}). Thus, it is necessary either to substitute a fluorescent purine or pyrimidine analog at the residue of interest or to attach a probe

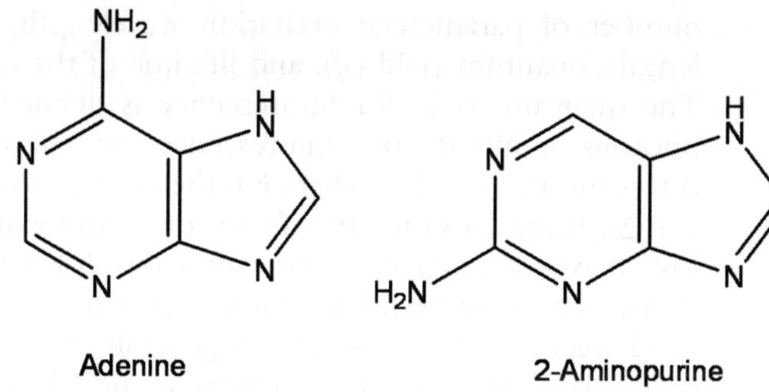

Fig. 8.3. Structures of adenine and 2-aminopurine.

molecule to the oligonucleotide. The adenine analog, 2-aminopurine (Fig. 8.3), has been applied in a large number of studies as a site-specific fluorescent reporter for polynucleotide studies. This fluorescent purine exhibits a number of desirable properties that render it a good probe molecule for polynucleotides. Among these are the following: (1) 2AP can form Watson–Crick H-bonds with thymidine, uracil, or cytidine, thus inducing little or no structural perturbation when substituted for adenine; (2) 2AP can be relatively inexpensively incorporated into oligonucleotides using common solid-phase synthetic methods; (3) The quantum yield of fluorescence is highly sensitive to environmental conditions and thus local changes in conformation; (4) Theoretical and practical aspects of the photochemistry of 2AP, especially in the context of oligonucleotide structure, are well established. In particular, 2AP fluorescence tends to increase with solvent exposure and to decrease with base stacking. The interested reader is referred to two excellent recent studies focusing on the application of site-specific 2AP probes of RNA folding (27, 28). These papers provide a comprehensive outline of the properties of 2AP in relation to oligonucleotide structure as well as a guide to recent studies involving 2AP-containing oligonucleotides.

In the following, we describe our use of 2AP as a probe of quadruplex folding. In these studies, we used the susceptibility of 2AP fluorescence to collisional quenching by acrylamide as a probe of adenine exposure present in loops or between tandem quadruplexes. Taken in conjunction with other studies, these fluorescence quenching experiments help distinguish between conformational models of a single quadruplex as well as tandem two-quadruplex models of human telomeric DNA.

1.3. Theoretical Background for Collisional Quenching of Fluorescence

Many excellent texts (for example (29)) provide good summaries of biological applications of fluorescence. Fluorescence is defined as the emission of a photon from a singlet excited state to the ground state subsequent to absorption of a photon by a fluorophore. The fluorescence properties of a system are characterized by a number of parameters: excitation wavelength, emission wavelength, quantum yield (q), and lifetime of the excited state (τ_0). The quantum yield for fluorescence is defined as the ratio of photons emitted by fluorescence to photons absorbed. Alternatively, $q = k_f/\Sigma k_i$, where k_f is the rate of fluorescence emission and Σk_i represents the sum of rate constants for all mechanisms of loss of excitation energy. This expression shows explicitly the link between quantum yield and fluorescence lifetime.

Direct determination of q is generally not convenient; however, it is often sufficient to determine the relative quantum yield F, which is the emission intensity of the system measured under defined conditions. F is a function of the instrument (intensity of excitation source, sensitivity of the detection system, etc.) as well as the conditions of the measurement (fluorophore concentration, temperature, and solution conditions). F is most accurately obtained from integration over the wavelengths of the emission envelope; however, it may be adequate to determine F as the emission intensity at the emission wavelength maximum. Depending on the instrument, F may be given in photons per second or in arbitrary units.

The excited state lifetime is the average time that a fluorophore spends in the excited state before emitting a photon. For a homogeneous system, fluorescence decay consists of a single exponential process. In general, for fluorophores in a heterogeneous environment, the decay of fluorescence consists of multiple exponentials as defined by the relationship $F(t)/F_0 = \Sigma\alpha_i \cdot \exp(-t/\tau_{0,i})$ where α_i is the amplitude of the fluorescence intensity for species i, and $\tau_{0,i}$ is the lifetime for that species. Fluorescence lifetimes are generally in the subnanosecond to nanosecond time range and require specialized instrumentation for determination.

For many fluorophores, the measurable characteristics of fluorescence such as emission maximum, q, and τ_0 are sensitive to the local environment of the fluorophore. However, for 2AP, it turns out that the emission maximum is not very sensitive to its environment, but q and τ_0 are quite sensitive to the environment. Often, the structural basis of these variations in τ_0 and q can be quite subtle. For example, 2AP fluorescence is enhanced by cooling (indicating the importance of conformational motion), quenched by base stacking, enhanced by energy transfer from neighboring adenine, and strongly quenched by collision with neighboring G residues. In addition, 2AP emission is quenched by solutes such as acrylamide, which quenches by collision with the excited state fluorophore. As described below, the action of collisional quenchers is generally described by the Stern–Volmer equation or modifications of it.

1.4. Stern–Volmer Quenching

Classical Stern–Volmer quenching (also referred to as dynamic quenching) can be explained by a simple mechanism in which the quenching agent Q collides with an excited fluorophore F*. In this type of quenching, the excited state energy is lost when the excited fluorophore collides with Q and forms a transient encounter complex as depicted in Eq. 8.1:

$$F^* + Q \leftrightarrow (F^* \cdots\cdots Q) \leftrightarrow (F \cdots\cdots Q) \rightarrow F + Q + \text{thermal energy.} \quad (8.1)$$

Formation of $(F^* \cdots\cdots Q)$ depends on the diffusion rate constant for F* and Q, their concentrations, and the efficiency of the quenching process. For a single fluorescent species in a homogeneous environment, the degree of quenching increases with increasing [Q] according to the Stern–Volmer relationship:

$$F_0 / F_Q = 1 + K_{SV}[Q] = 1 + \tau_0 k_q [Q] = 1 + \tau_0 f_Q k_0 [Q] \quad (8.2)$$

where F_0 is the fluorescence observed in the absence of quencher, F_Q is the fluorescence observed in the presence of quencher at concentration [Q], K_{SV} is the Stern–Volmer quenching coefficient, τ_0 is the lifetime of the excited state in the absence of Q, k_q is the bimolecular quenching constant, f_Q is the efficiency of quenching, and k_0 is the (diffusion-controlled) bimolecular collisional rate constant. From Eq. 8.2, a plot of F_0/F_Q vs. [Q] is linear with a slope K_{SV} from which f_0 can be extracted if τ_0 and k_0 are known. The former may be available in the literature or measured directly, and the latter can be estimated from the Smoluchowski equation (*see* Note 1). Under defined conditions for a given quenching agent and a homogeneous fluorescent system, f_0 is related to the accessibility of F* to Q. In the absence of known values of τ_0 and k_0, the relative accessibility of F* to Q can qualitatively be assessed by comparing values of K_{SV} measured for the same fluorophore under different solvent conditions. Hence, under ideal circumstances, collisional quenching can provide insight into the relative degree of exposure of a fluorophore to solvent.

Experience has shown that the Stern–Volmer plot is often not linear but exhibits either a concave or convex relationship with respect to [Q]. A nonlinear Stern–Volmer plot indicates a heterogeneous fluorescing system. This heterogeneity may arise either from fluorophores with different k_0, f_0, and/or τ_0 values, or from formation of static complexes with Q. In systems with a single fluorophore (such as the quadruplexes under consideration in this article), downward curvature is interpreted to indicate that the fluorophore experiences more than one microenvironment. Such a heterogeneous quenching system can in general be described by Eq. 8.3:

$$F_0 / F_Q = \sum_{i=1}^{n} \frac{f_i}{1 + K_{SV,i}[Q]}, \quad (8.3)$$

where $K_{SV,i}$ is the quenching constant for the *i*th species and f_i is the fractional contribution of the *i*th species to the total fluorescence. Explicitly for a two-component system:

$$F_0 / F_Q = \frac{f_1}{1 + K_{SV,1}[Q]} + \frac{1 - f_1}{1 + K_{SV,2}[Q]}. \quad (8.4)$$

Eftink (30) has shown that if $K_{SV,1} \geq 4 \times K_{SV,2}$, the F_0/F_Q plot will exhibit visible downward curvature. Fitting the data to Eq. 8.4 by nonlinear least squares allows determination of the optimal values of the parameters f_1, $K_{SV,1}$, and $K_{SV,2}$ (although see Note 2).

Upward curvature is generally interpreted to indicate formation of both static and dynamic (collisional) complexes (see Note 3). This case will not be treated here; the interested reader is referred to a standard text book for details regarding interpretation and quantitative analysis of upwardly curving Stern–Volmer plots.

1.5. Summary of Fluorescence Properties of 2AP

The fluorescence properties of 2AP in water are as follows: emission involves a $\pi^* \rightarrow \pi$ transition, with an excitation maximum at ~305 nm and emission maximum at ~370 nm, $q = 0.7$; fluorescence decay is mono-exponential with $\tau_0 = 12$ ns (31). The emission maximum is relatively insensitive to solvent polarity, and q is not affected by pH but is strongly dependent on temperature. In the context of an oligonucleotide, q is strongly influenced by nearest neighbors and by base stacking. Quenching of 2AP fluorescence by acrylamide gives a Stern–Volmer plot with slight upward curvature, which has been attributed to a weak static quenching component at high [acrylamide]. K_{SV} for acrylamide quenching of 2AP is ~28 M^{-1} at 10°C and increases to ~58 M^{-1} at 70°C.

2. Materials

1. A variety of different buffer systems have been used to study quadruplexes. For the studies described in Fig. 8.3, BPES (6 mM Na_2HPO_4, 2 mM NaH_2PO_4, 1 mM Na_2EDTA, 185 mM NaCl, pH 7.1) and BPEK (6 mM Na_2HPO_4, 2 mM NaH_2PO_4, 1 mM Na_2EDTA, 185 mM KCl, pH 7.1) buffers were used.
2. Concentrated acrylamide stock solutions (4–6 M) can be prepared by carefully adding the desired amount of high-quality acrylamide to a tared volumetric flask followed by addition of

the appropriate buffer to volume (see Note 4). Acrylamide solutions are susceptible to degradation and should be stored at 4 °C in a foil-wrapped container to protect from exposure to light. (see Note 4 regarding toxicity and disposal of acrylamide).

3. Desalted, lyophylized deoxyoligonucleotides with single dA residues substituted serially with 2AP were purchased from commercial sources: (Oligos Etc, http://www.oligosetc.com/index.phpor or IDT, http://www.idtdna.com/home/home.aspx). The oligonucleotides were used without further purification and were dissolved in a small volume of an appropriate buffer to give a stock solution of ~0.5–1 mM. Oligonucleotide concentrations were determined from the absorbance at 260 nm using absorption coefficients supplied by the manufacturer (see Note 5).

3. Methods

3.1. Acrylamide Titrations

1. The fluorescence measurements were carried out with a JASCO FP-6500 fluorometer equipped with a Peltier temperature controller and a magnetic mixer.
2. Oligonucleotides were excited at 305 nm and emission spectra were measured from 320 nm to 420 nm. Excitation and emission slits were set to 5 nm.
3. The fluorometer is allowed to stabilize for ~30 min after turning it on.
4. Add buffer to fluorescence cuvette; equilibrate with stirring for several minutes to achieve a constant temperature in the cuvette. We use a 1 × 1 cm quartz fluorescence cuvette that can accommodate 2.4–3.0 mL of sample.
5. Record and store fluorescence baseline of the buffer. Many fluorometers allow digital recording of spectra and signal averaging by recording multiple spectra to increase the signal-to-noise ratio if necessary.
6. Add oligonucleotide from the stock solution to give a reasonable signal at the emission maximum for 2AP of 372 nm. Depending on the fluorometer, the final oligo concentration is 0.2–2 μM in 2AP equivalents. Check the oligo concentration by determining its absorbance at 260 nm. The absorbance of the solution used for fluorescence measurements should be ≤0.1 to minimize inner filter effects (29).
7. Record and store the emission spectrum of 2AP–oligo without added acrylamide.

8. Add an aliquot of the stock acrylamide solution to give a final concentration of 20–100 mM. Ensure that mixing is adequate; record emission spectrum.
9. Continue serial addition of the stock acrylamide solution to give points encompassing a range of acrylamide concentrations between 0 and ~800 mM. Record the emission spectrum after each addition.

3.2. Data Analysis

1. Subtract buffer blank spectrum from each experimental spectrum.
2. Correct the spectra for dilution by multiplying by the appropriate dilution factor.
3. Plot F_0/F_Q determined at the emission maximum (~370 nm) vs. [acrylamide] (Stern–Volmer plot).
4. Fit the data points to an appropriate analytical expression: Eq. 8.1 if the Stern–Volmer plot is linear or Eq. 8.4 if the S–V plot shows downward curvature. We use the nonlinear least-squares module in the computer program Origin 7.0 (OriginLab, Northampton, MA; http://www.originlab.com/)

3.3. Interpretation of Data

Here we provide two examples of 2AP–acryamide quenching data used by our group to assess the degree of exposure of individual dA residues in two models of the human telomeric DNA sequence. In the first study, Li et al. (8) compared the acrylamide quenching curves of HuTel22 in sodium and potassium buffers. The purpose of this study was to compare the exposure of the loop adenines in solution with the predicted exposures assessed from the solution structure of HuTel22 in Na^+ (antiparallel basket) determined by NMR and the structure of the same oligonucleotide determined in K^+ by X-ray crystallography (all parallel propeller structure). The two topographical arrangements have very different loop geometries and hence different environments for the loop adenines. In the basket structure, A1, A13, and A19 are stacked upon G-tetrads; A7 is unstacked. This is consistent with the apparent quantum yields (F_0 values) of the 2AP substituted oligonucleotides in Fig. 8.4 and summarized in Table 8.1: A7 > A19 > A1 ≈ A13. It is well established that 2AP fluorescence is strongly quenched by nearby G residues, probably via an electron-transfer mechanism. The close proximity of A1, A13, and A19 to Gs in the crystal structure (1) may account for some or all of these quenching effects.

As described above, provided appropriate τ_0 values are known, collisional quenching data can provide information about environmental heterogeneity at the positions of substitution. According to the Stern–Volmer model (Eq. 8.1), a single fluorophore with a uniform microenvironment exhibits a linear relationship between

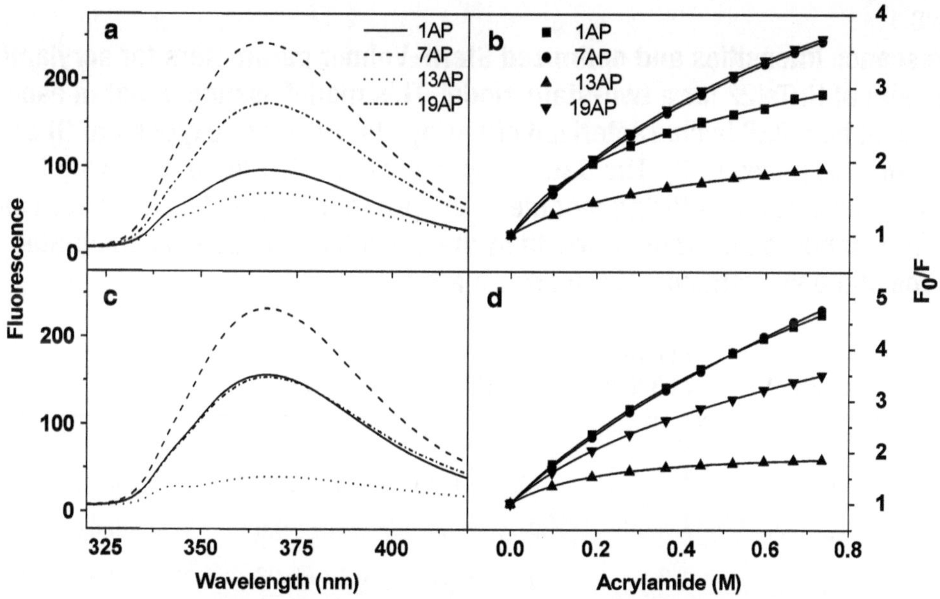

Fig. 8.4. Emission spectra (panels A and C) and fluorescence quenching curves (panels B and D) for HuTel22 measured in the presence of Na^+ or K^+ at 5°C. The lines were drawn in the Stern–Volmer plots using the optimized parameters $K_{SV,1}$, $K_{SV,2}$, and f_1 given in Table 8.1 and determined by nonlinear least-squares fitting of the quenching data to Eq. 8.4 as described in the text. The data are replotted from Ref. (8).

F_0/F and [Q]. Differences in K_{SV} between different single probe residues, each in a homogeneous microenvironment, may result either from alterations in k_0, τ_0, or both (Eq. 8.1). Clearly, without prior knowledge of τ_0 for different single 2AP substitutions, it is impossible to correlate K_{SV} directly with the accessibility of the reporter group to solvent. Kimura et al. (32) have determined that in Na^+ or K^+, the fluorescence decay curves of HuTel22 with serial dA→2AP substitutions consist of a single exponential with lifetimes (τ_0) of 0.54, 0.34, and 0.35 ns for AP7, AP13, and AP19, respectively (32, 33). This implies that, to a close approximation, the observed alterations in K_{SV} result from alterations in the k_0, which varies with the accessibility of the excited state fluorophore to the quenching agent at constant temperature and viscosity.

The quenching curves shown in Fig. 8.4b and d are clearly concave with respect to [acrylamide]. In the case of the Na^+ structure, the excited state 2AP must experience at least two distinct microenvironments: one that is relatively accessible to Q and the other that is shielded from Q. The structural basis of these differences is unclear; however, given that NMR indicates a single global fold (the basket structure), the structural heterogeneity probably results from highly localized, nanosecond fluctuations

Table 8.1
Fluorescence intensities and optimized Stern–Volmer parameters for acrylamide quenching of HuTel22 for a two-state model. This model assumes that in each oligonucleotide, 2AP is characterized either by state 1 (more exposed to Q) or state 2 (less exposed to Q). The data in Fig. 8.4 were fitted to Eq. 8.4 using the nonlinear least-squares fitting module of the program Origin 7.0. The error values show the standard deviation of the fit of the individual datasets as calculated from the diagonal elements of the error matrix

	SASA (Å^2)[a]	Relative F_0 at 370 nm[b]	$K_{SV,1}$ (M^{-1})	$K_{SV,2}$ (M^{-1})	f_1	$1-f_1$
Na$^+$ buffer						
1AP	102	2.4	12.84 ± 0.86	0.23 ± 0.06	0.68 ± 0.02	0.32
7AP	107	6.0	7.19 ± 0.09	−0.01 ± 0.03	0.86 ± 0.01	0.14
13AP	38	1.7	6.84 ± 0.41	0.12 ± 0.03	0.52 ± 0.02	0.48
19AP	25	4.3	7.57 ± 0.17	−0.02 ± 0.05	0.86 ± 0.01	0.14
K$^+$ buffer						
1AP	196	3.9	9.65 ± 0.17	0.18 ± 0.05	0.88 ± 0.01	0.12
7AP	98	5.8	9.07 ± 0.25	0.39 ± 0.10	0.88 ± 0.01	0.12
13AP	91	1.0	9.01 ± 0.19	−0.05 ± 0.01	0.55 ± 0.01	0.45
19AP	96	3.8	9.44 ± 0.20	0.29 ± 0.03	0.77 ± 0.01	0.23
2AP alone	269	192[c]	37.65 ± 0.37[d]	–	101.5 ± 2.2[e]	–

[a]SASA: solvent accessible surface area of the 2AP moiety calculated from the NMR structure of HuTel22 (Protein Data Bank entry 143D (1)) in Na$^+$ and the crystal structure in K$^+$ (PDB entry 1KF1 (23)).
[b]Normalized with respect to $F_0 = 40$ arbitrary units for 13AP in K$^+$.
[c]Determined for 2AP and 13AP in BPEK at 25 °C.
[d]Determined at 25 °C in 50 mM KOAc, 5 mM Mg(OAc)$_2$ by Ballin et al. (27).
[e]The Stern–Volmer plot showed upward curvature, indicative of static quenching (27).

in positioning of the adenines that are confined to the respective loop regions.

To put the analysis on a more quantitative basis, we fit the quenching curves to a two-state model (Eq. 8.3 with $n = 2$), even though we realized that such a model likely is an oversimplification. The resulting best fit parameters ($K_{SV,1}$, $K_{SV,2}$, and f_1) are summarized in Table 8.1. $K_{SV,1}$ varies from 6.8 M^{-1} to 12.8 M^{-1}; $K_{SV,2}$ varies from 0 to 0.2 M^{-1}; and f_1 varies between 0.52 and 0.86. This analysis suggests that in Na$^+$, AP7 and AP19 are nearly homogeneous with respect to their accessibility to acrylamide quenching ($f_1 \approx 0.86$), whereas a significant fraction of the AP1 and AP13 bases are relatively inaccessible to acrylamide quenching (32 and 42%, respectively). If the K_{SV} parameters in Table 8.1 in Na^+ are divided by the τ_0 values above, the bimolecular collisional constant $k_0 = 13.6$, 20.1, and 21.6 $M^{-1} ns^{-1}$ for AP7, AP13, and AP19, respectively, for the solvent-exposed residues, and approximately 100-fold less for the inaccessible fractions. The results are consistent with our earlier conclusions derived from the NMR structure of HuTel22 in Na^+ that A13 is packed within the diagonal loop at one end of the structure, whereas A7 and A19 are in more exposed regions in the lateral loops at opposite ends of the structure. However, the heterogeneity apparent in the f_1 values suggests some flexibility in the diagonal loop such that, on average, about half of the residues are relatively solvent accessible.

For K^+, all quenching curves also display downward curvature as shown in Fig. 8.4d, indicative of local environmental heterogeneity at the sites of substitution as described above with Na^+. When these data were analyzed by the two-state model (Eq. 8.4), only A13 exhibited nearly equal fractions of relatively accessible and inaccessible states ($f_1 = 0.55$ and $f_2 = 0.45$). For the other substitutions, only 10–20% of the 2AP residues are relatively inaccessible to acrylamide, suggesting a relatively homogeneous environment. Notably, the K_{SV} parameters associated with the accessible population of states are all approximately the same ($K_{SV} \approx 9.0$–$9.5\ M^{-1}$). Since the requisite τ_0 values are not available in K^+, it is not possible to directly relate these K_{SV} values to residue accessibility. However, f_1 values (for A13) that are markedly <1.0 indicate that a significant fraction of the residues reside in a microenvironment that is inaccessible to acrylamide. This leads to our previous conclusion that the structure of HuTel22 in K^+ solutions is not the propeller structure, in which all loop adenines (A7, A13, and A19) are in equivalent environments.

In a second example, Petraccone et al. (34) correlated emission intensities and acrylamide quenching data with solvent-accessible surface areas (SASAs) calculated for several different models of a 50-nt oligonucleotide d(TTAGGG)$_8$TT. This sequence models eight repeats of the TTAGGG human telomere sequence and consists of two tandem quadruplexes. Molecular dynamics was used to model various combinations of hybrid-1, hybrid-2, and propeller structures, and the SASA of each dA residue was calculated from the final 4 ns of the molecular dynamics trajectory. The authors compared the SASA with the observed F_0/F values at 0.6 M acrylamide for a series of oligonucleotides in which each of the eight loop dA residues was individually substituted with 2AP. They concluded that one combination of topologies (hybrid-1-hybrid-2) showed the best correlation between SASA and accessibility to acrylamide.

In summary, the two studies discussed above show how fluorescence emission and fluorescence quenching measurements can be used to distinguish unambiguously between closely related structural models of G-quadruplexes. In common with similar studies, it is possible to determine best fitting values of parameters such as K_{SV} which may be related to solvent accessibility and hence can be correlated with various structural models. However, because K_{SV} is a product of at least three independently variable physical parameters (τ_0, k_0, and f_0), and because of the difficulty of determining all of them independently from quenching curves, K_{SV} values should be interpreted cautiously with respect to specific structures. Nevertheless, others and we have demonstrated that even in the absence of a rigorous quantitative interpretation of these variables, when high-resolution model structures are available from which predictions regarding accessibility of dA residues can be made, the fluorescence properties of 2AP can be a valuable aid in model discrimination, especially when taken in conjunction with other biophysical measurements such as sedimentation.

4. Notes

1. The value of k_0, the diffusion-limited bimolecular collisional rate constant, can be estimated from the Smoluchowski equation: $k_0 = 4\pi DR_0N'$ where D and R_0 are the sum of the diffusion coefficients and molecular radii of the quencher and fluorophore and N' is Avogadro's number divided by 1000. The diffusion constants can be calculated from the Stokes–Einstein equation $D_i = k_B T/(6\pi R\eta)$ where k_B is the Boltzmann constant, T is the absolute temperature, R is the molecular radius of the species, and η is the solvent viscosity (30).
2. Nonlinear least-squares analysis of the fitting of $K_{SV,1}$, $K_{SV,2}$, and f_1 to Eq. 8.3 with $n=1$ or $n=2$ revealed that the optimized parameters are highly correlated (e.g., for the data sets shown, the covariance between any pair of parameters was ± ~0.9 as determined from the covariance matrix). Thus, the parameters may not represent unique values, but rather should be regarded as descriptive values that can reproduce the data within experimental error. In addition, even though the fit to the experimental data points is generally quite good as judged by the SSR and residual plots, this in itself is not sufficient to justify the use of the two-state model.
3. A modified Stern–Volmer equation has been used to describe quenching curves with upward curvature resulting from static quenching. This relationship is $F_0 / F = (1 + K_{SV}[Q])\exp(V[Q])$, where V is the volume sur-

rounding the fluorophore in which one or more quencher molecules reside at the instant of excitation (the so-called "active volume element") (30).

4. Concentrated acrylamide solutions are available commercially from a variety of laboratory suppliers. These solutions are generally prepared in water; thus the buffer concentration will change during the titration if such solutions are used for the titrations. Acrylamide is a known human neurotoxin and suspected carcinogen. It is readily absorbed through the skin and by inhalation. Please consult the appropriate material safety datasheet (http://www.jtbaker.com/msds/englishhtml/A1550.htm) for details on hazards, treatment for exposure, and procedures for cleanup of accidental spillage. Your institutional regulations should be consulted prior to disposal of acrylamide.
5. The absorption coefficient for 2AP at 305 nm has been reported to be 6000 $M^{-1} cm^{-1}$. For 2AP incorporated into an oligonucleotide, the absorption coefficient at 260 nm is 1700 $M^{-1} cm^{-1}$ (35, 36).

Acknowledgments

This work was supported by grant CA35635 from the National Cancer Institute.

References

1. Wang Y, Patel DJ (1993) Solution structure of the human telomeric repeat d[$AG_3(T_2AG_3)_3$] G-tetraplex. Structure 1:263–282
2. Davis JT (2004) G-quartets 40 years later: From 5′-GMP to molecular biology and supramolecular chemistry. Angew Chem Int Ed Engl 43:668–698
3. Burge S, Parkinson GN, Hazel P, Todd AK, Neidle S (2006) Quadruplex DNA: sequence, topology and structure. Nucleic Acids Res 34:5402–5415
4. Dai J, Carver M, Yang D (2008) Polymorphism of human telomeric quadruplex structures. Biochimie 90:1172–1183
5. Huppert JL (2008) Four-stranded nucleic acids: structure, function and targeting of G-quadruplexes. Chem Soc Rev 37: 1375–1384
6. Lane AN, Chaires JB, Gray RD, Trent JO (2008) Stability and kinetics of G-quadruplex structures. Nucleic Acids Res 36: 5482–5515
7. Neidle S, Parkinson GN (2008) Quadruplex DNA crystal structures and drug design. Biochimie 90:1184–1196
8. Li J, Correia JJ, Wang L, Trent JO, Chaires JB (2005) Not so crystal clear: the structure of the human telomere G-quadruplex in solution differs from that present in a crystal. Nucleic Acids Res 33:4649–4659
9. Gellert M, Lipsett MN, Davies DR (1962) Helix formation by guanylic acid. Proc Natl Acad Sci U S A 48:2013–2018
10. Williamson JR, Raghuraman MK, Cech TR (1989) Monovalent cation-induced structure of telomeric DNA: the G-quartet model. Cell 59:871–880
11. Wright WE, Tesmer VM, Huffman KE, Levene SD, Shay JW (1997) Normal human chromosomes have long G-rich telomeric overhangs at one end. Genes Dev 11:2801–2809
12. Paeschke K, Simonsson T, Postberg J, Rhodes D, Lipps HJ (2005) Telomere end-binding proteins control the formation of G-quadruplex

DNA structures in vivo. Nat Struct Mol Biol 12:847–854

13. Schaffitzel C, Berger I, Postberg J, Hanes J, Lipps HJ, Pluckthun A (2001) In vitro generated antibodies specific for telomeric guanine-quadruplex DNA react with *Stylonychia lemnae* macronuclei. Proc Natl Acad Sci U S A 98:8572–8577
14. Collado M, Blasco MA, Serrano M (2007) Cellular senescence in cancer and aging. Cell 130:223–233
15. Ruzankina Y, Asare A, Brown EJ (2008) Replicative stress, stem cells and aging. Mech Ageing Dev 129:460–466
16. Huppert JL, Balasubramanian S (2007) G-quadruplexes in promoters throughout the human genome. Nucleic Acids Res 35:406–413
17. Duquette ML, Handa P, Vincent JA, Taylor AF, Maizels N (2004) Intracellular transcription of G-rich DNAs induces formation of G-loops, novel structures containing G4 DNA. Genes Dev 18:1618–1629
18. Han H, Hurley LH (2000) G-quadruplex DNA: a potential target for anti-cancer drug design. Trends Pharmacol Sci 21:136–142
19. Teng Y, Girvan AC, Casson LK, Pierce WM Jr, Qian M, Thomas SD et al (2007) AS1411 alters the localization of a complex containing protein arginine methyltransferase 5 and nucleolin. Cancer Res 67:10491–10500
20. Shafer RH, Smirnov I (2000) Biological aspects of DNA/RNA quadruplexes. Biopolymers 56:209–227
21. Juskowiak B (2006) Analytical potential of the quadruplex DNA-based FRET probes. Anal Chim Acta 568:171–180
22. Alberti P, Bourdoncle A, Sacca B, Lacroix L, Mergny JL (2006) DNA nanomachines and nanostructures involving quadruplexes. Org Biomol Chem 4:3383–3391
23. Parkinson GN, Lee MP, Neidle S (2002) Crystal structure of parallel quadruplexes from human telomeric DNA. Nature 417:876–880
24. Ambrus A, Chen D, Dai J, Bialis T, Jones RA, Yang D (2006) Human telomeric sequence forms a hybrid-type intramolecular G-quadruplex structure with mixed parallel/antiparallel strands in potassium solution. Nucleic Acids Res 34:2723–2735
25. Luu KN, Phan AT, Kuryavyi V, Lacroix L, Patel DJ (2006) Structure of the human telomere in K^+ solution: an intramolecular (3+1) G-quadruplex scaffold. J Am Chem Soc 128:9963–9970
26. Xu Y, Noguchi Y, Sugiyama H (2006) The new models of the human telomere d[AGGG(TTAGGG)$_3$] in K^+ solution. Bioorg Med Chem 14:5584–5591
27. Ballin JD, Bharill S, Fialcowitz-White EJ, Gryczynski I, Gryczynski Z, Wilson GM (2007) Site-specific variations in RNA folding thermodynamics visualized by 2-aminopurine fluorescence. Biochemistry 46:13948–13960
28. Ballin JD, Prevas JP, Bharill S, Gryczynski I, Gryczynski Z, Wilson GM (2008) Local RNA conformational dynamics revealed by 2-aminopurine solvent accessibility. Biochemistry 47:7043–7052
29. Lakowicz JR (1999) Principles of fluorescence spectroscopy. Plenum, New York, NY
30. Eftink MR (1991) In: Lakowicz JR (ed.), Topics in fluorescence spectroscopy, vol 2. Plenum, New York, NY. pp. 53–126
31. Hardman SJ, Botchway SW, Thompson KC (2008) Evidence for a nonbase stacking effect for the environment-sensitive fluorescent base pyrrolocytosine-comparison with 2-aminopurine. Photochem Photobiol 84:1473–1479
32. Kimura T, Kawai K, Fujitsuka M, Majima T (2004) Fluorescence properties of 2-aminopurine in human telomeric DNA. Chem Commun:1438–1439
33. Kimura T, Kawai K, Fujitsuka M, Majima T (2004) Detection of the G-quadruplex-TMPyP4 complex by 2-aminopurine modified human telomeric DNA. Chem Commun 4:401–402
34. Petraccone L, Trent JO, Chaires JB (2008) The tail of the telomere. J Am Chem Soc 130:16530–16532
35. Johnson NP, Baase WA, Von Hippel PH (2004) Low-energy circular dichroism of 2-aminopurine dinucleotide as a probe of local conformation of DNA and RNA. Proc Natl Acad Sci U S A 102:3426–3431
36. Smagowicz J, Wierzchowski KL (1974) Lowest excited states of 2-aminopurine. J Lumin 8:210–232

Chapter 9

Assessing DNA Structures with ^{125}I Radioprobing

Timur I. Gaynutdinov, Ronald D. Neumann, and Igor G. Panyutin

Abstract

Iodine-125 radioprobing is based on incorporation of radioiodine into a defined position in a nucleic acid molecule. Decay of ^{125}I results in the emission of multiple, low-energy Auger electrons that, along with positively charged residual daughter nuclide, produce DNA strand breaks. The probability of such strand breaks at a given nucleotide is in inverse proportion to the distance from the ^{125}I atom to the sugar of that nucleotide. Therefore, conclusions can be drawn about the conformation or folding of a DNA or RNA molecule based on the distribution of ^{125}I decay-induced strand breaks. Here we describe in detail the application ^{125}I radioprobing for studying the conformation of quadruplex structures, and discuss the advantages and limitations of the method.

Key words: Iodine-125, Radioprobing, DNA structure, Telomere, Quadruplex

1. Introduction

Probing the chemical reactivity of DNA or RNA nucleotides with various agents is widely used in studies of the conformation of nucleic acids. Accessibility of the nucleotides to the agents is usually revealed by the analysis of the consequent strand breaks. Methods based on such probing provide valuable information on the conformation of nucleic acids and often complement and verify structural data obtained by X-ray crystallography and NMR. For example, the discovery of the quadruplex structure in G-rich oligonucleotides was based on the observation that the N7 position of guanines was protected from dimethylsulfate modification (1). Iodine-125 radioprobing is a similar methodology but different in approach (2). It is based on the analysis of nucleic acid strand breaks produced close to the site of decay of an ^{125}I atom. The majority of local breaks produced by ^{125}I decay result from the direct action of radiation. The probability of these breaks occurring depends mostly on the distance from the radionuclide

P. Baumann (ed.), *G-Quadruplex DNA: Methods and Protocols*, Methods in Molecular Biology, vol. 608
DOI 10.1007/978-1-59745-363-9_9, © Humana Press, a part of Springer Science+Business Media, LLC 2010

decay site to the corresponding nucleotides (3), and to a lesser extent on their accessibility to solution like in other probing methods that rely on a diffusible chemical reagent.

Iodine-125 belongs to a class of radioisotopes, called Auger-electron emitters, that decay by electron capture or internal conversion. Both of these processes create a vacancy in an inner atomic shell that results in a cascade of electron transitions accompanied by the emission of a number of orbital electrons, the so-called Auger electrons (4). Decay of ^{125}I produces on average 21 such electrons. The energy of the majority of these electrons is below 1 keV, which is considerably less than the energy of nuclear decay-generated beta particles. Auger electrons produce breaks in DNA or RNA strands either directly or by generation of OH radicals in the "bound" water. In addition, the nucleotides located next to that site are affected by the positively charged (+21 on average for ^{125}I) daughter nucleus, which, by stripping electrons from the neighboring bonds, may also contribute to the strand breakage (5).

Iodine-125 can be easily incorporated into DNA as [^{125}I]-iododeoxycytosine (^{125}I-dC) using a DNA polymerase and commercially available ^{125}I-dCTP as a precursor (6). Most of the breaks (90%) produced by decay of ^{125}I are located within approximately one helical turn from the ^{125}I-dC incorporation site (7). Analysis of strand breaks in DNA and DNA-protein complexes with known 3-D structure has confirmed experimentally that the probability of breaks depends primarily on the distance from ^{125}I to the sugar of a target nucleotide (7–9). The strand breaks are due to the energy deposition or radical attack in the sugar–phosphate moiety since damage to the DNA bases generally does not lead to strand scission. Therefore, radioprobing, like "low-resolution" NMR, allows one to obtain information on the interatomic distances and, in principle, to reconstruct the 3D structure of nucleic acids in complexes with proteins.

Unfortunately, due to the complex nature of DNA breaks and multiple mechanisms involved, currently there is no comprehensive theoretical model that would establish a relationship between the probability of breaks at a given nucleotide and the distance to its atoms. Therefore, the analysis of the strand breaks is limited to the establishment of the order, in terms of distance, in which the nucleotides are located relative to the ^{125}I decay site. However, even this simple analysis proved to be useful in providing unique information on the conformation of DNA and DNA–protein complexes studied so far (10–12). Here we describe application of radioprobing to the analysis of quadruplex DNA structures in oligonucleotides.

2. Materials

2.1. Oligodeoxyribonucleotides

Oligodeoxyribonucleotides (ODN) (primer and template, Fig. 9.1) were synthesized on an ABI394 DNA synthesizer (PE Applied Biosystems, Foster City, CA), and purified by denaturating polyacrylamide gel electrophoresis (PAGE) as described in (11). The concentration of single-stranded ODN was measured at 260 nm on an HP 8452A diode array spectrophotometer, and was calculated with an online oligonucleotide properties calculator (19).

2.2. Oligonucleotide Labeling

2.2.1. Telomeric ODN Labeling with [γ-^{32}P]-ATP and T4 PNK

1. T4 Polynucleotide kinase (T4 PNK) 10 U/μL, supplied with 10× reaction buffer: 500 mM Tris–HCl (pH 7.6 at 25°C), 100 mM $MgCl_2$, 50 mM dithiothreitol (DTT), 1 mM spermidine, and 1 mM (ethylenediaminetetraacetic acid) EDTA (Fermentas; Cat. # EK0031).
2. EasyTides [γ-^{32}P]-adenosine 5′-triphosphate ([γ-^{32}P]-ATP), specific activity: 6000 Ci/mmol, 10 mCi/mL, 50 mM Tricine pH 7.6 (Perkin Elmer; Cat. # NEG502Z500UC).
3. EDTA (0.5 M, pH 8.0): dissolve 18.61 g EDTA in 80 mL of water. Adjust pH to 8.0 with NaOH (approx 2 g NaOH). Bring the volume to 100 mL with water.
4. MicroSpin G-25 columns (GE Healthcare).

2.2.2. Incorporation of ^{125}I-dCTP by Primer Extension Reaction With Klenow Fragment of DNA Polymerase I

1. 5-[^{125}I]Iodo-2′-deoxycytidine-5′-triphosphate ([^{125}I]-dCTP), specific activity: 2200 Ci/mmol, 1.12 mCi/mL, in a solution containing methanol:water 3:1 (Perkin Elmer; Cat. # NEX074S).
2. Klenow fragment of DNA polymerase I (Klenow fragment) 10 U/mL, supplied with 10× reaction buffer: 500 mM Tris–HCl (pH 8.0 at 25°C), 50 mM $MgCl_2$, 10 mM DTT (Fermentas, Cat. # EP0051).
3. dNTP set – 100 mM aqueous solutions at pH 7.0 of each of dATP, dCTP, dGTP, and dTTP (Fermentas; Cat. # R0181).

2.2.3. PAGE Purification

1. Formamide stop solution (USB Corporation; Cat. # 70724).
2. SequaGel sequencing system consists of SequaGel concentrate (237.5 g of acrylamide, 12.5 g of methylene bisacrylamide,

```
                                     125I
             5'-ACGTTATATGAGGGTTAGGGTTAGGGTCAGGG    primer
       3'-CGATCCTATGCAATATACTCCCAATCCCAATCCCAGTCCC   template
```

Fig. 9.1. Sequences of oligodeoxyribonucleotides used. Nucleotides incorporated by the primer extension reaction are underlined.

and 450.5 g urea in 1 L of deionized aqueous solution), SequaGel diluent (7.5 M urea in deionized water), and SequaGel buffer: 200 mL bottles containing 0.89 M Tris-Borate, 20 mM EDTA pH 8.3 (10× TBE), and 7.5 M urea (National Diagnostics; Cat. # EC-833).

3. Tris-buffered saline (TBS): Prepare 10× stock with 1.37 M NaCl, 27 mM KCl, 250 mM Tris-HCl, pH 7.4. Dilute 100 mL with 900 mL water for use.
4. Spin-X centrifuge tube filter with 0.22 mm cellulose acetate (Corning Incorporated).

2.3. Required Equipment

1. Vacuum-centrifuge with a cold trap (Jouan RC 1010 or equivalent) – needed to dry [^{125}I]- dCTP from methanol/water mixture.
2. Power supply for sequencing PAGE, capable of providing up to 3000 V, 60 W.
3. Gel dryer.
4. Freezers (−20 and −80°C).
5. Water bath or incubator at 37°C.

3. Methods

A repeated noncoding DNA sequence d(TTAGGG)$_n$ is present in the telomeric ends of all human chromosomes. These repeats can adopt multiple inter- and intramolecular non-B-DNA conformations that may play an important role in biological processes. NMR spectroscopy and X-ray crystallography have solved several intramolecular structures of the telomeric oligonucleotides – antiparallel, parallel, and "mixed-type" hybrid type I and II structures containing both parallel and antiparallel strands (Fig. 9.2) (13–18).

We applied ^{125}I-radioprobing to determine the conformation in solution of the human telomeric quadruplex in dAGGG(TTAGGG)$_3$ (Fig. 9.1), the same sequence that was used in the NMR and crystallographic analyses. An oligonucleotide containing four human telomeric repeats was labeled with ^{125}I at a single internal position, was allowed to fold into quadruplexes under different ionic conditions, and then was frozen for decay accumulation. Probabilities of breaks at individual nucleotides were determined by sequencing gel electrophoresis. Analysis of the quadruplex fold was based on a comparison of break probability with the distance from ^{125}I to the corresponding nucleotides in different quadruplex models.

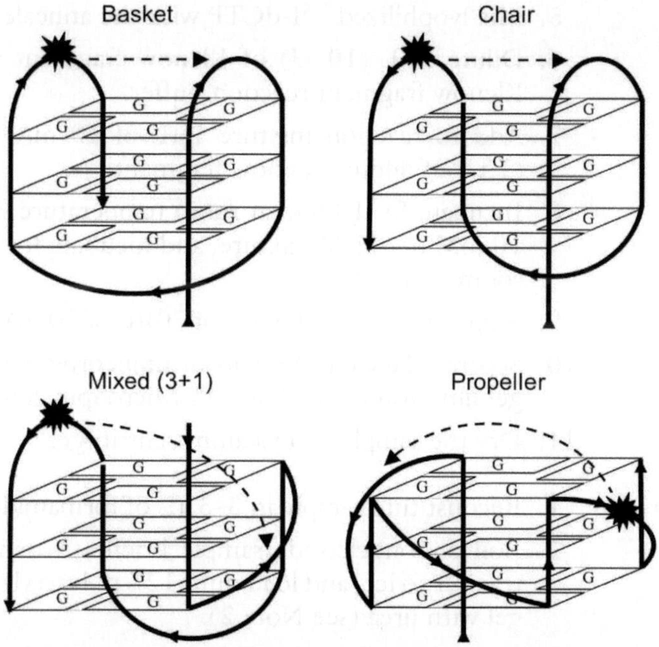

Fig. 9.2. The schematic diagram of four intramolecular conformations of human telomeric quadruplexes. Position of ^{125}I-C is shown by the mark. *Arrows* illustrate the distances between ^{125}I and T5–A7 loop: shorter in the antiparallel basket and chair conformations and longer in the mixed and parallel propeller conformations of G-4 quadruplex.

3.1. Oligonucleotide Labeling

3.1.1. Telomeric ODN Labeling With [γ-^{32}P]-ATP with T4 PNK

1. Take 20 pmol of G4-forming oligonucleotide (primer) and mix it with 2 μL of 10× T4 PNK buffer and 40–80 pmol of [γ-^{32}P]-ATP.
2. Add 1 μL (10 units) of T4 PNK. Adjust volume of the reaction mixture to 20 μL with water.
3. Incubate the sample at 37°C for 40 min.
4. Stop reaction by addition of 30 μL of 20 mM EDTA, pH 8.0.
5. Separate labeled DNA from unincorporated [γ-^{32}P]-ATP by gel filtration through a MicroSpin column.
6. Dry the sample in a vacuum centrifuge.

3.1.2. Incorporation of ^{125}I-dCTP by Primer Extension Reaction with Klenow Fragment of DNA Polymerase I

1. Vacuum-dry approximately 80–100 pmol of ^{125}I-dCTP (see Note 1).
2. Reconstitute dried oligonucleotide in the mixture of 1 μL of Klenow fragment reaction buffer and 6 μL of water.
3. Add 24 pmol of the complementary template oligonucleotide in 1 μL.
4. Incubate for 15 min at 37°C to form a duplex.

5. Mix lyophilized ^{125}I-dCTP with the annealed duplex.
6. Dilute 1 μL (10 U) of Klenow fragment with 9 μL of 1× Klenow fragment reaction buffer.
7. Add to reaction mixture 1 μL of 10 mM dATP and 1 μL (1 U) of diluted Klenow fragment.
8. Incubate for 10 min at room temperature, then add 1 μL of 10 mM of dNTP mixture, and incubate for another 5 min at room temperature.
9. Stop reaction by addition of 30 μL of 20 mM EDTA pH 8.0.
10. Separate labeled DNA from unincorporated nucleotides by gel filtration through a G25 MicroSpin column.
11. Dry the sample in a vacuum centrifuge.

3.1.3. PAGE Purification

1. Reconstitute sample in 3–5 μL of formamide stop solution.
2. Boil oligonucleotide sample briefly (1 min) in water bath, transfer to ice, and load onto 12% polyacrylamide sequencing gel with urea (see Note 2).
3. Run electrophoresis at 55 W for approximately 1 h.
4. Excise band from the gel, crush it in an Eppendorf tube, and elute with ~200 μL of 1× TBS for 1 h on shaking platform at room temperature.
5. Purify sample from acrylamide debris on 0.22-μm cellulose acetate Spin-X filter, and precipitate oligonucleotide with ethanol.
6. Wash the pellet with 70% ethanol stored at −20°C, and dry the pellet.
7. Re-suspend sample in 50 μL of water and desalt by gel filtration through a MicroSpin column (GE Healthcare or equivalent, packed with G-25 resin equilibrated in water).
8. Dry the sample in a vacuum centrifuge.

3.2. Analysis of Break Distribution

3.2.1. Sample Preparation

1. Take ~0.1–0.3 μCi to prepare mixtures with different counterions and/or quadruplex binding compounds (see Note 3).
2. Incubate for at least 1 h at 37°C to allow quadruplex structure formation.
3. Quickly freeze the samples in liquid nitrogen and store for 2 weeks at −80°C to accumulate breaks.

3.2.2. Analysis of Break Distribution

1. After 2 weeks, thaw the samples and analyze by 12% denaturing PAGE (Fig. 9.3).
2. Quantify DNA strand breaks by analyzing digitized autoradiograph using image analyzing software (see Note 4). The probabilities of breaks for the samples in lanes 1 and 2 in Fig. 9.3 are plotted in Fig. 9.4.

Higher probabilities of breaks at the nucleotides from A(1) to T12 in presence of 150 mM KCl reflect folding of telomeric ODN into a compact structure. The maximum of breaks is observed in the loop T5–A7 that is adjacent to ^{125}I-dC18 in the basket and chair antiparallel structures (Fig. 9.2). Therefore, folding of the telomeric DNA is detected by radioprobing, and distribution of probability of breaks is consistent with the presence of the antiparallel conformations in solution.

4. Notes

1. Consult your Radiation Safety Department on the local regulations regarding work with radioiodinated compounds. ^{125}I-dCTP is supplied in a methanol/water mixture. It is not volatile and does not require special iodination hood for handling. Once dried, it degrades rapidly. To dry such an amount usually takes 30–40 min. Plan ahead to minimize the time that dried ^{125}I-dCTP will spend on the bench.

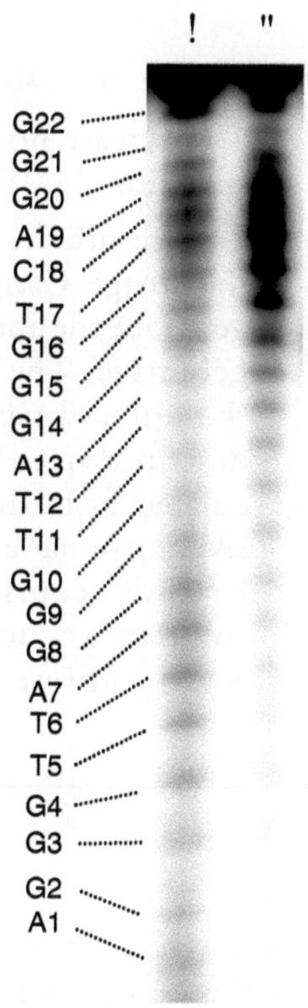

Fig. 9.3. Analysis of DNA strand breaks in 12% denaturing PAGE. Lane 1: telomeric ODN + 150 mM KCl; lane 2: duplex of telomeric ODN and complementary template + 150 mM KCl.

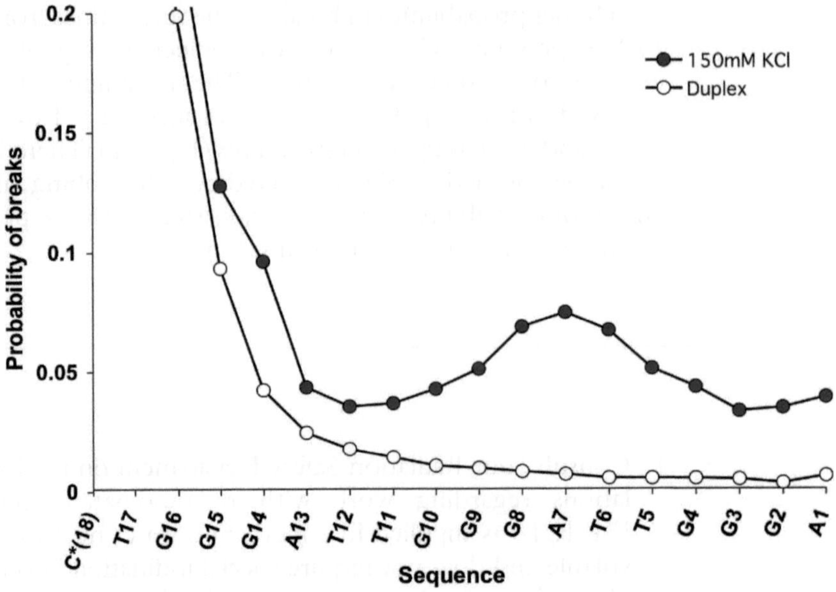

Fig. 9.4. Distribution of ^{125}I-induced strand breakage probability in the telomeric ODN calculated from the data on Fig. 9.3 in 150 mM KCl (-●-) and in duplex form (-○-).

2. We use 20 × 40 cm glass plates with 0.4 mm spacers.
3. To decrease nonspecific degradation of DNA, 10% dimethylsulfoxide (DMSO) may be used.
4. To measure the intensity of the individual bands, we are using the Image Gauge software (FUJI Medical Systems USA) to generate the intensity profile of each lane from the digitized gel image. The profile should be deconvoluted to a series of the Lorentz-type peaks corresponding to individual bands as described in detail in (20). The probability of breaks are calculated from the areas of the individual peaks using a recursive formula and assuming that probability of breaks at [^{125}I]-dC equals 1 (Fig. 9.4), as described in detail in (7). Another way to analyze digitized gel images is by using the SAFA software (21).

Acknowledgment

This research was supported by the Intramural Research Program of the NIH, Clinical Center.

References

1. Sen D, Gilbert W (1988) Formation of parallel four-stranded complexes by guanine-rich motifs in DNA and its implications for meiosis. Nature 334:364–366
2. Panyutin IG, Neumann RD (2005) The potential for gene-targeted radiation therapy of cancers. Trends Biotechnol 23:492–496
3. Martin RF, Haseltine WA (1981) Range of radiochemical damage to DNA with decay of iodine-125. Science 213:896–898
4. Kassis AI (2004) The amazing world of auger electrons. Int J Radiat Biol 80:789–803
5. Lobachevsky PN, Martin RF (2000) Iodine-125 decay in a synthetic oligodeoxynucleotide. II. The role of auger electron irradiation compared to charge neutralization in DNA breakage. Radiat Res 153:271–278
6. Panyutin IG, Neumann RD (1994) Sequence-specific DNA double-strand breaks induced by triplex forming 125I labeled oligonucleotides. Nucleic Acids Res 22:4979–4982
7. Lobachevsky PN, Martin RF (2000) Iodine-125 decay in a synthetic oligodeoxynucleotide. I. Fragment size distribution and evaluation of breakage probability. Radiat Res 153: 263–270
8. Panyutin IG, Neumann RD (1997) Radioprobing of DNA: distribution of DNA breaks produced by decay of 125I incorporated into a triplex-forming oligonucleotide correlates with geometry of the triplex. Nucleic Acids Res 25:883–887
9. Karamychev VN, Zhurkin VB, Garges S, Neumann RD, Panyutin IG (1999) Detecting the DNA kinks in a DNA-CRP complex in solution with iodine-125 radioprobing. Nat Struct Biol 6:747–750
10. Karamychev VN, Tatusov A, Komissarova N et al (2003) Iodine-125 radioprobing of E. coli RNA polymerase transcription elongation complexes. Methods Enzymol 371:106–120
11. He Y, Neumann RD, Panyutin IG (2004) Intramolecular quadruplex conformation of human telomeric DNA assessed with 125I-radioprobing. Nucleic Acids Res 32: 5359–5367
12. Malkov VA, Panyutin IG, Neumann RD, Zhurkin VB, Camerini-Otero RD (2000) Radioprobing of a RecA-three-stranded DNA complex with iodine 125: evidence for recognition of homology in the major groove of the target duplex. J Mol Biol 299:629–640
13. Wang Y, Patel DJ (1993) Solution structure of the human telomeric repeat d[AG3(T2AG3)3] G-tetraplex. Structure 1:263–282
14. Parkinson GN, Lee MP, Neidle S (2002) Crystal structure of parallel quadruplexes from human telomeric DNA. Nature 417:876–880
15. Luu KN, Phan AT, Kuryavyi V, Lacroix L, Patel DJ (2006) Structure of the human telomere in K+ solution: an intramolecular (3 + 1) G-quadruplex scaffold. J Am Chem Soc 128:9963–9970
16. Ambrus A, Chen D, Dai J, Bialis T, Jones RA, Yang D (2006) Human telomeric sequence forms a hybrid-type intramolecular G-quadruplex structure with mixed parallel/antiparallel strands in potassium solution. Nucleic Acids Res 34:2723–2735
17. Phan AT, Luu KN, Patel DJ (2006) Different loop arrangements of intramolecular human telomeric (3+1) G-quadruplexes in K+ solution. Nucleic Acids Res 34:5715–5719
18. Dai J, Carver M, Punchihewa C, Jones RA, Yang D (2007) Structure of the Hybrid-2 type intramolecular human telomeric G-quadruplex in K+ solution: insights into structure polymorphism of the human telomeric sequence. Nucleic Acids Res 35:4927–4940
19. Kibbe WA (2007) OligoCalc: an online oligonucleotide properties calculator. Nucleic Acids Res 35:W43–46
20. Panyutin IV, Luu AN, Panyutin IG, Neumann RD (2001) Strand breaks in whole plasmid dna produced by the decay of (125)I in a triplex-forming oligonucleotide. Radiat Res 156:158–166
21. Das R, Laederach A, Pearlman SM, Herschlag D, Altman RB (2005) SAFA: semi-automated footprinting analysis software for high-throughput quantification of nucleic acid footprinting experiments. RNA 11:344–354

Chapter 10

Monitoring the Temperature Unfolding of G-Quadruplexes by UV and Circular Dichroism Spectroscopies and Calorimetry Techniques

Chris M. Olsen and Luis A. Marky

Abstract

DNA oligonucleotides containing guanine repeat sequences can adopt G-quadruplex (GQ) structures in the presence of specific metal ions. We report on how to use a combination of spectroscopic and calorimetric techniques to determine the spectral characteristics and thermodynamic parameters for the temperature-unfolding of GQs. Specifically, we investigated the unfolding of d($G_2T_2G_2TGTG_2T_2G_2$), *G2*, and d($G_3T_2G_3TGTG_3T_2G_3$), *G3* by a combination of UV and circular dichroism (CD) spectroscopies, and differential scanning calorimetry (DSC).

Analysis of the UV and CD spectra of these GQs at low (100% helix) and high (100% random coil) temperatures yielded the optimal wavelengths to determine the melting curves. In addition, the CD spectra yielded the particular conformation(s) that each GQ adopted at low temperature. DSC curves yielded complete thermodynamic profiles for the unfolding of each GQ. We use these profiles to determine the thermodynamic contributions for the formation of a G-quartet stack.

Key words: Thrombin-binding aptamer, G-quartets, DNA quadruplexes, Cations, UV and CD melting curves, DSC thermograms, Thermodynamics, Heat and stability

1. Introduction

Controlling the formation and dissociation of quadruplex DNA structures in vivo is an important strategy for developing novel therapeutics for the treatment of cancer and other diseases. Knowledge concerning the stabilities of DNA quadruplex structures is also critical for understanding the formation, longevity, and resolution of these structures in vivo. This information is useful in predicting the stabilities of putative quadruplex structures that have been identified based on their sequence (1), and whether chemical approaches to manipulating quadruplex structures are

P. Baumann (ed.), *G-Quadruplex DNA: Methods and Protocols*, Methods in Molecular Biology, vol. 608
DOI 10.1007/978-1-59745-363-9_10, © Humana Press, a part of Springer Science+Business Media, LLC 2010

likely to be effective at disrupting or stabilizing G-quadruplex (GQ) structures in vivo.

Previously, we have investigated the thermodynamic folding (and unfolding) of intramolecular GQs in the presence of different cations (2–4). In this article, we report on how UV and circular dichroism spectroscopies and differential scanning calorimetry techniques are applied to investigate both the temperature-unfolding and the overall conformation of GQs, and how the analysis of the resulting curves yields the associated standard thermodynamic profiles. Furthermore, we also present complete thermodynamic unfolding profiles for two intramolecular GQ structures (Fig. 10.1). The experimental data yielded the thermodynamic contributions for the formation of a single G-quartet stack, which can be used to determine the unfolding/folding thermodynamics of a variety of biologically important GQs.

2. Materials

1. Oligonucleotide sequence (HPLC purified)
2. Chromatography equipment
3. Freeze dry and lyophilizer system
4. Filtration apparatus with nylon 0.45-μm filter paper
5. pH meter
6. HEPES buffer (10 mM, >99.5% purity)
7. Cesium hydroxide monohydrate (99.5% on basis of metal content)
8. Potassium chloride (>99% purity)
9. Double distilled high-purity water

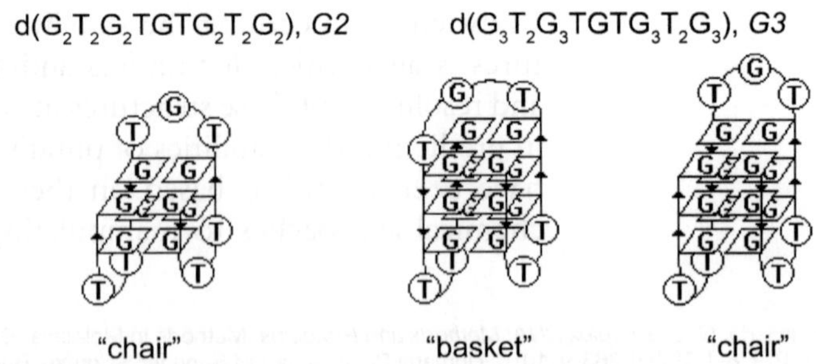

Fig. 10.1. Sequences of oligonucleotides, designations, and putative structures.

3. Methods

The methods described below outline sample and buffer preparation (Subheading 3.1), UV spectroscopy (Subheading 3.2), circular dichroism (CD) spectroscopy (Subheading 3.3), differential scanning calorimetry (DSC) (Subheading 3.4), analysis of spectroscopic and calorimetric melting curves (Subheading 3.5), experimental approach to obtain unfolding thermodynamics (Subheading 3.6), and thermodynamics of a G-quartet stack (Subheading 3.7).

3.1. Sample and Buffer Preparation

The oligonucleotides (*see* Note 1) (ODNs, and their designations, Fig. 10.1) $d(G_2T_2G_2TGTG_2T_2G_2)$, *G2*; and $d(G_3T_2G_3TGTG_3T_2G_3)$, *G3*, were synthesized by the Core Synthetic Facility of the Eppley Research Institute at University of Nebraska Medical Center (UNMC), purified by HPLC, and desalted by column chromatography using G-10 Sephadex exclusion chromatography. The concentrations of the oligomer solutions were determined at 260 nm and 80°C using a Perkin-Elmer Lambda-10 spectrophotometer and the following molar extinction coefficients for the single strands: 146 mM^{-1} cm^{-1} (*G2*), and 191 mM^{-1} cm^{-1}(*G3*). These values were obtained by extrapolation of the tabulated values for dimers and monomeric bases at 25°C (5, 6) to higher temperatures using procedures reported previously (7). Inorganic salts from Sigma were reagent grade, and used without further purification. All measurements were made in buffer solutions containing 10 mM Cs-HEPES at pH 7.5, adjusted to the desired salt concentration and pH with KCl or CsOH, respectively (see Note 2). Oligonucleotide solutions were prepared by dissolving the dry and desalted ODN in buffer. The Cs^+/K^+ ion exchange was done initially by heating the ODN solution to 100°C for 5 min and cooled to room temperature over 25 min.

3.2. UV Spectroscopy

The UV spectra (absorbance versus wavelength) of each oligonucleotide were measured in the 350–200 nm range in 1 nm increments, using a thermoelectrically controlled Aviv spectrophotometer Model 14DS UV-VIS (Lakewood, NJ). The reported spectra correspond to an average of at least three scans. To determine the optimum wavelengths of the melting curve, spectra were obtained at low and high temperatures corresponding to 100% formation of the quadruplex and to 100% random coil, respectively. The optimum wavelengths are selected where the absorbance changes are minimal i.e., at wavelengths of maximum absorbance, and where the largest absorbance difference is observed between the high- and low-temperature scans.

UV melting curves were performed at 297 nm using optical quartz cells with 0.1–1 cm path lengths. The temperature was scanned from 10 to 100°C at a heating rate of ~0.6°C/min. The cell chamber was continuously flushed with nitrogen gas to prevent water condensation at low temperatures. Shape analysis of the melting curves yielded transition temperatures (T_M), which are the midpoint temperatures of the helix–coil transitions, and model-dependent van't Hoff enthalpies (ΔH_{vH}) (7, 8). Transition molecularities were obtained from the dependence of the T_M over a tenfold range in oligonucleotide concentration (in strands). Formation of intramolecular complexes is obtained when the T_M is independent of strand concentration, i.e., the T_M remains constant with increasing strand concentration.

3.3. Circular Dichroism Spectroscopy (CD)

We used an Aviv Circular Dichroism Model 202SF spectrometer (Lakewood, NJ), equipped with a Peltier temperature control system, to obtain the CD spectra of the oligonucleotides at low and high temperatures. Spectra were obtained from 340 to 220 nm in 1 nm increments using free strained quartz cuvettes, and the reported spectra correspond to the average of at least two scans. The conformation of the complex was determined by inspection of the CD spectra at temperatures where the oligonucleotide is 100% helical. CD melting curves were determined by following the change in ellipticity as a function of temperature at 262 and/or 292 nm; these wavelengths were determined from the largest ellipticity difference between the spectra at high and low temperatures. ODN solutions were heated at a rate of ~0.9°C/min using free strained quartz cuvettes with a path length of 1 or 0.1 cm. Analysis of the CD melting curves, using standard procedures, yielded T_Ms, and model dependent van't Hoff enthalpies, ΔH_{vH} (7).

3.4. Differential Scanning Calorimetry (DSC)

The total heat required for the unfolding of each GQ was measured with a VP-DSC differential scanning calorimeter from Microcal (Northampton, MA). We used a temperature range of 10–100°C and a heating rate of 45°C/h. Analysis of the resulting thermograms yielded T_M, ΔH_{cal}, ΔS_{cal}, $\Delta G°(T)$, and ΔH_{vH}. These parameters are obtained with the following relationships: $\Delta H_{cal} = \int \Delta C_p(T)\ dT$; $\Delta S_{cal} = \int \Delta C_p(T)/T\ dT$, and the Gibbs equation, $\Delta G°(T) = \Delta H_{cal} - T\Delta S_{cal}$; where ΔC_p is the anomalous heat capacity of the oligonucleotide during the transition, ΔH_{cal} is the unfolding enthalpy, and ΔS_{cal} is the unfolding entropy. The ΔH_{vH} terms are also obtained from the DSC profiles using the software provided with the instrument.

3.5. Analysis of Spectroscopic Melting Curves

Spectroscopic, UV, and CD melting curves of a particular nucleic acid structure provide fundamental information about the helix–coil transition of the molecule. This includes its thermal stability or transition temperature and overall standard thermodynamic profiles: free energy, enthalpy, and entropy. All of these parameters are obtained from the analysis of the shape of the melting curve using standard thermodynamic relationships (7). The starting point of this analysis is the use of the van't Hoff equation: $\partial \ln K/\partial T = \Delta_{HvH}/RT^2$, where K is the equilibrium constant for the helix→coil transition of the oligonucleotide, T is the absolute temperature, ΔH_{vH} is the van't Hoff enthalpy, and R is the universal gas constant. Thus, the dependence of the equilibrium constant with temperature yields the energy or heat needed to disrupt a cooperative duplex unit into single strands. However, one should be careful in using this approach because ΔH_{vH} will often vary with T. This occurs when heat capacity effects are involved between the initial and final states (9). The procedure involves the evaluation of the equilibrium constant K as a function of temperature. For intramolecular transitions, K is expressed in terms of α, which is the fraction of strands in the helical state, according to the relationship, $K(T) = \alpha/(1 - \alpha)$. Substitution of $K(T)$ into the van't Hoff equation, differentiating, and evaluating the resulting expression at the midpoint of a transition, where $T = T_M$ and $\alpha = ½$, yields the relationship: $\Delta H_{vH} = 4RT_M^{\ 2}\ (\partial\alpha/\partial T)_{T=Tm}$. The T_M and the slope $(\partial\alpha/\partial T)_{T=T\,m}$ are determined from an α versus T profile, which is obtained from the experimental melting curve using the level rule throughout the transition, as outlined elsewhere (7). Furthermore, at the T_M, $K = 1$ ($\Delta G^{\circ} = 0$) and ΔG° at any temperature is given by $\Delta G^{\circ}(T) = \Delta H_{vH}(1 - T/T_M)$.

3.6. Experimental Approach to Obtain Unfolding Thermodynamics

Initially, we use UV spectroscopy to obtain the overall spectral characteristics of a GQ at low and high temperatures; typical spectra are shown in Fig. 10.2a. The spectra of *G2* and *G3* have absorbance maxima at ~255 nm; however, 297 nm is the optimum wavelength to measure melting curves because of its large hypochromic effect relative to the spectra at high temperature. Fig. 10.2b shows the UV melting curves of *G2* and *G3*. Both curves are sigmoidal, with *G2* unfolding in a monophasic transition (T_M of 51°C), and *G3* unfolding in a biphasic transition (T_M's of 54 and 76°C). We also obtained ΔH_{vH}s from shape analysis of the curves (7) (*see* Table 10.1).

We measure UV melting curves over a tenfold strand concentration (data not shown) to determine whether each GQ folds intramolecularly. Fig. 10.2c shows the T_M dependences on strand concentration. For each transition the T_M remains constant; therefore, we conclude that each GQ folds intramolecularly.

The next step is to use CD spectroscopy to determine the conformation of each GQ. To obtain reproducibility of the CD spectra of *G3*, the following temperature treatment was used: the *G3* solution was heated to 90°C, cooled to 60°C slowly, kept at 60°C for 15 min, slowly cooled to 2°C, and maintained at this temperature for 40 min. Fig. 10.3a shows the CD spectra of *G2*

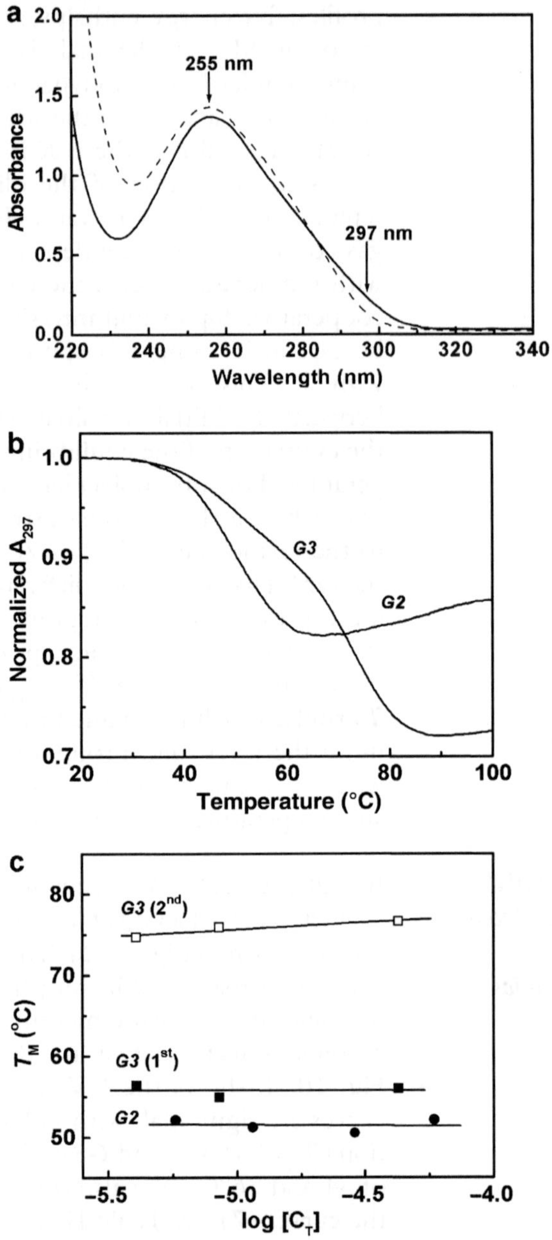

Fig. 10.2. UV spectroscopy data. All experiments were done in 10 mM Cs-HEPES buffer, 100 mM K^+ at pH 7.5. (**a**) UV spectra of *G2* at ~8 μM (in strands), at 15°C (*solid line*) and 80°C (*dashed line*); (**b**) UV-melting curves of *G2* and *G3*, (**c**) T_M-dependence on strand concentration: *G2* (*circles*), *G3* (first transition, *closed squares*), (second transition, *open squares*).

and *G3*. *G2* has maxima at 292 and 247 nm, while *G3* has maxima at 262 and 292 nm (shoulder). However, the largest ellipticity differences, relative to the spectrum of the random coil state (Fig. 10.3a), are observed at 262 and 292 nm; these two wavelengths are selected to obtain CD melting curves. Fig. 10.3b

Table 10.1
Van't Hoff folding enthalpies (kcal/mol)

	ΔH_{vH} (DSC)	ΔH_{vH} (CD)	ΔH_{vH} (UV)	Average ΔH_{vH}
G2	-40	-41	-42	-41
G3 (first)	-33	-24	-31	-29
G3 (second)	-64	-50	-68	-61

All experiments were performed in 10 mM Cs-Hepes buffer at pH 7.5 and 100 mM KCl. Experimental errors are shown in parentheses: ΔH_{vH} (±15%).

shows the CD melting curves of *G2* at 292 nm, and *G3* at 262 and 292 nm. All curves are sigmoidal and show that each G-quadruplex unfolds in monophasic transitions with T_{Ms} of 52°C (*G2*) and 58 and 70°C (*G3*), which are in fair agreement with the T_Ms obtained from UV melting curves. The associated ΔH_{vH}s, obtained from shape analysis of these curves (7), are shown in Table 10.1.

To assign the particular conformation(s) that *G2* and *G3* adopt, we used a pair of control GQs with known arrangements of guanines: d($G_4T_4G_4$) forms a bimolecular quadruplex with guanines in the antiparallel orientation (10, 11), while d(T_4G_4) forms a tetramolecular quadruplex with guanines in the parallel orientation (10, 12). Their spectra are shown in Fig. 10.3c; [d($G_4T_4G_4$)]$_2$ has a wavelength maximum at 292 nm, while the maximum of [d(T_4G_4)]$_4$ is at 262 nm. The comparison of the above spectra with the low-temperature spectra of each GQ shows that the spectral characteristics of *G2* are closer to those of the [d($G_4T_4G_4$)]$_2$, indicating that *G2* has an antiparallel arrangement of guanines at every face (2, 13, 14), adopting a "chair" conformation (15). *G3* contains spectral characteristics of both types of quadruplexes, parallel and antiparallel. This indicates that *G3* contains a mixture of two conformers, one assigned to the "chair" conformation, and the second assigned to the "basket" conformation in which the strand has flipped to the front of the structure, yielding a conformer with guanines in the parallel orientation (front and back), called "adjacent parallel" (Fig. 10.1) (16). To determine the percentages of each conformer present in a given spectrum of *G3* (Fig. 10.3c), a set of two equations and two unknowns was solved simultaneously at the maximum wavelengths, using the molar ellipticity of the control GQs as 100% of each conformation and assuming spectral additivity.

DSC was used to determine complete thermodynamic profiles for the temperature-unfolding of each GQ (*see* Fig. 10.4). To obtain reproducible curves for *G3*, the solution was temperature-treated as described in the CD section. The DSC curves show

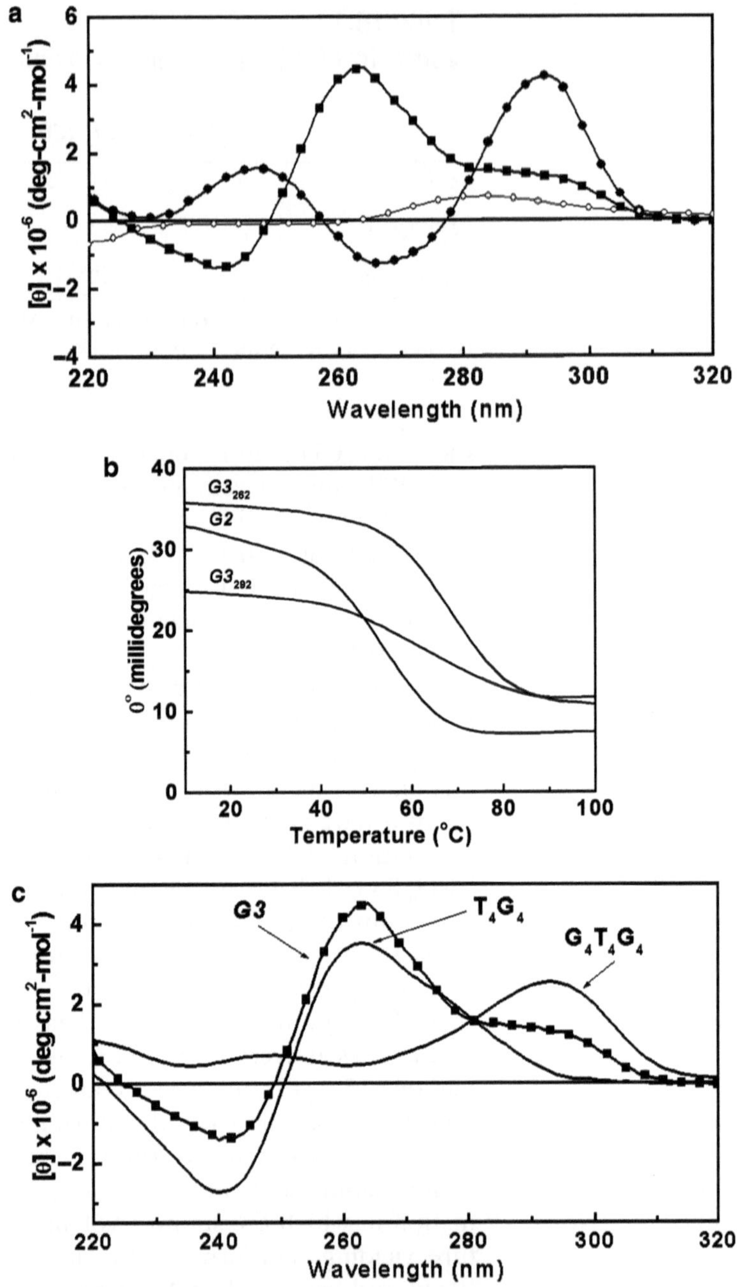

Fig. 10.3. CD Spectroscopy Data. All experiments were done in 10 mM Cs-HEPES buffer, 100 mM K^+ at pH 7.5. (**a**) CD spectra of *G2* at 6.85 μM (in strands), at 20°C (*closed circles*) and 90°C (*open circles*); *G3* at 5.24 μM and 20°C (*squares*); (**b**) CD melting curves of *G2* and for each conformer of *G3*; (**c**) Comparison of the CD spectra of *G3* (*squares*) and control oligonucleotides.

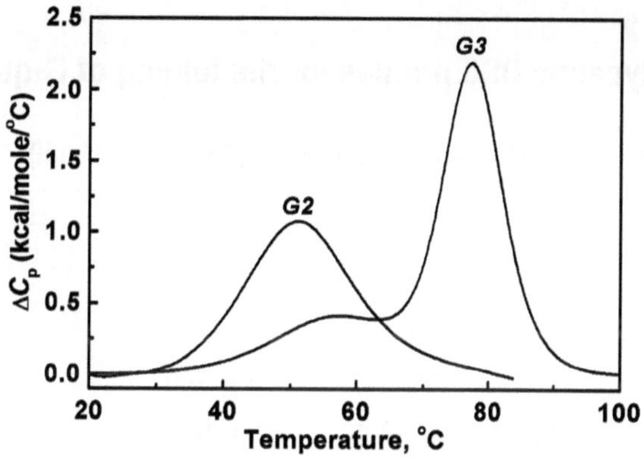

Fig. 10.4. DSC curves of G-quadruplexes in 10 mM Cs-HEPES buffer, 100 mM K^+ at pH 7.5. The oligonucleotide concentrations (in total strands) are 40 μM (*G2*) and 80 μM (*G3*).

that *G2* unfolds in a monophasic transition with a T_M of 52.6°C, while *G3* unfolds in a biphasic transition with T_M's of 57.3 and 77.5°C. The first transition for *G3* is at ~5°C above that of *G2*, consistent with *G3* having one additional G-quartet. The T_M comparison with those obtained from CD melts allows us to assign the first transition to the unfolding of the "chair" conformer of *G3* and second transition to the unfolding of the "basket" conformer. This assignment was confirmed on the basis of a more detailed analysis of the CD spectra for *G3* at several temperatures (data not shown). Complete thermodynamic profiles for the formation of *G2* and *G3* are reported in Table 10.2. The ΔH_{cal} values of *G3* were determined in terms of the percentage of each conformer present in solution. The unfolding of a GQ is accompanied by an unfavorable free energy contribution, which results from the typical unfavorable enthalpy–favorable entropy compensation. The unfavorable enthalpy contributions correspond to the energy needed to break G-quartet stacks and Hoogsteen hydrogen bonding, while favorable entropy contributions correspond to the increase in disorder of the system due to the formation of the random coil and the putative release of counterions and water molecules. For instance, the higher heat observed in the unfolding of *G3* (38.3 kcal/mol) is attributed to breaking one additional G-quartet stack, relative to *G2* (23.7 kcal/mol), since the stacking of the loops contribute equally (3).

A comparison of the ΔH_{vH} values obtained from these three techniques is listed in Table 10.1. All three methods have similar ΔH_{vH} values for a given GQ; however, the values obtained from the CD melting curves for *G3* are lower by ~20% over the DSC and UV values. This is probably due to the unfolding of each conformer contributing to the CD melting curve at each wavelength.

Table 10.2
Thermodynamic DSC profiles for the folding of G-quadruplexes at 20°C

Conformer	T_M (°C)	ΔH_{cal} (kcal/mol)	$T\Delta S_{cal}$ (kcal/mol)	$\Delta G°_{20}$ (kcal/mol)	ΔH_{vH} (kcal/mol)
G2					
"Chair"	52.6	−23.7	−21.6	−2.0	−41
G3					
"Chair"	57.3	−38.3	−34.0	−4.3	−29
"Basket"	77.5	−36.5	−30.5	−6.0	−61

All experiments were performed in 10 mM Cs-HEPES buffer at pH 7.5 and 100 mM KCl. Experimental errors are shown in parentheses: T_M (± 0.5°C), ΔH_{cal} (± 5%), $T\Delta S_{cal}$ (± 5%), $\Delta G°_{20}$ (± 7%), ΔH_{vH} (± 15%) are from Table 10.1.

However, the ΔH_{vH} determination has an experimental error in excess of 15%, and therefore we conclude that all transitions of these GQs are two-state i.e., without forming intermediate states.

3.7. Thermodynamics of a G-quartet Stack

G2 and *G3* have two and three G-quartets, respectively, and similar loops. Using the standard thermodynamic profiles for the folding of *G2* and the "chair" conformation of *G3*, we have determined the energetic contributions for the stacking of a single G-quartet by subtracting the thermodynamic parameters of *G2* from those of *G3* (Table 10.2). This exercise yields the following energetics for the stacking of two G-quartets at 20°C: $\Delta G° = -2.2$ kcal/mol, $\Delta H_{cal} = -14.6$ kcal/mol, and $T\Delta S_{cal} = -12.4$ kcal/mol (3).

4. Notes

1. The presence of three or more guanines in an oligonucleotide results in the formation of different aggregated states of the oligonucleotide. This constitutes a real problem in investigating the physical properties of GQs. To prevent the formation of undesirable aggregates and to obtain reproducible melting curves, a temperature treatment is recommended. This temperature treatment involves heating the GQ solution to high temperatures (achieving 100% of the random coil state), cooling slowly to ~1°C below the T_M where the temperature is

maintained for 15 min, then slowly cooling to 2°C where the temperature is maintained for 45 min (such as *G3*).

2. In general, to obtain quality melting curves in UV, CD, or DSC experiments, it is important to degas the buffer to prevent formation of bubbles, which interferes with the measurements.

To obtain highly reproducible DSC thermograms, utmost care should be taken to have a matching solvent system between sample and buffer solutions, and this is accomplished by using dialyzed samples: in other words, scanning the well-equilibrated solution of a dialysis experiment, i.e., the solution of the dialysis bag, against the external buffer solution. Special care needs to be taken in the loading of the DSC cells, to prevent introduction of air bubbles in the cells.

Acknowledgments

This work was supported by Grant MCB-0315746 from the National Science Foundation and a UNMC Graduate Fellowship to Chris Olsen.

References

1. Todd AK, Johnston M, Neidle S (2005) Highly prevalent putative quadruplex sequence motifs in human DNA. Nucleic Acids Res 33:2901–2907
2. Kankia BI, Marky LA (2001) Folding of the thrombin aptamer into a G-quadruplex with Sr^{2+}: stability, heat, and hydration. J Am Chem Soc 123:10799–10804
3. Olsen CM, Gmeiner WH, Marky LA (2006) Unfolding of G-quadruplexes: energetic, and ion and water contributions of G-quartet stacking. J Phys Chem B 110:6962–6969
4. Olsen CM, Gmeiner WH, Marky LA (2006) Interaction of Cd^{2+} with G-quadruplexes containing K^{+} or Sr^{2+}. J Biomed Nanotech 2:62–70
5. Cantor CR, Warshow MM, Shapiro H (1970) Oligonucleotide interactions. III. Circular dichroism studies of the confromation of deoxyoligonucleotides. Biopolymers 9:1059–1077
6. Marky LA, Blumenfeld KS, Kozlowski S, Breslauer KJ (1983) Salt-dependent conformational transitions in the self-complementary deoxydodecanucleotide d(CGCGAATT CGCG): evidence for hairpin Formation. Biopolymers 22:1247–1257
7. Marky LA, Breslauer KJ (1987) Calculating thermodynamic data for transitions of any molecularity from equilibrium curves. Biopolymers 26:1601–1620
8. Privalov PL, Potekhin SA (1986) Thermodynamic effects of mutations on the denaturation of T4 lysozyme. Methods Enzymol 131:4–51
9. van Holde KE (1985) Physical Biochemistry, 2nd edn. Prentice-Hall, Englewood Cliffs, NJ, pp 55–57
10. Guo Q, Lu M, Marky LA, Kallenbach NR (1992) Interaction of the dye ethidium bromide with DNA containing guanine repeats. Biochemistry. 31:2451–2455
11. Miyoshi D, Nakao A, Sugimoto N (2003) Structural transition from antiparallel to parallel G-quadruplex of $d(G_4T_4G_4)$ induced by Ca^{2+}. Nucleic Acids Res 4:1156–1163
12. Williamson JR, Raghuraman MK, Cech TR (1989) Mono-valent cation induced structure of telomeric DNA-The G-quartet model. Cell 59:871–880
13. Dapic V, Abodmerovic V, Marrington R, Peberdy J, Rodger A, Trent JO, Bates PJ (2003) Biophysical and biological properties of quadruplex oligonucleotides. Nucleic Acids Res 31:2097–2107

14. Lu M, Guo Q, Kallenbach NR (1993) Thermodynamics of G-tetraplex formation by telomeric DNAs. Biochemistry 32: 598–601
15. Macaya RF, Schultze P, Smith FW, Roe JA, Feigon J (1993) Thrombin-binding DNA aptamer forms a unimolecular quadruplex structure in solution. Proc Natl Acad Sci USA 90:3745–3749
16. Simonsson T (2001) G-quadruplex DNA structures-variations on a theme. Biol Chem 382:621–628

Chapter 11

Probing Telomeric G-Quadruplex DNA Structures in Cells with In Vitro Generated Single-Chain Antibody Fragments

Christiane Schaffitzel, Jan Postberg, Katrin Paeschke, and Hans J. Lipps

Abstract

Guanine-rich sequences have been shown to readily form parallel or antiparallel G-quadruplex DNA structures in vitro. All telomeric repeat sequences contain stretches of guanine residues that can form quadruplex structures. In order to demonstrate the occurrence of the quadruplex structure in vivo, we generated by ribosome display, scFv antibodies specific for quadruplex DNA structures formed by the telomeric sequence of the ciliate *Stylonychia*. The macronucleus of this hypotrichous ciliate contains 10^8 telomere-capped nanochromosomes and was stained with the antibody recognizing the antiparallel G-quadruplex DNA in indirect immuno-fluorescence assays. This antibody was also used as a specific probe to study the interaction of the telomere end-binding proteins with the G-quadruplex during different stages of the cell cycle.

Key words: Ciliate, Telomeres, Macronucleus, G-quadruplex-specific antibodies, Ribosome display, Radio-immunoassays, Indirect immunofluorescence staining, FISH, Gel shift experiments

1. Introduction

Much of the work on telomeric G-quadruplex DNA structure and the demonstration of its occurrence in vivo has been done in ciliated protozoa. This group of organisms was among the first model systems used for studying telomere biology, and many of the results obtained in these cells later proved to be true for other eukaryotic cells, including humans. The first telomere sequence was obtained from the rDNA of *Tetrahymena* (1). Later Klobutcher et al. (2) sequenced the macronuclear telomeres of four different species of ciliates and demonstrated the existence of a protrusion of the 3′-G-rich strand, a structural feature common to all telomeres (3–5). The first telomere end-binding proteins were characterized

P. Baumann (ed.), *G-Quadruplex DNA: Methods and Protocols*, Methods in Molecular Biology, vol. 608
DOI 10.1007/978-1-59745-363-9_11, © Humana Press, a part of Springer Science+Business Media, LLC 2010

in *Oxytricha* (6,7), and the enzyme telomerase was first described in *Tetrahymena* (8) and later purified from *Euplotes* (9).

The specific mechanism of gene amplification in these organisms offers unique possibilities to study telomere structure and function. Each cell contains two types of nuclei, small diploid micronuclei and at least one large macronucleus (Fig. 11.1a). While the

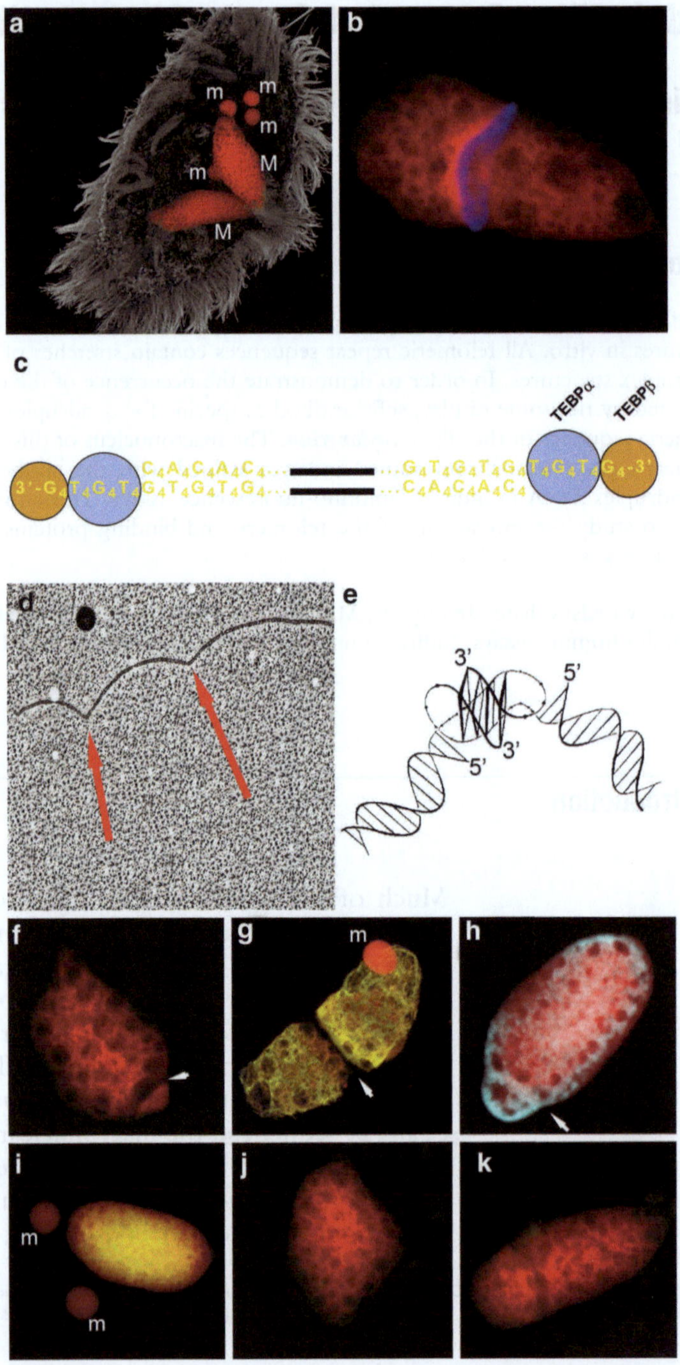

macronucleus encodes for all genes used for vegetative growth, the micronucleus is required during sexual reproduction, the process of conjugation. During this process the old macronucleus is degraded and micronuclei are exchanged between the two conjugation partners, fusing to form a diploid zygote nucleus. This nucleus divides mitotically and one of the daughter nuclei remains as a micronucleus while the other differentiates into a macronucleus (10). During the course of macronuclear differentiation, dramatic DNA excision, elimination, and fragmentation processes take place (for review, (11,12)). Although the extent of these DNA elimination and fragmentation processes varies between different ciliate species, common to all is the organization of macronuclear DNA in subchromosomal DNA-molecules, each terminating in telomeres. For example, the size of macronuclear

Fig. 11.1. G-quadruplex DNA structure in the macronucleus of *Stylonychia lemnae*. (**a**) *Stylonychia lemnae* cell containing micronuclei (m) and macronuclei (M). *Gray channel*: the cell was counterstained using an antibody directed against α-tubulin (Sigma). *Red channel*: micro- and macronuclear DNA was counterstained with To-Pro-3. (**b**) Details of a macronucleus with a replication band. Blue channel: to monitor replication in the replication band; newly replicated DNA was pulse-labeled with BrdU which was subsequently visualized using an anti-BrdU antibody (Sigma). *Red channel*: DNA was counterstained with To-Pro-3. (**c**) Schematic illustration of a nanochromosome from the *Stylonychia* macronucleus. A nanochromosome usually contains one ORF and is flanked at both sides by telomeres. The 16 nucleotide telomeric 3 -overhang is competent to bind a heterodimeric telomere end-binding protein (TEBP) consisting of both an α- and a β-subunit. (**d**) Electron micrograph of spread nanochromosomes from a *Stylonychia* macronucleus. The nanochromosomes interact end-to-end to form garland-like structures (15). (**e**) A model for telomeric end-to-end aggregations via antiparallel G-quadruplex DNA formation, proposed by Sundquist et al.(17). (**f**) Immunofluorescence using a scFv antibody directed against the parallel G-quadruplex DNA conformation reveals no signal. Red channel: DNA was counterstained with To-Pro-3. Arrow points to the replication band. (**g**) Immunofluorescence using a scFv antibody directed against the antiparallel G-quadruplex DNA conformation reveals a strong signal (*green*) distributed over the entire macronucleus. However, no signal could be observed in the replication band. Notably, a micronucleus (m) in this image remains unstained. Red channel: DNA was counterstained with To-Pro-3. Arrow points to the replication band. (**h**) Immunofluorescence using a telomeric oligonucleotide probe for FISH reveals a strong signal (cyan) distributed over the entire macronucleus and the replication band. Red channel: DNA was counterstained with To-Pro-3. Arrow points to the replication band. (**i**) Immunofluorescence using a scFv antibody directed against the antiparallel G-quadruplex DNA conformation reveals a strong signal (green) distributed over the entire macronucleus. Notably, micronuclei (m) in this image remain unstained. *Red channel*: DNA was counterstained with To-Pro-3. (**j**) Immunofluorescence using a scFv antibody directed against the antiparallel G-quadruplex DNA conformation reveals no signal when the expression of TEBPα is post-transcriptionally silenced by RNAi (20). *Red channel*: DNA was counterstained with To-Pro-3. (**k**) Immunofluorescence using a scFv antibody targeted to the antiparallel G-quadruplex DNA conformation reveals no signal, when the expression of TEBPβ is posttranscriptionally silenced by RNAi (20). *Red channel*: DNA was counterstained with To-Pro-3.

DNA in the ciliates *Oxytricha* or *Stylonychia* ranges between about 400 bp and 40,000 bp. In *Stylonychia*, each of the approximately 20,000 genes is amplified to an average copy number of 15,000 per macronucleus, resulting in more than 10^8 DNA molecules per nucleus. In the macronucleus of this group of ciliates, telomere length is tightly regulated to exactly 20 bp of double-stranded (G_4T_4) repeat with a 16-mer 3′-overhang (Fig. 11.1c) (2). This high concentration of telomeres facilitated their isolation and characterization, and analysis of their structural conformation in vivo. Replication of macronuclear DNA in these ciliates takes place in a morphological distinct region, the replication band (Fig. 11.1b) (13,14), providing an important internal control and allowing analysis of telomere conformation during replication.

The first observation that macronuclear telomeres may adopt an unusual DNA configuration in vitro was made with telomeric DNA sequences from *Stylonychia* (15) and later shown to be true in vivo (6). When the macronucleus was gently lysed, the DNA molecules did not exist as separate, individual entities, but formed long coherent linear structures (Fig. 11.1d). As it was later shown that macronuclear telomeres can adopt G-quadruplex DNA structures (16–18), the observed end-to-end aggregation could be explained by the existence of guanine quadruplexes formed by the guanine-rich 3′-telomeric overhang (Fig. 11.1e).

We used ribosome display (Fig. 11.2) to select the single chain Fv antibodies (scFvs) Sty49 and Sty3 against G-quadruplex DNA with the *Stylonychia* telomeric sequence from a naive antibody library (19). In the selection experiment, different families of antibodies recognizing various epitopes of the G-quadruplex DNA structure were obtained. The scFvs were characterized by radio-immunoassays and surface plasmon resonance with respect to their specificity and affinity. Two scFvs with high affinity and specificity for G-quadruplex DNA and for the telomeric sequence were used for indirect immuno-fluorescence staining of the nuclei of *Stylonychia lemnae*. The macronucleus – with exception of the replication band – was stained by the scFv Sty49, thus providing the first evidence for the existence of G-quadruplex DNA structure (Fig. 11.1f, g, i) at telomeres (Fig. 11.1h) in vivo.

Subsequently, RNA-interference was used to demonstrate the in vivo interaction of the *Stylonychia* telomere end-binding proteins (TEBPs) with the telomeric G-quadruplex DNA structure (Fig. 11.1j, k) (20). The formation of the G-quadruplex DNA structures was shown to be regulated in a cell-cycle dependent manner by phosphorylation of TEBPβ.

Here we describe the in vitro selection of quadruplex-specific scFvs by ribosome display, the testing of the selected scFvs for specific binding of the G-quadruplex DNA structure, and their application to eukaryotic nuclei.

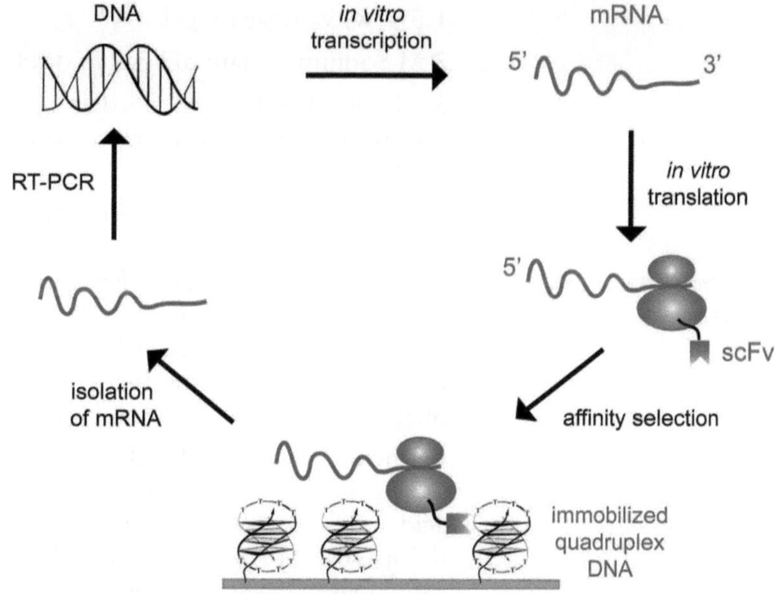

Fig. 11.2. The principle of ribosome display, a method of in vitro selection and evolution. Stable ternary complexes of ribosome, mRNA, and scFv antibody are formed due to deletion of the stop codon from the in vitro transcribed DNA pool encoding the scFv antibody library. Binders are isolated by selection against immobilized ligand such as the G-quadruplex DNA. Ribosome nascent chain complexes (RNCs) with nonbinding scFvs are washed away. The mRNA of bound RNCs is isolated, reverse transcribed, and PCR-amplified. Further diversification of the initial scFv library is achieved by using an error-prone DNA polymerase in the PCR step.

2. Material

2.1. Preparation of G-quadruplex DNA Structures and CD Analysis

1. Oligonucleotides: $d(G_4T_4G_4T_4G_4T_4G_4)$, $d(T_4G_4T)$, 5′-biotin-$d(T_5G_4T_4G_4T_4G_4T_4G_4)$
2. Bufferl: 25 mM Tris-HCl pH 8.0, 50 mM NaCl
3. Buffer2: 10 mM potassium phosphate pH 7.0, 1 M KCl
4. Superdex-75 column (GE Healthcare)
5. CD spectrophotometer

2.2. In vitro Selection of G-quadruplex DNA Specific Antibodies by Ribosome Display

2.2.1. In vitro Transcription and mRNA Purification

1. 70% Ethanol
2. 1 M Guanidinium isothiocyanate in ddH_2O
3. 6 M LiCl in ddH_2O
4. 5× Loading buffer: 50% (v/v) glycerol, 200 mM Tris-HCl pH 8.0, 100 mM acetic acid, 5 mM EDTA, bromophenol blue
5. NTPs: ATP, CTP, GTP, UTP (50 mM each), in ddH_2O
6. RNA denaturation buffer: 10 μL formamide, 3.5 μL formaldehyde, 2 μL of 10× MOPS buffer (0.2 M MOPS pH 7.0, 80 mM sodium acetate, 10 mM EDTA)

7. 1.5% (w/v) Agarose gel
8. 3 M Sodium acetate pH 5.2 in ddH_2O
9. 5× T7 RNA polymerase buffer: 1 M HEPES-KOH pH 7.6, 150 mM magnesium acetate, 10 mM spermidine, 0.2 mM DTT
10. TBE buffer: 90 mM Tris, 90 mM boric acid, 10 mM EDTA

2.2.2. Preparation of E. coli S30 Cell Extract

1. Incomplete rich medium (IRM; 8 L total): 5.6 g/L potassium dihydrogen phosphate (KH_2PO_4), 28.9 g/L dipotassium hydrogen phosphate (K_2HPO_4), 1 g/L yeast extract, 2 mg/L thiamine. Autoclave the medium first and then add 1 ml/L magnesium acetate (1 M) and 50 ml/L glucose (40% (w/v)).
2. Preincubation mix (10 mL): 3.75 ml 2 M Tris-acetate (pH 7.5 at 4°C), 71 µL 3 M magnesium acetate, 75 µL amino acid mix (10 mM of each of the 20 amino acids; Sigma), 0.3 mL 0.2 M ATP, 0.2 g phosphoenolpyruvate, 50 U pyruvate kinase (P-1506 Sigma). The preincubation mix should be prepared immediately before use.
3. S30 buffer: 10 mM Tris-acetate (pH 7.5 at 4 °C), 14 mM magnesium acetate, 60 mM potassium acetate. Store at 4 °C or chill the buffer solution before use.
4. *Escherichia coli* strain MRE600 (21).
5. Equipment: 5-liter baffled flask, dialysis tubing with a cutoff of 6,000–8,000 Da, French Press, refrigerated centrifuge (30,000×*g*), shaker for bacterial culture at 25 °C and 37 °C.

2.2.3. In vitro Translation

1. Anti-ssrAoligonucleotide(200 µMinddH$_2$O)(5′-TTAAGCT-GCTAAAGCGTAGTTTTCGTCGTTTGCGACTA-3′)
2. 100 mM Magnesium acetate in ddH_2O
3. 2 M Potassium glutamate in ddH_2O
4. PremixZ: 250 mM Tris-acetate (pH 7.5 at 4 °C), 1.75 mM of each amino acid, 10 mM ATP, 2.5 mM GTP, 5 mM cAMP, 150 mM acetylphosphate, 2.5 mg/mL *E. coli* tRNA, 0.1 mg/mL folinic acid, 7.5% PEG 8,000
5. Washing buffer (WBTH): 50 mM Tris-acetate (pH 7.5 at 4 °C), 150 mM NaCl, 50 mM magnesium acetate, 0.1% (v/v) Tween-20, 2.5 mg/mL heparin
6. 22 µM Protein Disulfide Isomerase (bovine PDI) (Sigma P-3818) in ddH_2O

2.2.4. Affinity Selection

1. Biotinylated ligand: 5′-biotin-d($T_5G_4T_4G_4T_4G_4T_4G_4$)
2. Elution buffer (EB): 50 mM Tris-acetate (pH 7.5 at 4 °C), 150 mM NaCl, 20 mM EDTA, 50 µg/mL *S. cerevisiae* RNA (Sigma)

3. Low-fat skimmed milk (12% in water) (sterilized)
4. PBS buffer: 137 mM NaCl, 2.7 mM KCl, 10 mM Na_2HPO_4, 1.8 mM KH_2PO_4
5. Washing buffer (WB, 10×): 0.5 M Tris-acetate (pH 7.5 at 4 °C), 1.5 M NaCl, 0.5 mM magnesium acetate
6. Washing buffer (WBT, 1×): 50 mM Tris-acetate (pH 7.5 at 4 °C), 150 mM NaCl, 50 mM magnesium acetate, 0.1% Tween-20
7. Avidin immobilized on agarose beads (Sigma)
8. Streptavidin magnetic particles (Roche Diagnostics)
9. Equipment: magnet, microtiter plate strips or plates (Nunc), panning tubes (Nunc), rocking table or shaker

2.2.5. mRNA Purification and RT-PCR

1. dNTPs (20 mM each)
2. 0.1 M Dithiothreitol (DTT)
3. High Pure RNA Isolation Kit (Roche Diagnostics)
4. 50 mM $MgCl_2$
5. Oligonucleotide primers need to be designed for the amplification of the 3′-terminal stem-loop (T3te) and for the reintroduction of the T7 promoter sequence and the 5′-terminal stem-loop (usually two long oligonucleotides are necessary: SDA and T7P)
6. QIAquick gel extraction kit (Qiagen)
7. 10 mM Tris-HCl, pH 8.5

2.3. Radioimmunoassays – Testing the Affinity and Specificity of scFvs

1. All the material for in vitro transcription and translation (see Subheadings 11.2.2.2 and 11.2.2.3), but omit methionine from the PremixZ.
2. Biotinylated ligand: 5′-biotin-d($T_5G_4T_4G_4T_4G_4T_4G_4$)
3. (^{35}S)methionine (10mCi/ml, 1175Ci/mmol; New England Nuclear)
4. PBS (1×): 10 mM Na_2HPO_4 (pH 7.4), 140 mM NaCl, 15 mM KCl
5. PBST (1×): PBS with 0.5% (v/v) Tween-20
6. Milk powder: 2% w/v in PBST
7. 4 μg/mL Neutravidin (Pierce), in PBS
8. Nucleic acids competitors/ligands: parallel G-quadruplex d(T_4G_4T); antiparallel G-quadruplex d($G_4T_4G_4T_4G_4T_4G_4$); salmon sperm DNA (GE Healthcare); poly(d(GC)) (GE Healthcare); poly(dA)·poly(dT) (GE Healthcare); *Stylonychia* telomeric sequence in a double-stranded B-DNA conformation (5′-CGCGAATCGCTTTTGGGGTACCCCAAAAG CGATTCGCG-3′); poly(dT) (GE Healthcare); DNA hairpin structure (5′-CGCGCGCGTTTTCGCGCGCG-3′); *E. coli* tRNA (Sigma).

9. SDS: 4%(w/v) in PBS
10. Equipment: microtiter plates; scintillation counter; and liquid scintillation cocktail 'OptiPhase2' (Wallac, Finland)

2.4. Expression, Refolding and Purification of scFv Antibody Fragments

2.4.1. Expression of the scFvs as Inclusion Bodies

1. BL21(DE3) *E. coli* strain (Stratagene)
2. 0.5 M EDTA pH 8.0
3. 1 M IPTG in ddH_2O
4. Resuspension buffer: 10 mM Tris-HCl, 2 mM $MgCl_2$, pH 8.0
5. SB medium: 20 g/L tryptone, 10 g/L yeast extract, 5 g/L NaCl, 45 mM K_2HPO_4, 0.1% (w/v)glucose, 120 μg/mL ampicillin
6. Wash bufferl: 100 mM Tris-HCl, 0.5 M guanidine HCl, 10 mM EDTA, pH 8.0
7. Wash buffer2: 100 mM Tris-HCl, 0.5% Triton X-100, 10 mM EDTA, pH 8.0

2.4.2. Refolding of the scFv

1. Refolding buffer: 200 mM Tris, 0.5 mM 6-amino-n-capronic acid, 0.8 M arginine, 0.5 mM benzamidine hydrochloride, 2 mM EDTA, 0.2 mM glutathione reduced (GSH), 1 mM glutathione oxidized (GSSG); adjust the pH with HCl to 9.0 at 4 °C
2. Solubilization buffer: 200 mM Tris-HCl, 6 M guanidine HCl, 10 mM EDTA, 50 mM DTT, pH 8.0 at 4 °C
3. Equipment: ultrafiltration cell and membrane (cutoff of 10 kDa)

2.4.3. Purification of the scFv

1. HBS buffer: 20 mM HEPES-KOH, 150 mM NaCl, pH 7.3
2. Wash buffer 1: 20 mM HEPES-KOH, 1 M NaCl, pH 7.3
3. Wash buffer 2: 20 mM HEPES-KOH, 150 mM NaCl, 10 mM imidazole, pH 7.3
4. Elution buffer: 20 mM HEPES-KOH, 150 mM NaCl, 100 mM imidazole, pH 7.3
5. NiNTA superflow column (Qiagen)
6. Superdex75 column (GE Healthcare)

2.5. In situ Antibody Staining with Antibodies Directed Against G-quadruplex Structure

2.5.1. Preparation and Fixation of Nuclei

1. 1% and 4% (w/v) solutions of paraformaldehyde in PBS pH 7.4
2. PBS pH 7.4: 137 mM NaCl, 2.7 mM KCl, 10 mM Na_2HPO_4, 1.8 mM KH_2PO_4
3. 0.5% (v/v) Triton X-100 in PBS
4. 4% (w/v) BSA in PBS/0.1% Triton X-100

2.5.2. In situ Antibody Staining and Detection

1. Goat anti-6-his (Bethyl)
2. Chicken anti-goat antibody labeled with fluorescein-5-isothiocyanate (FITC) (Bethyl). Antibodies were dissolved in PBS and diluted according to the manufacturer's recommendation.
3. PBS pH7.4: 137 mM NaCl, 2.7 mM KCl, 10 mM Na_2HPO_4, 1.8 mM KH_2PO_4
4. 0.1 μg/mL DAPI
5. 1 μM To-Pro-3 (Molecular Probes)
6. Vectashield antifade medium (Vector Laboratories)
7. Nail polish

2.6. FISH (fluorescence in situ hybridization) Analyses Using a Telomeric Probe

2.6.1. Labeling of Probes

1. digoxigenin-dUTP (Roche)
2. Cy3-dUTP (GE Healthcare)
3. *E. coli* DNA (AppliChem)
4. Hybridization mixture: 0.9 M NaCl, 20 mM Tris-HCl, pH 7.2, 0.01% SDS

2.6.2. Pretreatments Prior to In situ Hybridization

1. 20× SSC: 3 M NaCl, 0.3 M Na citrate pH 7.2
2. 1% (w/v) and 4% (w/v) solutions of paraformaldehyde in PBS pH 7.4
3. PBS pH 7.4: 137 mM NaCl, 2.7 mM KCl, 10 mM Na_2HPO_4, 1.8 mM KH_2PO_4
4. 0.5% Triton X-100 in PBS
5. 0.1 N and 0.01 N HCl
6. 0.002% Pepsin (Fermentas) in 0.01 N HCl
7. 200 μg/mL Rnase A (Fermentas)
8. 2× SSC: from 20× SSC stock solution
9. 50% formamide in 2× SSC (from 20× SSC stock solution) pH 7.0
10. 70% formamide in 2× SSC (from 20× SSC stock solution) pH 7.0
11. 70% and 100% Ethanol

2.6.3. Hybridization and Visualization

1. Rubber cement (Marabu)
2. 2× SSC: from 20× SSC stock solution
3. 0.1× SSC: from 20× SSC stock solution
4. 4% (w/v) BSA in PBS/0.1% Triton X-100
5. PBS: 137 mM NaCl, 2.7 mM KCl, 10 mM Na_2HPO_4, 1.8 mM KH_2PO_4 pH 7.4
6. Mouse antidigoxigenin monoclonal antibody (Sigma)

7. Goat anti-mouse Alexa Fluor 488™ (Molecular Probes)
8. 0.1 µg/mL DAPI
9. 1 µM To-Pro-3 (Molecular Probes)
10. Vectashield antifade medium (Vector Laboratories)
11. Nail polish

2.7. In vitro Demonstration of Telomeric G-quadruplex DNA Structure Using Specific Antibodies

2.7.1. In vitro Assembly of the Telomere–Protein Complex

1. 20% denaturing polyacrylamide gel: 19% (w/v) polyacrylamide 1% (w/v) bisacrylamide, 8 M urea, 1× TBE (0.089 M Tris, 0.089 M boric acid, 2 mM EDTA pH 8.3)
2. 0.1 M KCl
3. 60% (v/v) methanol
4. 50 mM Tris-HCl pH 8.0, 125 mM KCl, 5 mM DTT, 10% (v/v) glycerol
5. Superdex-200 column
6. *E. coli* DNA (AppliChem)
7. NiNTA agarose (Qiagen)
8. Sep-pak cartridge (Qiagen)
9. T4 polynucleotide kinase (Fermentas)

2.7.2. Demonstration of G-quadruplex DNA Structure by DMSO Modification

1. 20% Denaturing polyacrylamide gel: 19% (w/v) polyacrylamide 1% (w/v) bisacrylamide, 8 M urea, 1× TBE (0.089 M Tris, 0.089 M boric acid, 2 mM EDTA pH 8.3)
2. Piperidine (Merck)
3. DMSO (Sigma)

2.7.3. Electromobility Shift Assay (EMSA)

1. Agarose
2. 0.25 M TBE
3. (^{32}P)-γATP (GE Healthcare)
4. DE81 paper

3. Methods

3.1. Preparation of G-quadruplex DNA Structures and CD Analysis

The formation of G-quadruplex DNA requires neutral pH and alkaline cations (>50 mM K^+ or Na^+), which are an integral part of the G-quadruplex DNA structure. The formation of the G-quadruplex DNA structure can be easily monitored by CD spectroscopy as the spectra of four-stranded DNA are very different from the spectra of A-DNA, B-DNA, Z-DNA, or triple helices. Furthermore, the CD spectra of parallel and antiparallel G-quadruplex DNA are easy to discriminate, because the peaks are nearly symmetrical but different in sign.

1. For the generation of an antiparallel G-quadruplex DNA structure, the oligonucleotide $d(G_4T_4G_4T_4G_4T_4G_4)$ is dissolved to a concentration of 10 μM or less in 25 mM Tris-HCl pH 8.0, 50 mM NaCl (22) and incubated at room temperature for 24 h.
2. Parallel G-quadruplex DNA is formed by dissolving the oligonucleotide $d(G_4T_4G_4T_4G_4T_4G_4)$ or $d(T_4G_4T_4)$ to ≥100 μM in 10 mM potassium phosphate pH 7.0, 1 M KCl. The DNA is heated for 30 min to 90 °C and then cooled slowly to room temperature over 12 h (23). If the oligonucleotide is not completely dissolved, remove aggregates by centrifugation.
3. The G-quadruplex DNA structures are further purified to conformational homogeneity by gel filtration with a Superdex75 column using the respective buffer (see above). The G-quadruplex DNA is in equilibrium with single-stranded DNA. However, as this equilibrium is very slow, the four-stranded structure is stable for at least a week at 4 °C.
4. CD spectra of the individual peaks are recorded from 320 nm to 220 nm. Parallel G-quadruplex DNA has a strong, characteristic maximum at 264 nm and a minimum at 243 nm. The antiparallel conformation has a maximum at 295 nm and a strong, characteristic minimum at 265 nm, with a smaller maximum at 243 nm.

3.2. In vitro Selection of G-quadruplex DNA Specific Antibodies by Ribosome Display

Preparation of the ribosome display construct

The construct used for ribosome display contains a T7 promoter, a ribosome-binding site, and the sequence encoding the scFv antibody library, followed by a spacer sequence which is in frame with the scFv antibody. The spacer tethers the nascent scFv to the ribosome and keeps the structured part of the scFv outside the ribosomal tunnel, thus allowing folding and interaction of the scFv with ligands. At both ends of the mRNA, the ribosome display construct should include 5′- and 3′-stem-loops for stabilization. The generation of the antibody library for ribosome display has been described elsewhere (24–26).

3.2.1. In vitro Transcription and mRNA Purification

1. For the in vitro transcription reaction, thaw and mix the following reagents on ice: 40 μL 5× T7 RNA polymerase buffer, 28 μL NTPs, 8 μL T7 RNA polymerase (50 U/μL), 4 μL RNasin, and 0.5–1 μg DNA template encoding the scFv antibody library. Add RNase-free water to a volume of 200 μL and incubate the reaction mix for 2–3 h at 37 °C (see Note 1).
2. Add 200 μL RNase-free water and 400 μL 6 M LiCl to the 200 μL in vitro transcription mixture and retain on ice for 30 min. Centrifuge the mixture 20–30 min at full-speed in a microcentrifuge (4 °C) and discard the supernatant. Wash the pellet once with 500 μL 70% ethanol and allow to dry.

3. Resuspend the pellet in 200 μL RNase-free water and centrifuge for 5 min at full-speed in a microcentrifuge at 4 °C. Mix the supernatant (approximately 180 μL) with 18 μL 3 M sodium acetate and 500 μL 97% ethanol. Incubate 30 min on ice and centrifuge for 20–30 min at full-speed in a microcentrifuge at 4 °C. Wash the pellet with 500 μL 70% ethanol, air-dry and resuspend it in 40 μL RNase-free water. The mRNA should be stored at −20 °C.
4. Determine the concentration of mRNA by measuring the OD at 260 nm (an OD_{260nm} of 1 corresponds to 40 μg/mL). Also run an analytical agarose gel (1.5% (w/v)): Add 10 μL of RNA denaturation buffer to 1 μg mRNA and incubate for 10 min at 70 °C. Subsequently chill the samples on ice, mix with 2 μL of gel loading buffer, and separate by 1.5% (w/v) agarose gel electrophoresis in TBE buffer and in the presence of 20 mM guanidinium isothiocyanate.

3.2.2. Preparation of E. coli S30 Cell Extract

1. Grow a 100 ml starter culture of *E. coli* MRE600 in incomplete rich medium overnight at 37 °C with shaking.
2. The next day, inoculate 1 L of incomplete rich medium in a 5 L baffled shaker flask with 10 mL of the overnight culture and grow at 37 °C. Harvest the cells at an OD_{600nm} of 1.0, when they are still in the early exponential growth phase, by centrifugation for 15 min at 3,500×*g* (4 °C). Discard the supernatant and wash the pellet three times with 50 mL ice-cold S30 buffer per liter of culture. The cell pellet can be frozen at −80 °C or in liquid nitrogen and stored for a maximum of two days.
3. Thaw the cell pellet on ice and wash it again with S30 buffer. Weigh the cell pellet and resuspend it in ice-cold S30 buffer at a ratio of 1.27 mL of buffer per gram of wet cells.
4. Lyse the cells by one passage through a French Press using a chilled French Press cell at 6,000 psi. Centrifuge the lysed cells immediately for 30 min at 30,000×*g* at 4 °C. Transfer the supernatant to a clean centrifuge tube and centrifuge again at 30,000×*g* at 4 °C for 30 min.
5. Transfer the supernatant of the second centrifugation to a clean flask and add 1 mL of preincubation mix for each 6.5 mL of S30 extract. This solution is slowly shaken for 1 h at 25 °C.
6. Transfer the S30 cell extract to dialysis tubing and dialyze three times at 4 °C against a 50-fold volume of chilled S30 buffer. Change the dialysis solution hourly. Subsequently, centrifuge the cell extract for 10 min at 4,000×*g* (4 °C) and freeze the supernatant in aliquots of 100–500 μL in liquid nitrogen. Store samples at −80 °C (see Note 2).

3.3. In vitro Translation

1. Chill all solutions and mix the translation reaction on ice in the indicated order: 11 μL 2 M potassium glutamate, 7.6 μL 0.1 M magnesium acetate (the final concentration of magnesium ions should be optimized within the range 11–16 mM), 2 μL anti-ssrA oligonucleotide, 22 μL PremixZ (ice-cold; thaw on ice and vortex before pipetting), 40 μL *E. coli* S30 cell extract (see above), 2 μL PDI , 10 μL library mRNA (10 μg; thaw the mRNA just prior to use and immediately freeze the remainder), and RNase-free water to 110 μL. Incubate the translation reaction at 37 °C for 6–15 min (the translation time should be optimized).
2. Stop the translation with 400 μL ice-cold WBTH, vortex briefly and gently, and place the tube on ice. Centrifuge the translation mix for 5 min at 14,000×*g* (4 °C) and transfer the supernatant to a new ice-cold tube. Continue immediately with the affinity selection step.

3.3.1. Affinity Selection

1. Preparation of the tubes: block 5 mL panning tubes with 4% (w/v) milk in PBS for 1 h by end-over-end rotation. Wash the panning tubes three times with PBS, three times with washing buffer WBT, and finally fill the tubes with WBT.
2. Prepare biotin-depleted low-fat milk: incubate 1 mL sterilized 12% (w/v) milk powder with 100 μL streptavidin-coated magnetic beads for 1 h at room temperature. Remove the streptavidin-coated beads with a magnet and discard. Transfer the milk to a new tube and store on ice.
3. Wash 100 μL streptavidin-coated magnetic beads four times with ice-cold washing buffer WBT and resuspend them in 100 μL WBT.
4. Empty the panning tubes (from step 1) and add 60 μL sterilized, biotin-depleted skimmed low-fat milk to 1–2% (w/v) final concentration. Add the in vitro translation mix (from Subheading 11.3.2.4) and 10 pmol biotinylated ligand (G-quadruplex DNA (see Subheading 11.3.1)) to the panning tubes. Seal the tubes and rotate end-over-end for 1 h in the cold room.
5. Add 100 μL streptavidin-coated magnetic beads and rotate end-over-end on ice for 15 min in the cold room. Use the magnet to remove all the solution and wash the beads five times with WBT.
6. For elution, add 200 μL ice-cold elution buffer EB for 5 min on ice and shake the tubes gently. Immediately continue with the mRNA purification.

3.3.2. mRNA Purification and RT-PCR

1. Isolate mRNA using the RNA isolation kit according to the manufacturer's instructions. Elute the purified mRNA in 35 μL of RNase-free water. Denature the mRNA at 70 °C for 10 min. Subsequently, chill the mRNA samples on ice for 1–2 min.

2. For reverse transcription, prepare a premix on ice: 0.25 μL primer T3te (100 μM), 0.5 μL dNTP (20 mM each), 0.5 μL RNasin (40 U/μL), 0.5 μL Superscript reverse transcriptase (200 U/μL), 4 μL 5× Superscript first-strand synthesis buffer, 2 μL DTT (0.1 M). Add 12.25 μL denatured mRNA to the premix, mix, and centrifuge briefly at 4 °C. Incubate 1 h at 50 °C.
3. Set up the PCR reaction on ice (50 μL total volume): 0.125 μL primer T3te (100 μM), 0.125 μL primer SDA (100 μM), 0.5 μL dNTP (20 mM each), 0.25 μL Taq Polymerase (5 U/μL), 5 μL 10× PCR-reaction buffer, 2.5 μL DMSO, 1.55 μL $MgCl_2$ (50 mM), 32.45 μL H_2O, 7.5 μL DNA template (from the reverse transcription reaction). The PCR reaction is carried out for 4 min at 94 °C, followed by 20 cycles (or more if necessary) of 30 s at 94 °C, 30 s at 50 °C and 2.5 min at 72 °C. The reaction is finished by 10 min at 72 °C.
4. Purify the PCR product by agarose gel electrophoresis using the gel extraction kit.
5. The gel-extracted PCR product is the template for the second PCR amplification using the same reaction set-up (as above) with the primers T3te and T7P. After initial denaturation for 4 min at 94 °C, 10–15 PCR cycles are performed with 30 s at 94 °C, 30 s at 60 °C, 2.5 min at 72 °C, and the reaction is finished by 10 min at 72 °C.

3.4. Radioimmunoassays – Testing the Affinity and Specificity of Selected Antibodies

1. Coat the microtiter plate wells overnight at 4 °C with neutravidin (100 μL per well, 4 μg/mL in PBS).
2. Wash the plate three times with PBS. To each well, add 50 pmol of biotinylated ligand (G-quadruplex DNA) in 100 μL buffer and incubate with gentle shaking for 30 min at 25 °C.
3. After washing with PBST, block the microtiter plate wells with 4% (w/v) skimmed low-fat milk in PBS for 1 h.
4. Carry out an in vitro transcription reaction using 1 μg DNA (expression plasmid with the gene encoding the scFv or PCR product with T7 promoter). Carry out the subsequent in vitro translation with the following modifications to protocol 3.2.4: add 2 μL (^{35}S)methionine (0.3 μM, 50 μCi/mL final) but no cold methionine to the translation mix. Translate for 30 min at 37 °C.
5. Dilute the translation reaction mixture fourfold with PBST and centrifuge for 5 min at 14,000×*g*.
6. Dilute the supernatant with the same volume of 4% (w/v) milk in PBST containing either no ligand or, for inhibition studies, different concentrations of unbiotinylated ligand (G-quadruplex DNA) of competitor RNA or DNA respectively. Incubate for 1 h at room temperature, and then add 100 μL of this reaction mixture to the microtiter well.

7. Allow the binding reaction with the immobilized ligand (G-quadruplex DNA) to proceed for 30 min at room temperature, with gentle shaking.
8. Wash five times with PBST and elute with 4% (w/v) SDS in PBS.
9. To 5 mL scintillation fluid, add the eluted fraction and quantify the radioactivity in a scintillation counter.

3.5. Expression, Refolding, and Purification of scFv Antibody Fragments

3.5.1. Expression of the scFv Antibody as Inclusion Bodies

1. Transform the plasmid pTFT74_Sty49 (pTFT74_Sty3 respectively) (19) into BL21(DE3) (Stratagene).
2. Inoculate 25 mL SB medium (20 g/L tryptone, 10 g/L yeast extract, 5 g/L NaCl, 45 mM K_2HPO_4, 0.1% (w/v) glucose), with the addition of 120 μg/mL ampicillin, with a single colony and grow the preculture overnight at 37 °C with gentle shaking.
3. Inoculate 1 L SB medium (with 120 μg/mL ampicillin) with 20 mL preculture and grow at 37 °C, 160 rpm in a 5 L flask to an OD of 1–2 at 550 nm.
4. Induce with 500 μL 1 M IPTG and allow the culture to grow for 4 h.
5. Harvest the cells by centrifugation (e.g. 9,500×*g*, 5 min, GS3 rotor), resuspend the pellet in 30 mL resuspension buffer and transfer to a 50 mL falcon tube.
6. Add 1–2 mg of DNaseI (Sigma) and lyse the cells by two passages through a French press cell.
7. Incubate the cell extract for 15 min at room temperature with gentle shaking. Then, add 0.6 mL Triton X-100 and incubate for 15 min at room temperature with gentle shaking.
8. Add 0.6 ml 0.5 M EDTA, pH 8.0, and centrifuge for 10 min at 20,000×*g* in an SS34 rotor at 4 °C.
9. Resuspend the pellet in wash buffer 2 by pipetting and centrifuge for 10 min at 20,000×*g* in an SS34 rotor.
10. Repeat **step 9** until the pellet loses its dark yellow color.
11. Resuspend the pellet once in wash buffer 1 by pipetting and centrifuge 10 min at 20,000×*g* in an SS34 rotor at 4 °C.
12. Completely resuspend the pellet in wash buffer 2 and centrifuge for 10 min at 20,000×*g* in an SS34 rotor at 4 °C.
13. Determine the wet weight of the pellet (typically the yield is 500–1000 mg protein) and store the inclusion body pellet at –20 °C.

3.5.2. Refolding of the scFv Antibody

1. Resuspend the inclusion body pellet in 10–25 mL solubilization buffer by stirring until completely dissolved.

2. Dialyze the solubilized inclusion body suspension against 2×250 mL solubilization buffer at 4 °C (use at least a tenfold excess of solubilization buffer).
3. Centrifuge for 20 min at 49,000×*g* in an SS34 rotor, 4 °C.
4. Retain the supernatant for refolding: slowly stir 400 mL refolding buffer in the cold room and add the solubilized inclusion body suspension in small drops. Do not add more than 4 mL or more than 1 μM scFv (final concentration in the refolding buffer) during one 24 h period.
5. Allow the refolding solution to stir for 24 h. Then, add the next 4 mL of solubilized inclusion body suspension.
6. Repeat step 5 several times until the final concentration of scFv in the refolding solution is approximately 3 μM.
7. Concentrate the refolding solution to 20 mL using an ultrafiltration cell with a membrane cutoff of 10 kDa.

3.5.3. Purification of the scFv Antibody

1. Dialyze the refolded scFv against HBS buffer and centrifuge 15 min at 49,000×*g* in an SS34 rotor at 4 °C.
2. Load the scFv onto a NiNTA superflow column (Qiagen) (2 mL), equilibrated with HBS buffer.
3. Wash the column with 20 mM HEPES-KOH, 1 M NaCl, pH 7.3.
4. Wash the column with 20 mM HEPES-KOH, 150 mM NaCl, 10 mM imidazole, pH 7.3. Then, elute the scFv with 20 mM HEPES-KOH, 150 mM NaCl, 100 mM imidazole, pH 7.3.
5. Load the NiNTA eluate onto a Superdex75 column (GE Healthcare), equilibrated with HBS buffer. The scFv antibody elutes in several peaks. The scFv monomer elutes at approximately 30 kDa. Typically, scFv dimers, trimers, and higher molecular complexes of the scFv are also observed.
6. Determine the concentration of the scFv by measuring the OD at 280 nm. The extinction coefficients are 53,010 $M^{-1}cm^{-1}$ for scFv Sty49, and 49,170 $M^{-1}cm^{-1}$ for Sty3. The purified scFv antibody can be stored for several months at 4 °C. Alternatively, it can be frozen with 10% (v/v) glycerol and stored at −80 °C.

3.6. In situ Antibody Staining with Antibodies Directed Against G-quadruplex DNA Structure (see Note 3)

3.6.1. Preparation and Fixation of Nuclei

1. Isolate nuclei according to standard protocols. Spin down and then fix with 4% (w/v) paraformaldehyde in PBS pH 7.4 for 10 min at room temperature. Wash once in PBS and resuspend in PBS.
2. Immobilize nuclei on poly-L-lysine coated microscopic slides or coverslips (see Note 4). Transfer approximately 250 μL nuclei in PBS onto a coverslip and leave for at least 60 min.
3. Remove excess liquid and wash once in PBS. Permeabilize nuclei by treatment with 0.5% (v/v) Triton X-100 in PBS for 10 min at room temperature. Wash three times in PBS.

4. Incubate nuclei in 200 μg/mL RNase A for 30 min at 37 °C. Wash three times with PBS, and post-fix with 1% (w/v) paraformaldehyde in PBS pH 7.4.
5. Block samples with 4% (w/v) BSA in PBS/0.1% Triton X-100 for 30 min at room temperature prior to detection.

3.6.2. In situ Antibody Staining and Detection

1. For detection of G-quadruplex DNA structure use the his-tagged scFv antibody as a primary layer, followed by an antibody specific for the his-tag as a secondary layer, e.g. goat anti-6-his (Bethyl). As a third layer use a fluorochrome-conjugated antibody targeted to the secondary antibody, e.g. chicken anti-goat antibody labeled with fluorescein-5-isothiocyanate (FITC) (Bethyl).
2. Dilute antibodies according to the manufacturer's recommendations in blocking solution. Incubate successively for 90 min at room temperature or overnight at 4 °C. After each incubation, wash three times in PBS.
3. Counterstain nuclei with either 0.1 μg/mL 4′,6-diamidino-2-phenyl-indole (DAPI) or 1 μM To-Pro-3 for 10 min at room temperature. Wash in PBS. Mount samples in Vectashield anti-fade medium (Vector Laboratories) and seal with nail polish.
4. Excitation wavelengths are 310 nm for DAPI, 488 nm for FITC, 633 nm for To-Pro-3.

3.7. FISH (fluorescence in situ hybridization) Analyses Using a Telomeric Probe

3.7.1. Labeling of Probes

1. Label telomeric (or any other sequence) oligonucleotide probes (see Note 5) at their 3′-ends with hapten or fluorochrome labeled dUTPs (e.g. digoxigenin-dUTP or Cy3-dUTP) according to the manufacturer's recommendations.
2. For probe DNA calculate an amount between ~1 ng/μL (for repetitive sequences) or 50 ng/μL as a rule of thumb (see Note 6). Use up to a 50-fold excess of competitor DNA, such as *E. coli* DNA.
3. Dissolve the probe and *E. coli* competitor DNA in hybridization mixture (0.9 M NaCl, 20 mM Tris-HCl pH 7.2, 0.01% SDS).

3.7.2. Pretreatments Prior to In situ Hybridization

1. Isolate nuclei according to standard protocols. Spin down, fix with 4% (w/v) paraformaldehyde in PBS pH 7.4 for 10 min at room temperature. Wash once in PBS and then resuspend in PBS.
2. Immobilize nuclei on poly-L-lysine coated microscopic slides or coverslips. Transfer approximately 250 μL nuclei in PBS onto a coverslip and leave for at least 60 min.
3. Remove excess liquid and wash once in PBS. Permeabilize nuclei by treatment with 0.5% Triton X-100 in PBS for 10 min at room temperature. Wash three times in PBS.

4. Incubate nuclei in 0.1 N HCl for 5–8 min at room temperature. Wash three times with PBS.
5. Digest nuclei with 0.002% pepsin in 0.01 N HCl for 8 min at 37 °C, wash three times with PBS.
6. Incubate nuclei in 200 µg/mL RNase A for 30 min at 37 °C. Wash three times with PBS, and post-fix subsequently with 1% paraformaldehyde in PBS pH 7.4.
7. Equilibrate coverslips with nuclei in 50% formamide in 2× SSC pH 7.0 for 1–2 h at room temperature. Incubation in this solution for 1–2 weeks at 4 °C improves hybridization efficiency (see Note 7).
8. Denature the probe in a boiling water bath. Denature DNA by incubation of coverslips in 70% formamide/2× SSC pH 7.0 at 73 °C for 3 min, wash in cold PBS, wash in 70% and 100% ethanol, and wash once in PBS.

3.7.3. Hybridization and Visualization

1. Use 15 µL labeled probe for one 22 × 22 mm coverslip. Drop the probe onto a microscope slide and cover with the coverslip containing the immobilized nuclei. Seal with rubber cement.
2. Hybridize for at least overnight or longer (2–3 days) at 42 °C.
3. Wash coverslips three times in 2×SSC at 42 °C and then three times in 0.1× SSC at 60 °C.
4. Block samples with 4% BSA in PBS/0.1% Triton X-100 for 30 min at room temperature prior to detection.
5. *Proceed directly to* step 6 *when using probes labeled with Cy3-dUTP or another fluorochrome.* Dilute antibodies targeted to hapten-labeled probes in blocking solution according to the manufacturer's recommendations (e.g. for detection of digoxigenin-tailed oligonucleotides use mouse anti-digoxigenin monoclonal antibody (Sigma) as the primary layer, followed by goat anti-mouse Alexa Fluor 488™ conjugate (Molecular Probes) as the secondary layer). Incubate for 90 min at room temperature or overnight at 4 °C. Wash three times in PBS.
6. Counterstain nuclei with either 0.1 µg/mL 4′,6-diamidino-2-phenyl-indole (DAPI) or 1 µM To-Pro-3 for 10 min at room temperature. Wash in PBS. Mount samples in Vectashield antifade medium (Vector Laboratories) and seal with nail polish.
7. Excitation wavelengths are 310 nm for DAPI, 488 nm for Alexa Fluor 488™, 514 nm for Cy3, 633 nm for To-Pro-3.

3.8. In vitro Demonstration of Telomeric G-quadruplex DNA Structure Using Specific Antibodies

3.8.1. In vitro Assembly of the Telomere–Protein Complex

1. Telomeric oligonucleotides: TelG 5′-GCTACACTTGCCAT $_4G_4T_4G_4T_4G_4$-3′ and TelC 5′-C_4A_4TGGCAAGTGTAGC-3′ are purified by denaturing polyacrylamide gel electrophoresis (19% (w/v) acrylamide, 1% (w/v) bisacrylamide, 8 M urea, 1× TBE, run for 3–4 h at 50 mA). Extract the corresponding bands and filter off the acrylamide using a syringe and a 0.45 micron filter. Add 0.1 M KCl and measure OD_{260}. Activate the sep-pak cartridge (Qiagen) by washing with 10 mL acetonitrile, followed by 10 mL Millipore-filtered water. Apply the DNA solution, wash away unbound oligonucleotides and urea with 20 mL water, and then elute oligonucleotides with 2 mL 60% (v/v) methanol. Label G-rich oligonucleotides with T4 polynucleotide kinase using (^{32}P)-γATP. Mix both oligonucleotides in equal molarity in 50 mM Tris-HCl pH 8.0, 125 mM KCl, 5 mM DTT, and 10% glycerol, boil for 5 min and cool down slowly (over 1.5 h).
2. Preparation of recombinant telomere end-binding proteins: Proteins are his-tagged, purified using NiNTA agarose, and further purified on a Superdex-200 column. Fractions containing the proteins identified on an SDS-acrylamide gel are pooled and digested by adding appropriate amounts of TEV9 (between 1:50 and 1:100 by mass) to remove the his-tag. His-tags are bound to NiNTA following cleavage. Proteins are dialyzed against 10 mM Tris pH 8.0, 100 mM KCl, 1 mM DTT, and 10% (v/v) glycerol.
3. Assembly of the telomere–protein complex: Incubate labeled DNA (10 nM) with proteins (20 nM each) in 50 mM Tris pH 8.0, 125 mM KCl, 5 mM DTT, and 10% (v/v) glycerol. Use 10 μg/mL sheared *E. coli* DNA and 10 μg/mL BSA as competitors. Incubate at 4 °C overnight.

3.8.2. Demonstration of G-quadruplex DNA Structure by DMSO Modification

Following a standard protocol, separate samples on a 20% (w/v) denaturing polyacrylamide gel.

3.8.3. Electromobility Shift Assay (EMSA)

1. Assemble various telomere-protein complexes (*see* Subheading 11.3.7.1), separate on a 0.4% (w/v) native agarose gel in 0.25 M TBE. Dry gel onto DE81 paper (Whatman) and expose.
2. Add the appropriate antibody to the various complexes using different concentrations of antibody, incubate for 2 h at room temperature, and then analyze on a 0.4% (w/v) native agarose gel (Fig. 11.3).

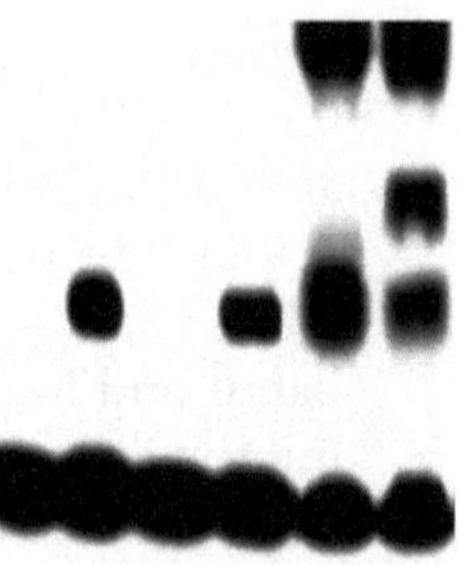

Fig. 11.3. Both TEBPs are required for G-quadruplex DNA formation in vitro. Electrophoretic mobility shift assay. Lane 1: telomeric oligo, lane 2: telomeric/TEBPα complex, lane 3: telomeric oligo plus G-quadruplex specific antibody, lane 4: telomeric/TEBPα complex plus G-quadruplex specific antibody, lane 5: telomeric/TEBPα/β complex lane 6: telomeric/TEBPα/β complex plus G-quadruplex specific antibody. Addition of the G-quadruplex antibody results in a complex of altered electrophoretic mobility when the telomeric DNA is bound by the TEBPα/β heterodimer (lane 6). In this experiment the concentration of the telomeric DNA is 0.1 μM due to the limiting concentration of the G-quadruplex antibody that was available. Analysis was carried out on a 0.4% (w/v) native agarose gel (20).

4. Discussion

The scFvs Sty49 and Sty3 were the first antibodies specific for guanine quadruplex DNA structures (19). For a number of reasons, ribosome display is particularly well-suited for the selection of specific, high-affinity antibodies targeted against this unusual DNA structure. In ribosome display, the selection is performed in the presence of all proteins and nucleic acids of the *E. coli* cytoplasm. Heparin is also added for the affinity selection as it competes together with the cytoplasmic RNA and DNA for the binding of less specific anti-DNA antibodies. In addition, antibodies can evolve during the ribosome display selection with respect to affinity and specificity due to PCR mutations which introduce additional diversity in the antibody pool. This can lead to the selection of specific, high-affinity antibodies, which were not in the initial library.

We anticipate that ribosome display will be equally successful in the generation of quadruplex-specific antibodies formed by the human telomere sequence and other G-rich sequences implicated to form a four-stranded structure. Considering the low number of telomeres in mammalian cells, antibodies suitable for the detection of G-quadruplex DNA in human telomeres must have an even higher affinity than the Sty49 scFv antibody ($K_D = 5$ nM; (19)), and the same high specificity for the G-quadruplex DNA structure and the respective telomeric sequence.

Most interestingly, the homologs of the ciliate telomere end-binding proteins (TEBPα and β) have also been identified in mammalian cells, indicating a much higher evolutionary conservation of the ciliate model system than anticipated (27,28). POT1 (the human TEBPα homolog) has been shown to trap the single-stranded form of telomeric DNA and thus shifts the equilibrium away from the G-quadruplex DNA structure (29). In complex with TPP1 (the recently identified human TEBPβ homolog), POT1 stimulates the activity of telomerase (28) and WRN and BLM helicases on telomeric substrates (30). Both helicases have been reported to dissolve G-quadruplex DNA (31) and mutations in one of these genes, as in Bloom's and Werner's syndrome, lead to inherent genomic instability. It remains to be investigated whether TPP1 can promote G-quadruplex formation as reported for TEBPβ (20). The presence of these proteins acting on telomeric quadruplex DNA indicates the existence of the G-quadruplex DNA structure in mammalian cells. However, a biological role for the G-quadruplex DNA structure in telomeres has yet to be established. Specific probes such as G-quadruplex-specific antibodies are essential tools for this task.

5. Notes

1. The in vitro transcription and translation reaction can be easily scaled-up. Care should be taken that the diversity of the DNA library is conserved, i.e. the number of molecules of DNA template used should be several times higher than the diversity of the scFv antibody library.
2. The *E. coli* S30 cell extract can be stored for several months at −80 °C without losing activity. Activity will start to decrease after more than two freeze-thaw cycles.
3. Although this method is described for antibodies directed against G-quadruplex DNA structure, this technique works equally well for any other antibody.
4. Poly-L-lysine coated and other adhesive microscope slides are commercially available from a number of suppliers. However, to our knowledge, poly-L-lysine coated coverslips have to be made in-house. Incubate coverslips for at least 1 h in 1 mg/mL poly-L-lysine hydrobromide, molecular weight 150,000–300,000 (Sigma). A drop of Triton X-100 reduces the surface tension, resulting in a better coverage of the coverslips with poly-L-lysine. Allow coverslips to dry at room temperature.
5. Telomeric oligonucleotides must reflect the in vivo structure of telomeres.

6. This concentration can be increased further when trying to detect single copy sequences.
7. Specimens on coverslips can be stored for up to three months in 50% (v/v) formamide.

Acknowledgment

This work was funded by the Deutsche Forschungsgemeinschaft.

References

1. Blackburn EH, Gall JG (1978) A tandemly repeated sequence at the termini of the extra-chromosomal ribosomal RNA genes in *Tetrahymena*. J Mol Biol 120:33–53
2. Klobutcher LA, Swanton MT, Donini P, Prescott DM (1981) All gene-sized DNA molecules in four species of hypotrichs have the same terminal sequence and an unusual 3′ terminus. Proc Natl Acad Sci U S A 78:3015–3019
3. Henderson ER, Blackburn EH (1989) An overhanging 3′ terminus is a conserved feature of telomeres. Mol Cell Biol 9:345–348
4. Zakian VA (1989) Structure and function of telomeres. Annu Rev Genet 23:579–604
5. Makarov VL, Hirose Y, Langmore JP (1997) Long G tails at both ends of human chromosomes suggest a C strand degradation mechanism for telomere shortening. Cell 88:657–666
6. Lipps HJ, Gruissem W, Prescott DM (1982) Higher order DNA structure in macronuclear chromatin of the hypotrichous ciliate Oxytricha nova. Proc Natl Acad Sci U S A 79:2495–2499
7. Gottschling DE, Zakian VA (1986) Telomere proteins: specific recognition and protection of the natural termini of Oxytricha macronuclear DNA. Cell 47:195–205
8. Greider CW, Blackburn EH (1985) Identification of a specific telomere terminal transferase activity in *Tetrahymena* extracts. Cell 43:405–413
9. Lingner J, Cech TR (1996) Purification of telomerase from *Euplotes aediculatus*: requirement of a primer 3′ overhang. Proc Natl Acad Sci U S A 93:10712–10717
10. Grell KG (1973) Protozoology. Springer Verlag, Berlin, Heidelberg New York
11. Prescott DM (1994) The DNA of ciliated protozoa. Microbiol Rev 58:233–267
12. Juranek SA, Lipps HJ (2007) New insights into macronuclear development in ciliates. Int Rev Cyt 262:219–251
13. Olins AL, Olins DE, Franke WW, Lipps HJ, Prescott DM (1981) Stereo-electron microscopy of nuclear structure and replication in ciliated protozoa (Hypotricha). Eur J Cell Biol 25:120–130
14. Postberg J, Alexandrova O, Cremer T, Lipps HJ (2005) Exploiting nuclear duality of ciliates to analyse topological requirements for DNA replication and transcription. J Cell Sci 118:3973–3983
15. Lipps HJ (1980) In vitro aggregation of the gene-sized DNA molecules of the ciliate *Stylonychia mytilus*. Proc Natl Acad Sci U S A 77:4104–4107
16. Sen D, Gilbert W (1988) Formation of parallel four-stranded complexes by guanine-rich motifs in DNA and its implications for meiosis. Nature 334:364–386
17. Sundquist WI, Klug A (1989) Telomeric DNA dimerizes by formation of guanine tetrads between hairpin loops (see comments). Nature 342:825–829
18. Williamson JR, Raghuraman MK, Cech TR (1989) Monovalent cation-induced structure of telomeric DNA: the G-quartet model. Cell 59:871–880
19. Schaffitzel C, Berger I, Postberg J, Hanes J, Lipps HJ, Pluckthun A (2001) In vitro generated antibodies specific for telomeric guanine-quadruplex DNA react with *Stylonychia lemnae* macronuclei. Proc Natl Acad Sci U S A 98:8572–8577
20. Paeschke K, Simonsson T, Postberg J, Rhodes D, Lipps HJ (2005) Telomere end-binding proteins control the formation of G-quadruplex DNA structures in vivo. Nat Struct Mol Biol 12:847–854

21. Wade HE, Robinson HK (1966) Magnesium ion-independent ribonucleic acid depolymerases in bacteria. Biochem J 101:467–479
22. Wang Y, Patel DJ (1993) Solution structure of the human telomeric repeat d(AG3(T2AG3)3) G- tetraplex. Structure 1:263–282
23. Jin R, Gaffney BL, Wang C, Jones RA, Breslauer KJ (1992) Thermodynamics and structure of a DNA tetraplex: a spectroscopic and calorimetric study of the tetramolecular complexes of d(TG3T) and d(TG3T2G3T). Proc Natl Acad Sci U S A 89:8832–8836
24. Hanes J, Pluckthun A (1997) In vitro selection and evolution of functional proteins by using ribosome display. Proc Natl Acad Sci U S A 94:4937–4942
25. Knappik A, Ge L, Honegger A et al (2000) Fully synthetic human combinatorial antibody libraries (HuCAL) based on modular consensus frameworks and cdrs randomized with trinucleotides. J Mol Biol 296:57–86
26. Hanes J, Schaffitzel C, Knappik A, Pluckthun A (2000) Picomolar affinity antibodies from a fully synthetic naive library selected and evolved by ribosome display. Nat Biotechnol 18:1287–1292
27. Baumann P, Cech TR (2001) Pot1, the putative telomere end-binding protein in fission yeast and humans. Science 292:1171–1175
28. Wang F, Podell ER, Zaug AJ et al (2007) The POT1-TPP1 telomere complex is a telomerase processivity factor. Nature 445:506–510
29. Zaug AJ, Podell ER, Cech TR (2005) Human POT1 disrupts telomeric G-quadruplexes allowing telomerase extension in vitro. Proc Natl Acad Sci U S A 102:10864–10869
30. Opresko PL, Mason PA, Podell ER et al (2005) POT1 stimulates RecQ helicases WRN and BLM to unwind telomeric DNA substrates. J Biol Chem 280:32069–32080
31. Mohaghegh P, Karow JK, Brosh RM Jr, Bohr VA, Hickson ID (2001) The Bloom's and Werner's syndrome proteins are DNA structure-specific helicases. Nucleic Acids Res 29:2843–2849

Chapter 12

Detection of G-Quadruplexes in Cells and Investigation of G-Quadruplex Structure of $d(T_2AG_3)_4$ in K^+ Solution by a Carbazole Derivative: BMVC

Ta-Chau Chang and Cheng-Chung Chang

Abstract

Verification of the existence of quadruplex structure in native human telomeres and determination of the major structure of $d(T_2AG_3)_4$ (H24) in K^+ solution are the major questions regarding the structure of human telomeres. We have synthesized a fluorescent probe of 3,6-bis(1-methyl-4-vinylpyridinium)carbazole diiodide (BMVC) that has a very high binding affinity for G-quadruplex H24. BMVC stabilizes quadruplex structures and acts as a sensitive probe to the local environment. Although the circular dichroism patterns of H24 are different in Na^+ and K^+ solutions, similar binding behaviors of BMVC to H24 in these solutions led us to suggest that the major G-quadruplex structure of H24 in K^+ solution is very likely similar to that in Na^+ solution. Of particular interest is the fluorescent band detected at ~575 nm in quadruplex H24 and at ~545 nm in duplex DNA. In addition, the intensity of BMVC fluorescence increases by two orders of magnitudes upon interaction with either duplex or G-quadruplex DNA. BMVC has a greater binding preference for G-quadruplex H24 than for duplex DNA. Analyzing the BMVC fluorescence at the ends of metaphase chromosomes and other regions of chromosomes allowed us to verify the presence of G-quadruplex structure in human telomeres for the first time. Using fluorescence lifetime imaging microscopy, the longer decay time of BMVC in G-quadruplex H24 than in duplex DNA allowed us to map the G-quadruplex structure in human metaphase chromosomes.

Key words: G-quadruplex, Fluorescence, Telomeres, Carbazole, BMVC

1. Introduction

Telomeres, the ends of chromosomes, are essential for the stability of eukaryotic chromosomes (1, 2). It is believed that telomere length is closely related to "cancer" and "aging" (3, 4). Telomeres progressively shorten with each cell cycle in somatic cells, a process that eventually results in cellular senescence (4–7). In contrast, telomere length maintenance by telomerase in tumor cells

P. Baumann (ed.), *G-Quadruplex DNA: Methods and Protocols*, Methods in Molecular Biology, vol. 608
DOI 10.1007/978-1-59745-363-9_12, © Humana Press, a part of Springer Science+Business Media, LLC 2010

contributes to cellular immortalization (8–11). It was further proposed that molecules that stabilize the structure of telomeres have the potential to interfere with telomere replication and can therefore serve as antitumor agents (3, 12, 13). Therefore, verification of the telomere structure is not only for its own importance but also for its potential use as a therapeutic application.

Telomeres generally consist of many tandem repeats of guanine-rich (G-rich) motifs, for example, the hexameric repeats of TTAGGG/CCCTAA in vertebrate telomeres (14). Of interest is that a short 3 -overhang of G-rich single-stranded sequence could adopt G-quadruplex (G4) structures in vitro, especially in the presence of monovalent cations (15). The G4 structure of telomeric sequence $d[AG_3(T_2AG_3)_3]$ (H22) in Na^+ solution has been determined by NMR (16). However, the NMR spectrum of the H22 in K^+ solution showed a broad envelop with some fine lines, implying the presence of multiple conformational isomers (17, 18). Investigation of the G4 structures of the human telomeric repeats in K^+ solutions has recently received much attention (17–25). To this day, the G4 structure of H22 in K^+ solution remains open. Because K^+ predominates in cells and therefore is more biologically relevant than Na^+, it is of importance to elucidate the major G4 structure of human telomeric sequences in K^+ solution.

Although the G4 structure is easily formed in vitro, it is not easy to verify the existence of G4 structure in native telomeres. Schaffitzel et al. (26) used in vitro generated antibodies to show that the cilitate *Stylonychia* forms G4 structures in vivo. In addition to telomeres, the G4 structure was also found in the c-*myc* promotor sequence (27). Recently, Boussin et al. (28) used a specific G4 ligand to interact with the terminal ends of human chromosomes to support the presence of G4 structures at telomeres of human cells. We have applied the idea of single molecule spectroscopy to verify the presence of G4 structure of native human telomeres through a novel fluorescent probe (29, 30).

Considering the existence of a very small amount of telomeric G-quadruplexes compared to the vast excess of double-stranded DNA in chromosomes, it is essential to design either a fluorescent probe that has a higher binding preference for G4 structure than for duplex DNA, or a molecule that has distinct spectral features upon interaction with G4 structures. We have designed and synthesized a new fluorescent probe, 3,6-bis(1-methyl-4-vinylpyridinium) carbazole diiodide (BMVC) (31), with higher binding affinity for the quadruplex structure of human telomere sequence, $d(T_2AG_3)_4$ (H24), than for a linear duplex $[d(GCATATGGCCATATGC)]_2$ (LD16) (32). As BMVC binds to the G4 structure of H24 through external stacking to the end surface of the G-quartet (25, 33), the sensitivity of BMVC to the local environment reflected in its binding behaviors allowed us to elucidate the major G4 structure of H24 in K^+ solution. Moreover, the fluorescence of BMVC increases significantly in the presence of DNA and the emission peak is around 575 nm for quadruplex H24 and 545 nm for duplex LD16. In addition, the fluorescence decay of BMVC bound to LD16 is faster than that of H24. We capitalized on these properties of BMVC to verify the presence of G4 structure in native human telomeres (29, 30). Table 12.1 lists the DNA sequences studied in our work. The chemical structure of BMVC is shown in Scheme 12.1.

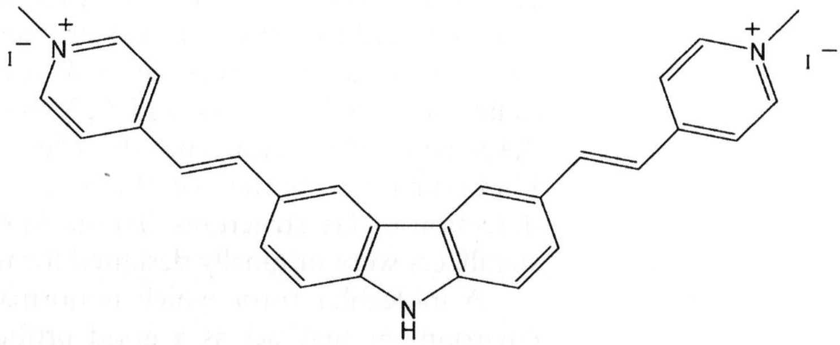

Scheme 12.1 Chemical structure of 3,6-bis(1-methyl-4-vinylpyridinium) carbazole diiodide (BMVC)

Table 12.1
Oligonucleotides used in this work

	Sequence	ε_{260} (M^{-1} cm^{-1})	Abbreviation
1.	5′-TTAGGGTTAGGGTTAGGGTTAGGG	244,600	H24
2.	5′-TTAGGGTTTGGGTTAGGGTTAGGG	238,700	H24-T9
3.	5′-TTAGGGTTAGGGTTTGGGTTAGGG	238,700	H24-T15
4.	5′-TTAGGGTTAGGGTTGGGTTAGGG	230,600	M23
5.	5′-TTAGGGTTAGGGTTTTGGGTTAGGG	246,800	M25
6.	5′-GGGTTAGGGTTAGGGTTAGGG	215,000	H21
7.	5′-$(TTAGGG)_9$	550,100	H54
8.	5′-$(TTAGGG)_{13}$	794,500	H78
9.	5′-AGGGTTAGGGTTAGGGTTAGGG	228,500	H22
10.	5′-TTAGGGTTAGGGTTAGGGTTAGGGTT	261200	H26

2. Design of BMVC and its Photophysical Properties

Fluorescent probes have attracted our attention because of their broad application, particularly in biotechnology, such as detection, identification, quantization, and characterization of biological systems. In order to verify the G4 structure in human telomeres, a fluorescent probe with a binding preference for G4 over duplex structure is critical. We began to search for a suitable probe among the documented G4 stabilizers. The first G4-targeted molecule of 2,6-diamidoanthraquinone derivative was documented by Neidle, Hurley, and coworkers (34). Since then, a number of dye molecules were reported to stabilize the G4 structure. Of these molecules, we have studied two dyes of 3,3′-diethyloxadicarbocyanine iodide (DODCI) (35, 36) and *N,N′*-bis-(2-(1-piperidino)ethyl)-3,4,9,10-perylenetetracarboxylic acid diimide (PIPER) (37, 38). Unfortunately, neither of them is a sensitive probe for the detection of G4 structures. To our knowledge, none of the G4 stabilizers were originally designed for use as fluorescent probes.

A molecular rotor which is normally sensitive to the local environment may act as a good probe (39–42). For example, auramine shows weak fluorescence in low-viscosity solvents but a strong fluorescence in DNA or in viscous solvents (43). The quantum yield of auramine in different environments depends upon its intramolecular charge transfer (ICT) in the excited state (44). Castex et al. (45) have reported that an ICT state can be formed in carbazole derivatives when electron-acceptor groups are linked to the electron-donor carbazole moiety. If a designed molecular rotor has different binding modes to the G4 and the duplex structures, it is possible to have distinct spectral features upon interaction with different DNA structures.

We therefore have synthesized a molecular rotor of BMVC with the electron-donor of a tri-cyclic molecule of carbazole as a core moiety together with the electron-acceptor of a pyridinium cation bridged by a vinyl group. Here we briefly introduce its photophysical property (46). Figure 12.1a shows the absorption and fluorescence spectra of 20 μM BMVC dissolved in different solvents at room temperature. Of particular interest is the significant increase of fluorescence quantum yield (Φ_F) of BMVC upon interaction with DNA. The fluorescence quantum yield of BMVC in glycerol is very similar to those of its complexes with H24 and LD16 DNA. Figure 12.1b shows the absorption and fluorescence spectra of BMVC in glycerol/water mixtures at room temperature. It appears that the viscosity plays a major role in determining the fluorescence quantum yield (Φ_F) of BMVC. Increasing the viscosity of the local environment could suppress nonradiative transitions because it could hinder the torsional motion of the fluorophore, and thus increase the quantum yield. In addition,

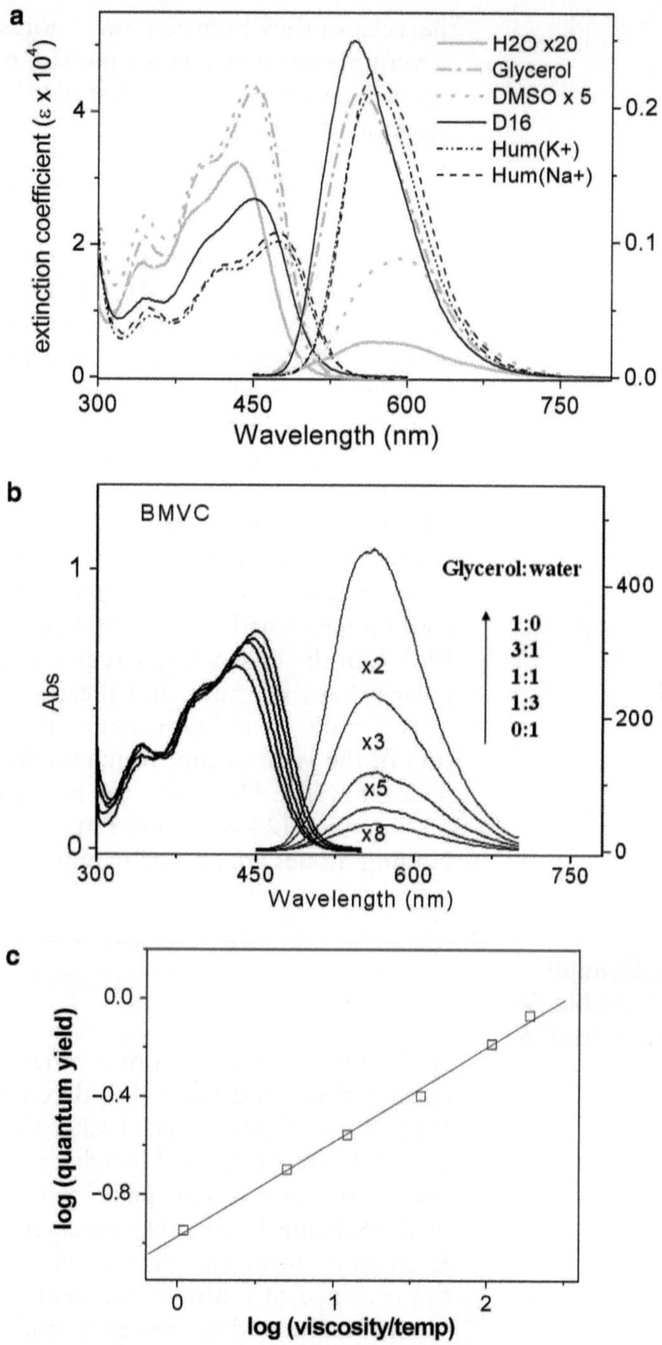

Fig. 12.1. (a) Absorption and Fluorescence spectra of BMVC in water, glycerol, DMSO, LD16, Hum24 (K^+), and Hum24 (Na^+) solvents. (**b**) Absorption and fluorescence spectra of BMVC in glycerol/water mixtures at room temperature. (**c**) The plots of log(Φ_F) vs. log (η/T) of BMVC in glycerol.

the relationship between the fluorescence quantum yield and solvent viscosity (η) as a function of temperature (T) in high viscosity solvents can be described by the Förster–Hoffmann equation (44, 47),

$$\Phi_F = B(h/T)^x,$$

where B and x are constants. Figure 12.1c shows the plots of log (Φ_F) vs. log (η/T) of BMVC in glycerol. The linear correlation between them suggests that BMVC exhibits a rotation-dependent nonradiative decay. Our data show that BMVC is sensitive to the viscosity and rigidity of the local environment. Thus, we suggested that the intramolecular twist of the vinyl groups in BMVC plays a major role in determining the fluorescence yield (46).

The marked increase of the fluorescence quantum yield of BMVC from ~0.002 to ~0.23 upon binding to DNA can be rationalized on the basis of the ICT processes perturbed by the rotation motion of the vinyl group. The positive charge of BMVC could interact with the negative charge of DNA. Immersion of BMVC in the hydrophobic region of DNA protects it against the polar water molecules, and therefore could affect the formation of ICT state. In addition, steric effects of DNA hinder reorientation of the vinyl group, which could also affect the fluorescence quantum yield. Moreover, the fluorescent emissions of BMVC at ~575 nm in H24 and ~545 nm in LD16 resulting from different binding modes are especially important to our study.

3. Investigation of the Possible G4 Structures of H24 by BMVC

G-rich sequences can adopt a variety of G4 structures that often coexist. Wang and Patel (16) determined the NMR structure of H22 d[$AG_3(T_2AG_3)_3$] as a basket-type G4 structure in Na^+ solution (Scheme 12.2a) and Neidle et al. (48) documented the x-ray crystal structure as a propeller-type G4 structure in the presence of K^+ (Scheme 12.2b). However, the NMR spectrum of H22 in K^+ solution showed a broad envelop with some fine lines, implying the presence of multiple conformational isomers (17, 18). The sedimentation and fluorescence studies suggested that the crystal structure of H22 cannot be the major structure in K^+ solution (20). Very recently, a number of groups have modified the original sequences to stabilize a single structure in K^+ solution. Yang et al. (17), Sugiyama et al. (23), and Patel et al. (49) independently suggested a mixed-I type G4 structure (Scheme 12.2c) for the H22 in K^+ solution. Although the modified structures could provide an approximation to the human telomeric structure, the native structure of H22 in K^+ solution remains unknown (22, 50).

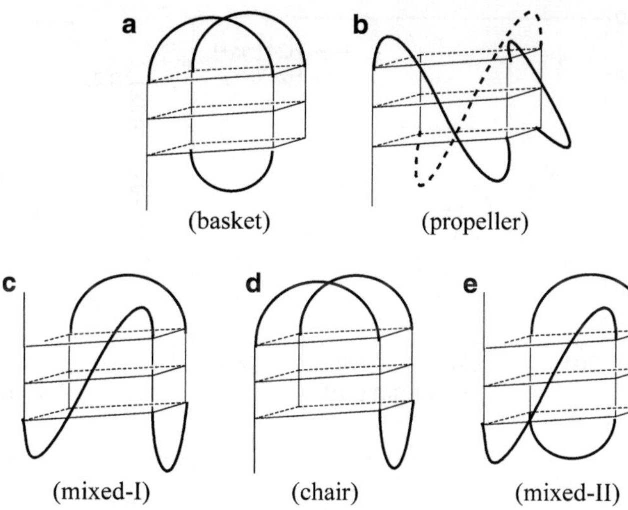

Scheme 12.2 Folding topologies of G-quadruplex DNA. Depicted are (**a**) "basket", (**b**) "propeller", (**c**) "mixed-I", (**d**) "chair", and (**e**) "mixed-II" structures

3.1. Circular Dichroism (CD) Spectra

CD spectra have been extensively applied to study the G4 structures (19, 20, 51–55). It is well known that linear parallel G4 structures, such as the propeller form, give a positive band at ~265 nm and a negative band at ~240 nm, while antiparallel G4 structures, such as basket and chair forms, show two positive bands at ~295 and ~240 nm and a negative band at ~265 nm. These spectral features are mainly attributed to the specific guanine stacking in various G4 structures. Figure 12.2a shows the CD spectra of H24 in solutions containing 150 mM of Na^+ or K^+ cation. The CD pattern of H24 in Na^+ solution is quite different from that in K^+ solution. According to the finding of multiple conformations from NMR analysis (17, 18), the positive CD band at ~290 nm associated with a positive shoulder at ~270 nm in the CD spectrum of H24 in K^+ solution can be attributed to the combination of several components.

3.2. Binding Characters

As different G4 structures can have different ligand binding sites, one may distinguish the structural isomers of quadruplexes by monitoring the signal from ligand binding. We have conducted spectral titrations to compare the binding ratio and binding affinity of BMVC to H24 in Na^+ and in K^+ solutions. Figure 12.2b shows the job plots of BMVC to H24 in Na^+ and in K^+ solutions to determine the binding ratio of BMVC to H24. A job plot is obtained by taking the intensity difference between the free and bound BMVC to the molar ratio of BMVC/H24. Our results reveal a ~2:1 binding ratio for BMVC to H24 in both Na^+ or K^+ solutions. It is likely that both ends of the G-quartet are binding sites of H24 for BMVC in both solutions.

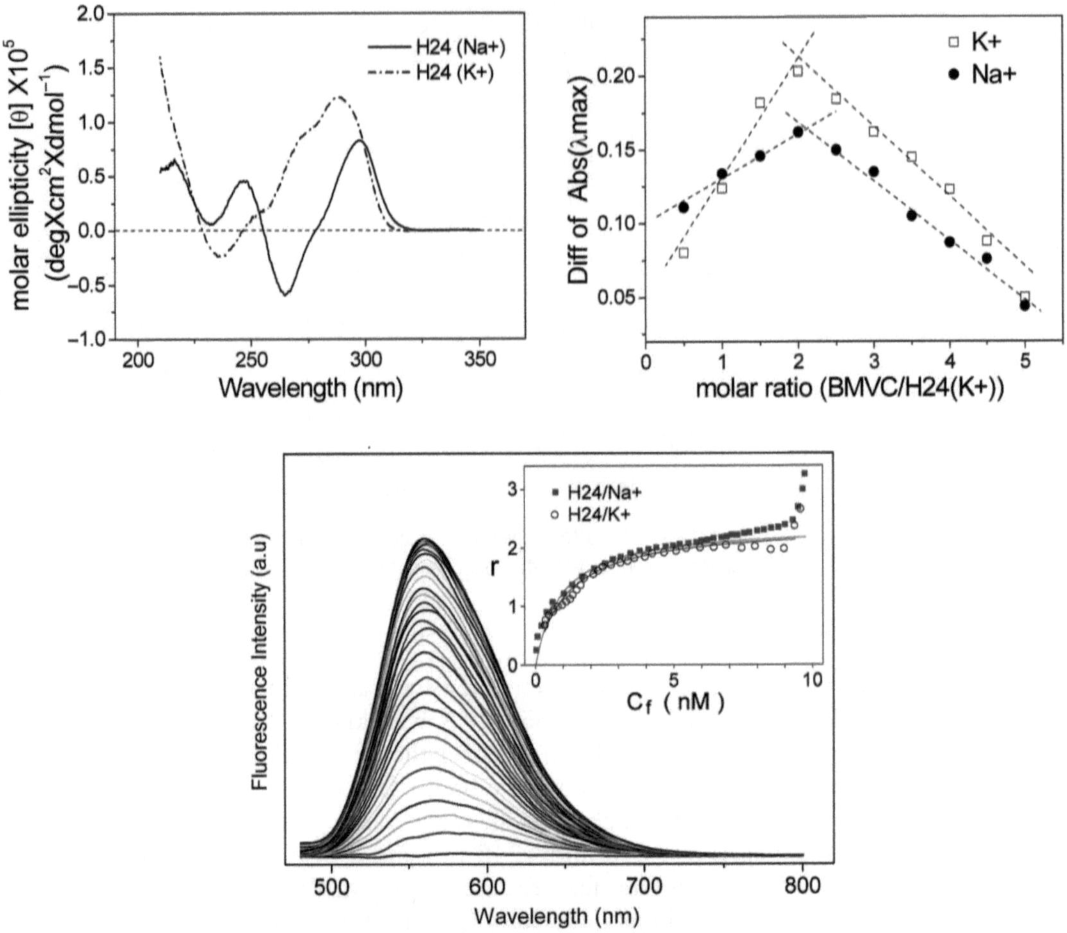

Fig. 12.2. (**a**) CD spectra of 10 μM H24 DNA in solutions containing 150 mM of Na^+ or K^+ cation. (**b**) The job plots of BMVC to the H24 in Na^+ and K^+ solutions obtained from absorption spectra of free BMVC and its complexes with H24. (**c**) Fluorescence titration of 10 nM BMVC by adding 0.25–12 nM H24 in K^+ solution. The inset shows the binding plots of γ vs. C_f for the titration. The fitting parameters to the equation, $\gamma = nKC_f/(1+KC_f)$, are $K = 1.12 \times 10^9$ with $n \sim 2.4$ in Na^+ solution, and $K = 1.07 \times 10^9\ \text{mol}^{-1}$ with $n \sim 2.3$ in K^+ solution.

To measure the binding affinity of BMVC to the G4 structure of H24, Fig. 12.2c shows the fluorescence titration of 10 nM BMVC by adding 0.25–12 nM H24 in K^+ solution. The titration data applied to construct the binding plots of γ vs. C_f are shown in the inset. The binding ratio γ is defined as C_b/C_{DNA}, where C_f, C_b, and C_{DNA} are the molar concentrations of free BMVC, bound BMVC, and DNA, respectively. The difference between C_t and C_b gives the magnitude of C_f, where C_t is the total concentration of BMVC. The curve of the binding plots again indicates that the binding is a complex process. Binding parameters can be obtained by fitting the plots with a multiple-equivalent-site model (56):

$$\gamma = nKC_f/(1 + KC_f),$$

where K is the equilibrium binding constant, and n represents the average number of ligands bound per DNA structure. Here the K values for BMVC to the H24 are 1.12×10^9 and $1.07\times10^9\,mol^{-1}$ with n ~2.4 and ~2.3 in Na^+ and K^+ solutions, respectively. The large value of γ at very low concentrations of H24 is probably due to nonspecific binding. Our data showed that the binding characteristics of BMVC to H24 are essentially the same in both Na^+ and K^+ solutions.

3.3. Binding Modes

As the loops and tails of the G4 structures could play an important role in ligand binding (53–55), substituting a base or varying the length of the loops may allow us to distinguish structural isomers and to determine the binding modes of different G4 structures. Figure 12.3a and b shows the CD spectra of H24, H24-T9, H24-T15, M23, M25, and H26 and the induced CD spectra of BMVC upon interaction with these G4 structures at their molar

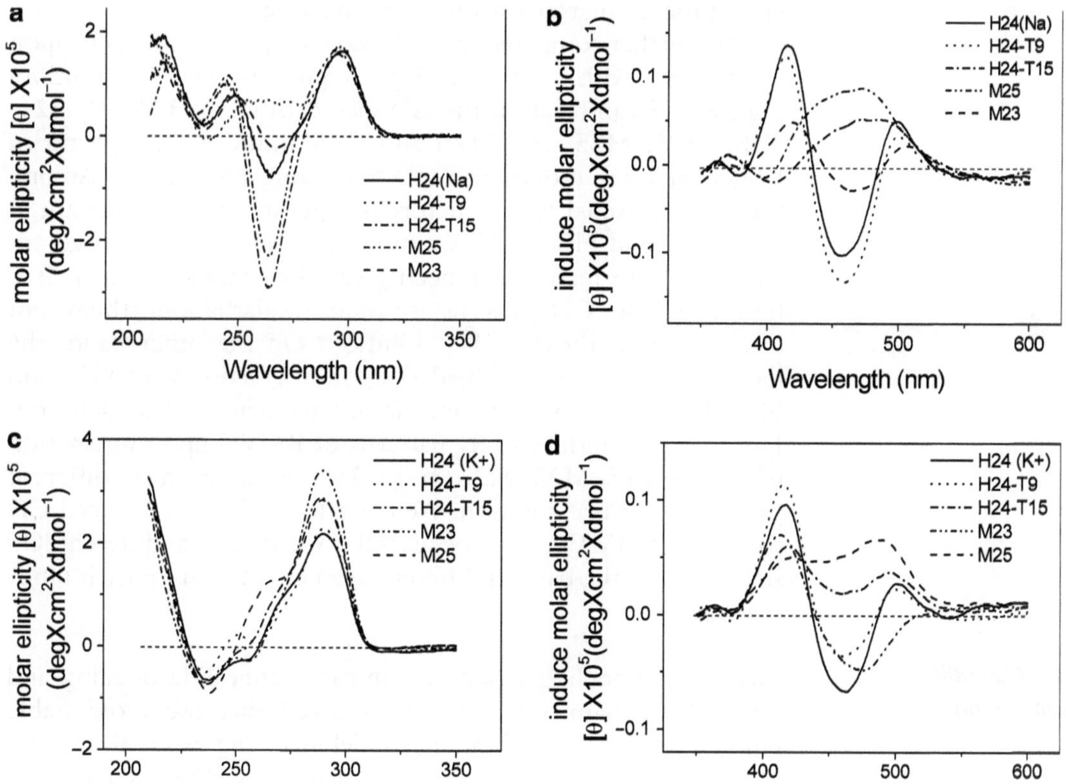

Fig. 12.3. (**a**) CD spectra of H24, H24-T9, H24-T15, M25, M23, and H26 and (**b**) the corresponding induced CD spectra of 20 μM BMVC upon interaction with these G4 structures at their molar ratio 1:1 in Na^+ solution. (**c**) CD spectra of H24, H24-T9, H24-T15, M25, M23, and H26 and (**d**) the corresponding induced CD spectra of 20 μM BMVC upon interaction with these G4 structures at their molar ratio 1:1 in K^+ solution.

ratio of 1:1 in Na^+ solution, respectively. It is found that substitution of the A-base in a TTA lateral loop by a T-base, such as H24-T9, could substantially change the CD spectra. On the other hand, both diagonal loops of the TTT in H24-T15 and the TTTT in M25 mainly enhance the negative CD band at ~265 nm. Our data clearly show no appreciable difference in the ~295 nm positive CD band, but show some changes to the ~265 nm and ~240 nm CD bands upon loop modification.

In contrast to very different CD patterns between H24 and H24-T9, Fig. 12.3b shows similar induced CD patterns of BMVC in H24 and H24-T9. It implies that the binding is not appreciably perturbed by substituting the A-base in the lateral TTA loops by a T-base. On the other hand, the induced CD pattern changes a lot in H24-T15 when substituting the A-base in the diagonal TTA loop by a T-base. The induced CD patterns of BMVC in H24-T15 and M25 characterized by a broad positive band around 470 nm are mainly attributed to the interaction with the TTT or the TTTT diagonal loops (57, 58). It appears that the induced CD pattern is dominated by the interaction of BMVC with the end of the G-quartet with the diagonal loop.

We further compared the induced CD spectra of BMVC upon interaction with these G-rich sequences in K^+ solution. Figure 12.3c and d show the CD spectra of H24, H24-T9, H24-T15, M23, M25, and H26 and their induced CD spectra of BMVC at a 1:1 molar ratio in K^+ solution, respectively. Among them, the mixed-I type structure is suggested to be dominant in both H24 and H26 in K^+ solution (17). To our knowledge, no structural information has been given for other sequences. It is found that the CD patterns are quite similar among them, but different from those in Na^+ solution. On the other hand, the induced CD patterns of BMVC upon interaction with H24 and H24-T9 in K^+ solution are similar to that in Na^+ solution. However, the induced CD patterns of BMVC upon interaction with H24-T15, M25, M23, and H26 in K^+ solution are different from those in Na^+ solution. We considered that the induced CD patterns of BMVC may be useful to distinguish different G4 structures on the basis of different loop interactions in their binding sites.

3.4. Molecular Simulation

Molecular modeling based on simulated annealing docking and molecular dynamics simulations showed that the most stable binding mode is a 2:1 binding model involving external stacking of BMVC to both ends of the G-quartet of H22, as shown in Fig. 12.4a (33). The molecular dynamics simulations further illustrated that the carbazole moiety, together with one pyridinium ring of BMVC, is sandwiched between the loop bases and the end surface of the G-quartet. The other pyridinium ring is bent to interact with the loop of G4 structure (33). BMVC stabilizes

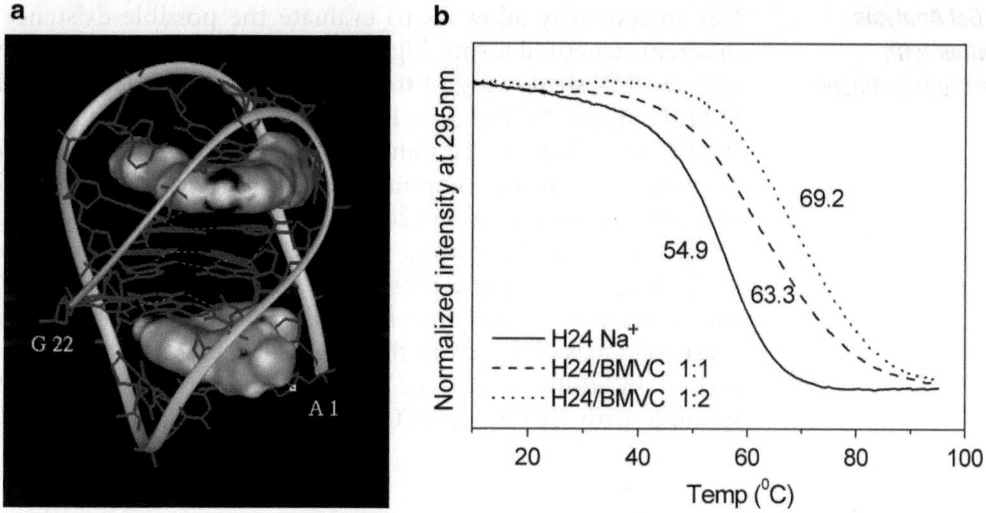

Fig. 12.4. (**a**) Plot of the structure of the 2:1 model where the each BMVC stacks to the end of G-quartet. (**b**) CD signal at 295 nm for the measurement of melting temperature of H24 and its complexes of BMVC as a function of temperature.

G4 quadruplexes via the π–π stacking interaction and the ionic interaction between the positive charge of pyridinium and the negative charge of phosphate groups. The binding energy of BMVC stacking to the end surface of G-quartets with parallel loops is −84.52 kcal/mol, to the end surface of G-quartets with diagonal loops is −69.06 kcal/mol, and to both end surfaces is −116.59 kcal/mol. The binding preference of BMVC to the end surface with a diagonal loop over parallel loops found in the experiments may be explained by the fast kinetics. Figure 12.4b shows the CD intensity at 295 nm of 10 μM H24 and its complexes with 10 μM and 20 μM BMVC in Na^+ solution as a function of temperature. The simulated result is consistent with the melting temperature measurement. This finding confirmed that both ends of the G-quartet are major binding sites for BMVC. In addition, the increased melting temperature of H24 upon BMVC binding clearly indicated that BMVC is a potent G4 stabilizer.

Apparently, this type of binding mode cannot be found in BMVC/LD16 complexes. The longer fluorescent wavelength of BMVC at ~575 nm in H24 than at ~545 nm in LD16 may be due to one pyridinium ring conjugated with the carbazole moiety on top of the G-quartet. Different binding modes between BMVC and various DNAs could perturb the intramolecular twist of the vinyl group and result in the enhancement of fluorescent yield with distinct fluorescent emissions. Nevertheless, distinct fluorescent emissions are especially useful to distinguish different DNA structures.

3.5. Gel Analysis Together with Fluorescence Decay

Gel analysis may allow us to evaluate the possible existence of different structural forms. Figure 12.5a and b show the prestained gels and UV shadowing of the gels of 20 μM each DNA of H21, H24, H54, and H78 (Lanes 1–4) and their complexes with 20 μM BMVC prestained for 10 min (Lanes 5–8) in the presence of Na^{+}, respectively. A major component is found in each lane, migrating to a very similar position in both the free DNA and its prestained DNA in the UV shadowing, implying that the major component is the free DNA in the prestained gels. However, the level of the fluorescence band due to BMVC/DNA complexes in the prestained gels differs from the level of the band in the UV shadowing. In addition, Fig. 12.5c shows the same gels after poststaining with 10 μM BMVC for 10 s. The band in the UV

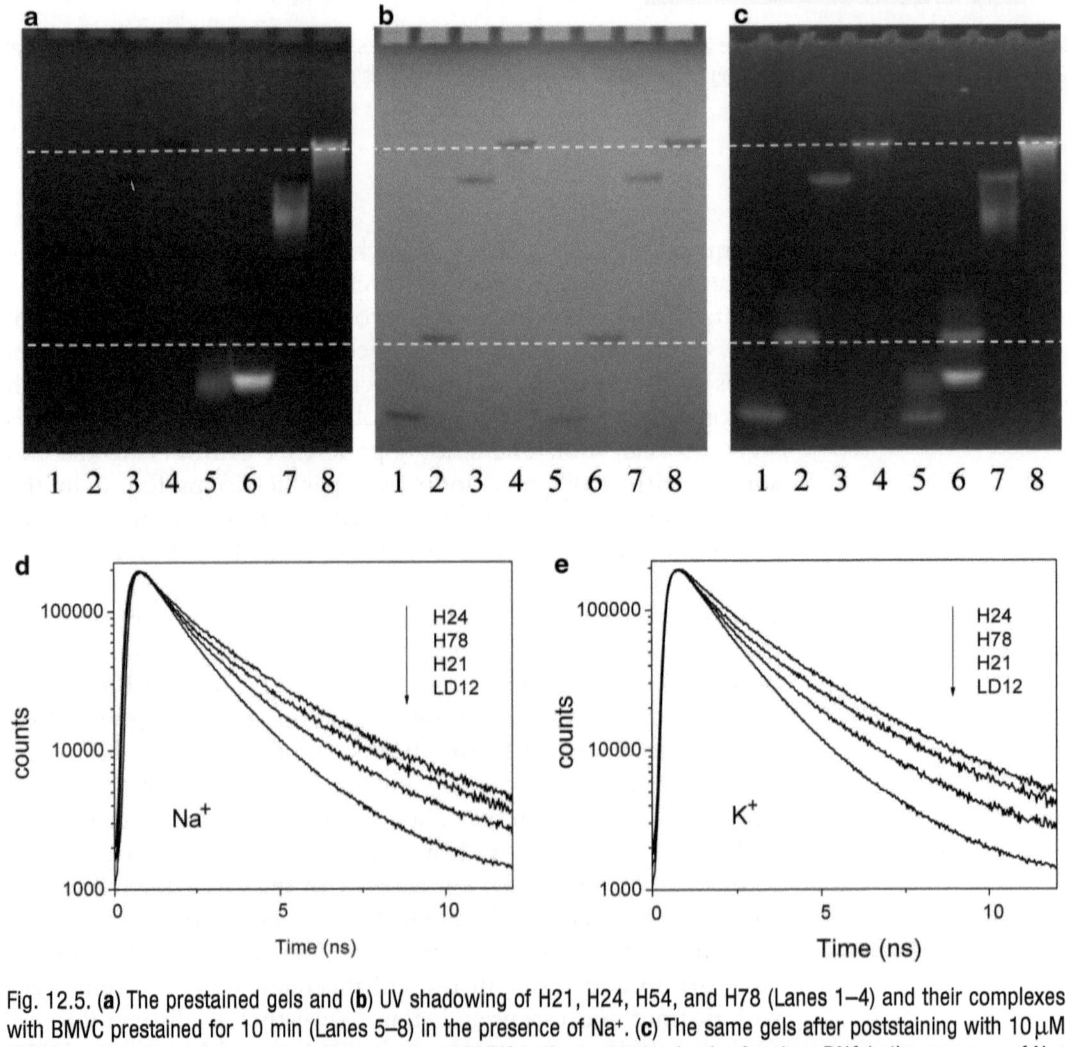

Fig. 12.5. (**a**) The prestained gels and (**b**) UV shadowing of H21, H24, H54, and H78 (Lanes 1–4) and their complexes with BMVC prestained for 10 min (Lanes 5–8) in the presence of Na^{+}. (**c**) The same gels after poststaining with 10 μM BMVC for 10 s. (**d**, **e**) Fluorescence decay curves of BMVC in the poststained gels of various DNA in the presence of Na^{+} and K^{+}, respectively.

shadowing resulting from the free DNA is clearly revealed by BMVC fluorescence in the poststained gels. Of particular interest is that the BMVC-bound H24, H54, and H78 complexes migrate faster than their free forms. In contrast, the BMVC-bound H21 complex migrates slower than its free form. The difference in migration between the BMVC/H24 and BMVC/H21 complexes and their free forms indicates that BMVC bound to the end of the G-quartet with a diagonal loop is stronger than that bound with two TTA loops. It further supports the involvement of the 5 -TTA sequence in the BMVC binding. Moreover, very similar results were found in the gel assays of H21, H24, H54, and H78 in K^+ solutions.

Different binding modes may be characterized by their distinct decay times. We have further applied fluorescence lifetime imaging microscopy (FLIM) to measure the fluorescence decay curves of BMVC in both prestained and poststained gels in Na^+ and K^+ solutions. Figure 12.5d and e show the semi-log decay curves of BMVC in the poststained gels of LD16, H21, H24, and H78 in Na^+ and K^+ solutions, respectively. The decay curve was measured by using time-correlated single photon counting (TCSPC) equipped with a picosecond diode laser at 470 nm. Although the decay curve cannot be fitted well by using a single exponential decay time, it is clear that the fluorescence decay of BMVC is faster in LD16 than in H21 and much faster than in H24 and H78. The faster decay of BMVC in H21 than in H24 again confirmed that the 5′-TTA sequence plays an essential role in BMVC binding. In addition, very similar decay curves were also detected in K^+ solution, implying the presence of similar binding modes of H24 in Na^+ and K^+ solutions.

3.6. The Major G4 Structure of H24 in K^+ Solution

It is well known that the basket form is dominant in the G4 structure of H24 in Na^+ solution (16), while multiple conformational isomers are suggested for the G4 structures of H24 in K^+ solution (17, 18, 23). Recently, three groups have determined the folding topology of H22 in K^+ solution. Yang et al. (17) found that the mixed-I form (Scheme 12.2c) is the major structure to a modified sequence, d(AAAG$_3$(T$_2$AG$_3$)$_3$AA) (M26), by NMR analysis. As the CD and NMR patterns of H26 are very similar to that of M26, they strongly suggested that the mixed-I form is also the major structure of wild-type H26. Sugiyama et al. (23) substituted the guanines in H22 with 8-bromoguanine and found, using CD spectroscopy, that H22 exists as a mixture of mixed-I form (Scheme 12.2c) and chair form (Scheme 12.2d). Patel et al. (49) studied the modified d[TTGGG(T$_2$AG)$_3$]A (M24) sequence by NMR analysis and concluded that the major structure is the mixed-I form (Scheme 12.2c, about 95 %). They all suggested that the mixed type form is a major structure for the four-repeated human telomeric sequences in K^+ solution. It should be noted that these G-rich sequences are all modified from the wide telomeric sequences.

Recently, Neidle et al. (50) argued that the mixed-I form of the sequences of M26 (17) and M24 (49) based on NMR analysis may not be valid to the G4 structure of H22, as both sequences have been slightly altered at the termini. Note that the flanking sequence plays a critical role for the folding of the G4 structure in K^+ solution (17, 59, 60). Furthermore, Patel et al. (49) showed that the extra flanking residues are stacked one on each end of the core of G-tetrads, and help to stabilize this particular topology. However, such a fold has not been observed with H22, which cannot form such base pairs. They concluded that the precise nature of all the species of H22 in K^+ solution deserves further study by fine structure methods (50). Nevertheless, the flanking sequence deserves further study as chromosomes contain a 50–200 nucleotide single-stranded telomeric sequence. We will discuss the G4 structure of a longer telomeric sequence in the next section.

In our work, we did observe different CD patterns of H24 in Na^+ and K^+ solutions. This difference is likely due to the coexistence of different types of G4 structures. Sugiyama et al. (23) suggested that the chair form and the mixed-I form are the major G4 structures of H24 in K^+ solution. At present, we are not able to distinguish the basket form from the chair form because they have very similar CD patterns. However, very similar binding characteristics, induced CD spectra, and gel analysis of BMVC upon binding to H24 in both Na^+ and K^+ solutions suggested that the G4 structures of H24 should have very similar binding sites in these solutions. In addition, the binding strength of BMVC to the end surface of the G-quartet is as follows: (5′-TTA tail + diagonal loop) ~ (5′-TTA tail + lateral loops) > (diagonal loop) ~ (two lateral loops) > (one lateral loop) (25). Note that a G-quartet with a 5′-TTA tail and a lateral loop at one end and only a lateral loop in the mixed-I form at the other end (Scheme 12.2c) is quite different from a G-quartet with a 5′-TTA tail and a diagonal loop at one end and a pair of lateral loops in the basket form at the other end (Scheme 12.2a). As the basket form of H24 is predominant in Na^+ solution, similar binding behaviors in these solutions do not favor the mixed-I form to be a major structure of H24 in K^+ solution.

On the other hand, several antiparallel duplex DNAs of various sequences, such as $[d(C_4G_4)]_2$, also show a CD pattern with a positive band at 265 nm and a negative band at 240 nm, which is very similar to the CD spectra of the parallel G4 structure (61). Moreover, similar CD patterns have been found for a single strand $d(GA)_{10}$ and for homoduplexes of $d(GA_3)_5$ and $d(GA_2)_7$ under certain conditions (62, 63). These results open a possibility that distinct CD spectra of H24 in Na^+ or K^+ solutions and the different

CD spectra of H24 and H24-T9 in Na^+ solution may be due to different loop base stacking and various intramolecular hydrogen bonding effects. If that is the case, the basket and the chair forms can be the main components of H24 in K^+ solution (21).

Alternatively, we anticipated that the mixed-II form (Scheme 12.2e) may be a candidate to coexist with the basket form and/or the chair form for H24 in K^+ solution (25). Note that the mixed-II form characterized by a diagonal loop at one end and a lateral loop with a 5′-TTA tail at the other end of the G-quartet can have two major binding sites, which are relatively comparable to the binding sites of the basket form of H24 in Na^+ solution. It is possible that the mixed form coexists with the basket form in K^+ solution. Nevertheless, we considered that the basket G4 structure is likely a major component of H24 in K^+ solution.

4. The G4 Structure of a Longer Sequence $d(T_2AG_3)_{13}$ (H78)

Considering the 50–200 bases in a single-stranded telomeric sequence and the possible effect of the flanking sequences, it is of interest to examine the structural units of G4 in longer sequences, such as $d(T_2AG_3)_9$ (H54) and $d(T_2AG_3)_{13}$ (H78). Figure 12.6a shows the CD spectra of H54 and H78 in Na^+ and in K^+ solutions. The CD patterns of H54 and H78 are almost identical to the patterns of H24 in Na^+ or in K^+ solutions, indicating the presence of G4 structure. In order to determine the unit number of the G4 structure in a longer sequence, Fig. 12.6b shows the absorption spectra of free BMVC and its complexes with H78 in Na^+ solution as a function of BMVC concentration. The job plots are shown in the inset. Our results reveal a binding ratio of ~5 for BMVC to H78 in Na^+ solution. Although the detailed structure of the whole H78 is not clear at present, the CD spectra together with the job plots suggest that H78 could form two units of G4 structures in Na^+ solution. This finding illustrates that a longer single-stranded telomeric sequence could form more than one unit of G4 structure.

Considering the presence of more complexes in the local environment around the G4 structure of H78, we have compared the induced CD spectra of BMVC/H78 with BMVC/H24. Figure 12.6c shows the induced CD spectra of BMVC/H78 in Na^+ solution upon BMVC titration. An induced CD positive band of BMVC/H78 complexes at ~415 nm increases, associated with a relative weak positive band at ~495 nm switching to a negative band at ~460 nm, upon BMVC titration up to a molar ratio of 12. Nevertheless, the induced CD patterns of BMVC/H78 at a molar ratio around 8 are similar to that of BMVC/H24 at a molar ratio around 2.

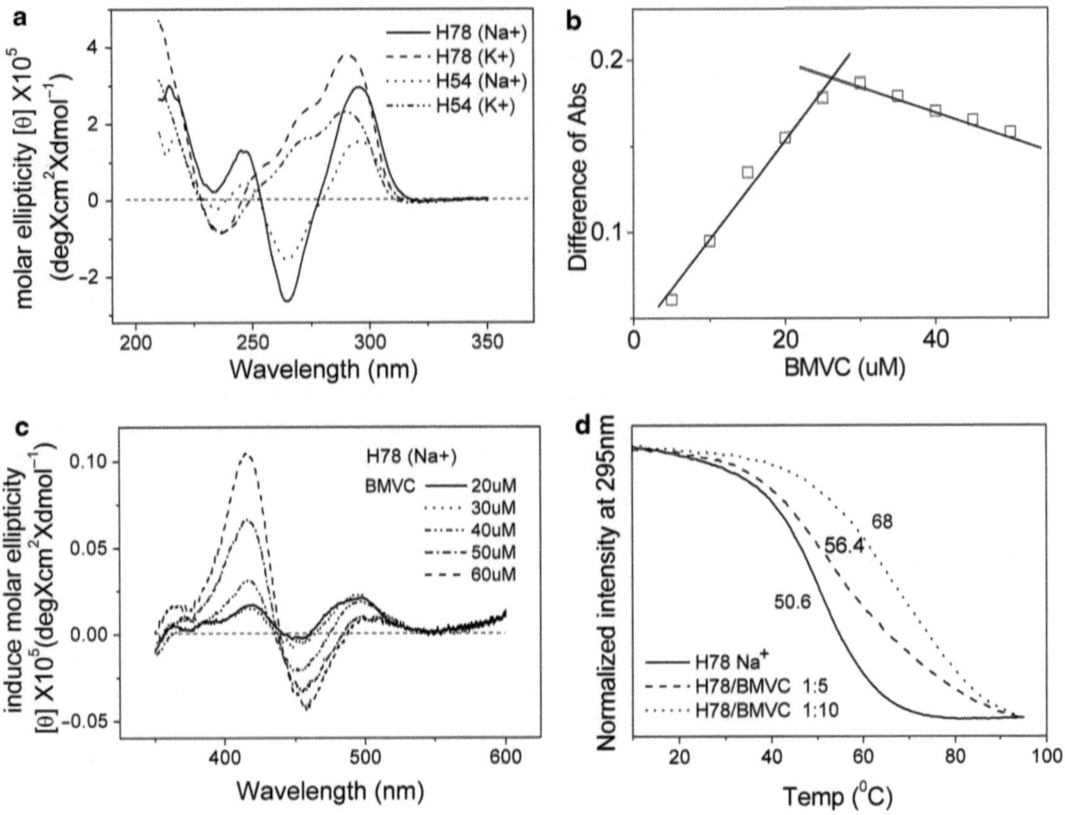

Fig. 12.6. (**a**) CD spectra of 10 μM H54 and 10 μM H78 DNA in the solutions containing 150 mM of Na^+ or K^+ cation. (**b**) Absorption spectra of free BMVC and its complexes with H78 in Na^+ solution upon BMVC titration. The inset shows the job plots to determine the binding ratio. (**c**) Induced CD spectra of BMVC/H78 in Na^+ solution upon BMVC titration. (**d**) CD signal at 295 nm for the measurement of melting temperature of H78 and its complexes of BMVC as a function of temperature.

In addition, it is of interest to evaluate whether the G4 structure of H78 can be stabilized by BMVC. Figure 12.6d shows the CD intensity at 295 nm of H78 and its complexes with various BMVC concentrations as a function of temperature. The melting temperature of H78 in Na^+ solution increases by ~17 °C upon interaction with BMVC at a molar ratio of 10, which is comparable to the melting temperature of H24 in Na^+ solution which increases by ~18°C upon interaction with BMVC at a molar ratio of 5. Together with the similar decay parameters of BMVC in the gel assays of H24 and Hu8, our spectral results suggested that the two units of the G4 structure in H78 are likely independent of each other. This further implies that the study of the G4 structure of H24 is valid for the long human telomeric sequences.

5. Detection of Quadruplex DNA Structures in Human Telomeres

We considered that a fluorescent probe of BMVC with distinct fluorescent emissions and decay times upon interaction with H24 and LD16 may be able to verify the presence of G4 structure in native human telomeres. For this purpose, we have further compared the binding preference of BMVC to quadruplex H24 and duplex LD16. Figure 12.7 shows competition analysis under high (2 μM) and low (0.2 μM) concentrations of BMVC (prestain, lanes 1–6). Visualization of BMVC-bound DNA molecules was achieved under a UV light. It is clear that BMVC is capable of binding to both H24 and LD16 under both BMVC concentrations (lanes 1, 2, 4, and 5). However, when a limiting amount of BMVC was incubated with an equal amount of H24 and LD16, most of the BMVC was bound to H24 (lanes 3 and 6). As a control, the same gels were then poststained with 10 μM of BMVC to visualize the position and level of the DNA loaded (lanes 7–12). Under our electrophoresis conditions, the BMVC-bound H24 complex migrates faster than free H24. Thus, both BMVC-bound and free H24 are revealed in the poststained gel. We found that a small amount of H24 was prestained when using 0.2 μM BMVC and most of the H24 was poststained by BMVC (lanes 10 and 12). Using 2 μM BMVC, the prestained H24 became dominant, as shown in lanes 7 and 9. More significantly, the BMVC fluorescence was only detected in quadruplex DNA when a limiting amount of BMVC was used. Thus, it is clear that BMVC has a binding preference toward quadruplex H24 over duplex LD16. A similar binding preference was also found in the presence of K^+ ions (30).

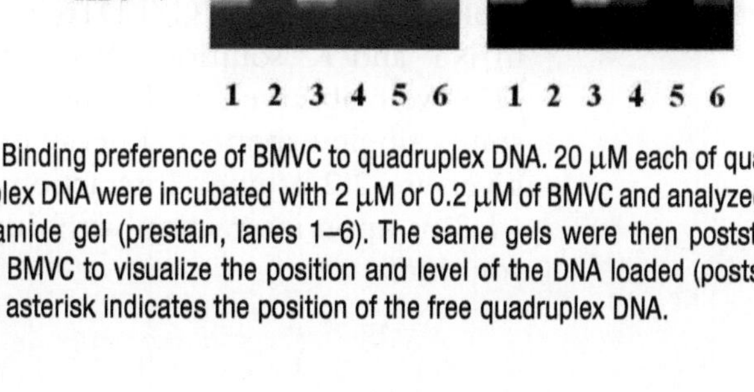

Fig. 12.7. Binding preference of BMVC to quadruplex DNA. 20 μM each of quadruplex or linear duplex DNA were incubated with 2 μM or 0.2 μM of BMVC and analyzed on a 20% polyacrylamide gel (prestain, lanes 1–6). The same gels were then poststained with 10 μM of BMVC to visualize the position and level of the DNA loaded (poststain, lanes 7–12). An asterisk indicates the position of the free quadruplex DNA.

If the G4 structure indeed existed in the telomeres of chromosomes, the small amount of quadruplex DNA would be completely overwhelmed by the amount of duplex DNA present. Conventional techniques based on ensemble methods can only measure the average properties. As a result, some important features with a very small number of events are normally averaged out. Moerner and Carter (64) introduced a statistical fine structure method to measure the absorption of individual molecules. We took advantage of the distinct fluorescent properties of BMVC and used confocal microscopy to compare the BMVC fluorescence at the ends of metaphase chromosomes and other regions of chromosomes. Verification of the existence of G4 structures of human telomeres is very important in supporting the recent development of telomerase inhibitors for antitumor agents.

Figure 12.8a shows fluorescence microscopy of BMVC in metaphase chromosomes of CL1-1 cells. Metaphase spreads prepared from CL1-1 cells on microscope slides were incubated with 50 nM BMVC for 10 min at room temperature. In order to compare the fluorescence spectra at 500–620 nm from the BMVC staining at the ends and other regions of chromosomes, the fluorescence spectra of telomere-proximal regions and other regions of chromosomes were obtained from an average of 60 individual chromosomes, as shown in Fig. 12.8b. Telomere-proximal regions emitted fluorescence at around 565 nm whereas the other chromosomal regions showed fluorescence emission at around 545 nm. As we were unable to completely eliminate the possible contribution from the interaction with duplex DNA in the telomere-proximal regions, the red shift of the fluorescence band to ~565 nm was likely due to the interference from the interaction of BMVC with duplex DNA. It appears that the fluorescence emission at around 565 nm at the ends of chromosomes and at around 545 nm at other regions of chromosomes could reveal the presence of G4 structures in human telomeres. The statistical fine structure allowed us for the first time to verify the existence of G4 structure in native human telomeres (29).

In addition, we have measured the fluorescence decays of BMVC upon interaction with LD16 and H24 in Na^+ and K^+ solution. The sample was dropped on a cover slip, and the decay curve was measured by using TCSPC equipped with a diode laser at 473 nm. Figure 12.8c shows typical semi-log decay curves of 1:1 molar ratios of BMVC/LD16 in Na^+ solution and BMVC/H24 in Na^+ and K^+ solutions. Very similar decay curves of BMVC/H24 were observed in Na^+ and K^+ solutions. For an easy comparison, a single exponential decay parameter used for the best fit is given as ~2.2 ns upon interaction with G4 structure of H24, and ~1.3 ns upon interaction with duplex LD16 in both Na^+ and K^+ solutions.

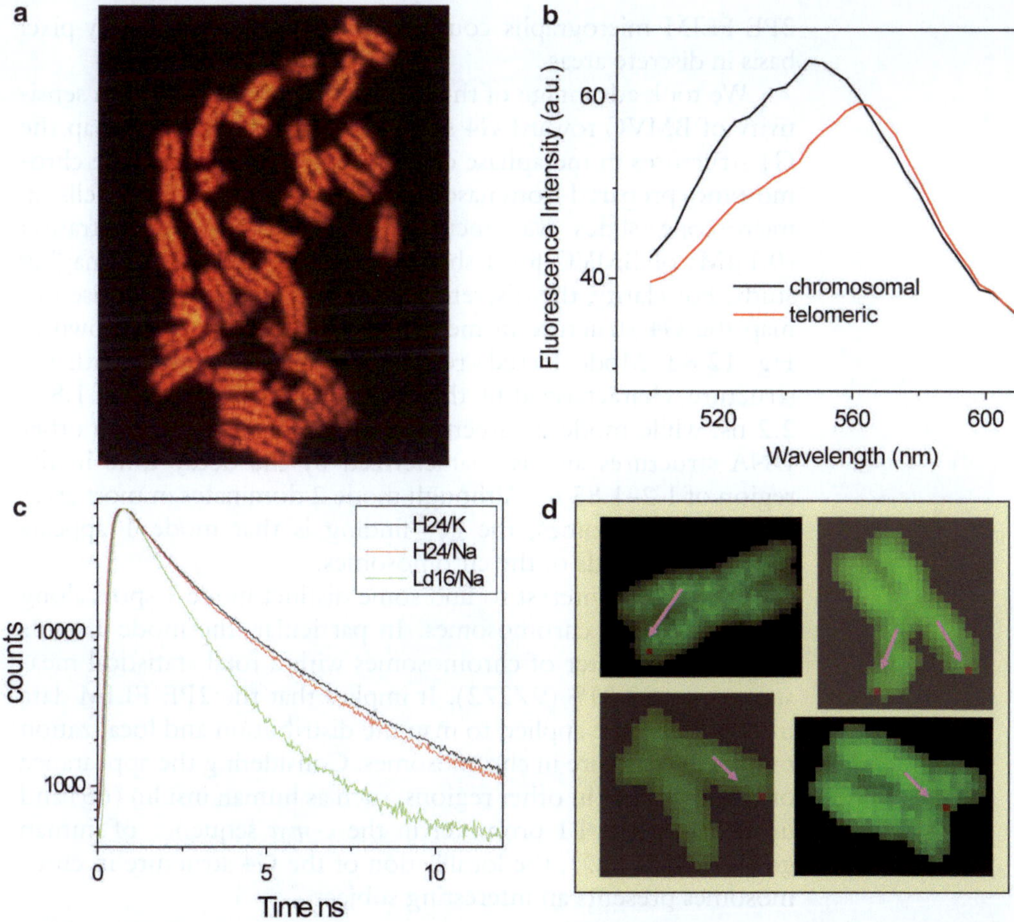

Fig. 12.8. (**a**) Fluorescence microscopy of BMVC-stained metaphase chromosomes of CL1-1 cells. (**b**) Fluorescence emissions of BMVC at 500–620 nm were collected at both the ends of chromosomes (telomeric) and at other regions of chromosomes (chromosomal) of human CL1-1 cells by scanning 3 nm wavelength windows for each image. The spectra were obtained from the average of 60 individual CL1-1 chromosomes. (**c**) Fluorescence decay curves of BMVC upon interaction with LD16 in Na^+ solution and H24 in Na^+ and K^+ solutions. (**d**) Four images of individual metaphase chromosomes of cancer cells. Here red is designated as mode 1, resulting from the interaction with G4 structure, characterized by the lifetime in the region of 1.85–2.2 ns, while green is designated as mode 2, reflecting the interaction with other DNA structures, characterized by the lifetime in the region of 1.2–1.85 ns.

Two photon-excitation (2PE) fluorescence lifetime imaging microscopy (FLIM) with submicron spatial resolution was further applied to verify the existence of the G4 structures in human telomeres and possibly map the localization of the G4 structures in metaphase chromosomes (30). The raster scanning mapping mechanism is achieved by a modified laser-scanning microscope (FV300, Olympus) equipped with a femtosecond laser pulse at 810 nm. The fluorescence of BMVC followed a quadratic dependence of the excitation intensity. A synchronized TCSPC module (SPC830, Becker and Hickle) was used to collect the fluorescent signal that is used to construct the 2PE-FLIM images. The lifetime analysis on

2PE-FLIM micrographs could be measured on a pixel-by-pixel basis in discrete areas.

We took advantage of the binding preference and high sensitivity of BMVC toward G4 structures over duplexes to map the G4 structures in metaphase chromosomes. The metaphase chromosomes prepared from nasopharyngeal carcinoma KJ-1 cells on microscope slides was incubated with a low concentration (0.1 μM) of BMVC for a short time (90 s) before the imaging study. For clarity, the discrete time mode at 1.85 ns was used to map the G4 structure in metaphase chromosomes, as shown in Fig. 12.8d. Mode 1 (red) results from the interaction with G4 structure, characterized by the decay time in the region of 1.85–2.2 ns, while mode 2 (green) reflects the interaction with other DNA structures and is characterized by the decay time in the region of 1.2–1.85 ns. Although mode 2 dominates in most areas of the chromosomes, the key finding is that mode 1 appears mostly at the ends of the chromosomes.

It is also of interest to find some distinct mode 1 spots along the edge of the chromosomes. In particular, the mode 1 spots occur at the center of chromosomes with a total statistical measurement of 3.31% (9/272). It implies that the 2PE-FLIM data of BMVC can be applied to map the distribution and localization of the G4 structure in chromosomes. Considering the appearance of G4 structures in other regions, such as human insulin (65) and upstream of the P1 promoter in the c-*myc* sequence of human genes (27, 66, 67), the localization of the G4 structure in chromosomes presents an interesting subject.

6. Summary and Future Prospects

The antitumor properties of fluorescent probes hold great promise for the future of chemotherapy. A BMVC molecule is a sensitive fluorescent probe with a very high binding preference for G4 structures. Similar binding behaviors of BMVC to the telomeric sequence H24 in Na^+ and K^+ solutions suggest that the basket G4 structure predominant in Na^+ solution is also likely a major G4 structure in K^+ solution. However, the distinct fluorescent properties of BMVC upon interaction with different DNA structures are of great value. The analysis of BMVC fluorescence on metaphase chromosomes allowed us to verify for the first time, the presence of the G4 structure in native human telomeres using statistical fine structure and fluorescence lifetime imaging microscope methods. Moreover, our results showed that BMVC is a promising lead molecule for the development of a second generation.

In addition, the diversity of structures may contribute to their biological functions (23). G-rich sequences could adopt a variety of structures, which may be convertible. For instance, telomerase cannot extend quadruplex-containing telomeres, because the enzyme only copies linear strands of DNA and therefore does not recognize them as a substrate. The folding of telomeric sequences into G4 structures could inhibit telomerase activity in cancer cell growth. Further studies to better understand the folding and unfolding of quadruplexes and the conformational flexibility are important not only for the biological role of G-quadruplexes, but also as a new direction in pharmaceutical therapy.

Considering a number of nontelomeric G-rich sequences documented in the human genome, investigation of the natural role and biological function of the nontelomeric G4 structures is starting to be explored, particularly for fundamental research and drug development in cancer research. It is essential to first identify the nontelomeric G4 structures and the possible structural morphology of nontelomeric G-rich sequences. Of particular interest is to map the localization of these G4 structures in metaphase chromosomes, particularly in normal cells vs. cancer cells. It would be great if one could determine the role of the G4 structures in cancer proliferation as many G-rich genes are possibly involved in cancer regulation.

Acknowledgments

This work was supported by Academia Sinica (AS-95-TP-AB2) and the National Science Council of the Republic of China (Grant NSC-93-2113-M001-044, NSC-94-2113-M-001-047, and NSC-95-2113-M-001-047). The authors would like to thank Chih-Wei Chien, Chi-Chih Kang, Jen-Fei Chu, I-Chun Kuo, Yi-Hsueh, Lin, Yu-Lin Tsai, Ting-Yuan Tseng, Cheng-Hao Chien, Chih-Chien Cho, and Wei-Chun Huang for their participation in this work. T.-C. is especially indebted to Pei-Jen Lou, Jing-Jer Lin, Chin-Tin Chen, Dah-Yen Yang, Sheh-Yi Sheu, and Fu-Jen Kao for their collaboration in this work.

References

1. Blackburn EH, Greider CW (1996) Telomeres. Cold Spring Harbor Laboratory Press, New York, NY
2. Williamson JR (1994) G-quartet structures in telomeric DNA. Annu Rev Biophys Biomol Struct 23:703–730
3. Mergny JL, Hélène C (1998) G-quadruplex DNA: a target for drug design. Nat Med 4:1366–1367
4. Harley CB, Futcher AB, Greider CW (1990) Telomeres shorten during ageing of human fibroblasts. Nature 345:458–460
5. Lundblad V, Szostak JW (1989) A mutant with a defect in telomere elongation leads to senescence in yeast. Cell 57:633–643
6. Sandell LL, Zakian VA (1993) Loss of a yeast telomere: arrest, recovery, and chromosome loss. Cell 75:729–739

7. Harley CB, Villeponteau MP (1995) Telomeres and telomerase in aging and cancer. Curr Opin Genet Dev 5:249–255
8. Greider CW, Blackburn EH (1987) The telomere terminal transferase of tetrahymena is a ribonucleoprotein enzyme with two kinds of primer specificity. Cell 51:887–898
9. Feng J, Funk WD, Wang S-S, Weinrich SL, Avilion AA, Chiu C-P, Adams RR, Chang E, Allsopp RC, Yu J, Le S, West MD, Harley CB, Andrews WH, Greider CW, Villeponteau B (1995) The RNA component of human telomerase. Science 269:1236–1241
10. Hahn WC, Counter CM, Lundberg AS, Beijersbergen RL, Brooks MW, Weinberg RA (1999) Creation of human tumor cells with defined genetic elements. Nature 400:464–468
11. Kim NW, Piatyszek MA, Prowas KR, Harley CB, West MD, Ho PL, Coviello GM, Wright WE, Weinrich SL, Shay JW (1994) Specific association of human telomerase activity with immortal cells and cancer. Science 266:2011–2015
12. Kerwin SM (2000) G-quadruplex DNA as a target for drug design. Curr Pharm Des 6:441–471
13. Han H, Hurley LH (2001) G-quadruplex DNA: a potential target for anti-cancer drug design. Trends Pharmacol Sci 21:136–142
14. Morin GB (1989) The human telomere terminal transferase enzyme is a ribonucleoprotein that synthesizes TTAGGG repeats. Cell 59:521–529
15. Gellert M, Lipsett MN, Davies DR (1962) Helix formation by guanylic acid. Proc Natl Acad Sci U S A 48:2013–2016
16. Wang Y, Patel DJ (1993) Solution structure of the human telomeric repeat d[$AG_3(T_2AG_3)_3$] G-tetraplex. Structure 1:262–283
17. Ambrus A, Chen D, Dai J, Bialis T, Jones RA, Yang D (2006) Human telomeric sequence forms a hybrid-type intramolecular G-quadruplex structure with mixed parallel/antiparallel strands in potassium solution. Nucleic Acids Res 34:2723–2735
18. Phan AT, Patel DJ (2003) Two-repeat human telomeric d(TAGGGTTAGGGT) sequence forms interconverting parallel and antiparallel G-quadruplexes in solution: distinct topologies, thermodynamic properties, and folding/unfolding kinetics. J Am Chem Soc 125:15021–15027
19. Rujan IN, Meleney JC, Bolton PH (2005) Vertebrate telomere repeat DNAs favor external loop propeller quadruplex structures in the presence of high concentrations of potassium. Nucleic Acids Res 33: 2022–2031
20. Li J, Correia JJ, Wang L, Trent JO, Chaires JB (2005) Not so crystal clear: the structure of the human telomere G-quadruplex in solution differs from that present in a crystal. Nucleic Acids Res 33:4649–4659
21. Redon S, Bombard S, Elizondo-Riojas MA, Chottard JC (2003) Platinum cross-linking of adenines and guanines on the quadruplex structure of the $AG_3(T_2AG_3)_3$ and $(T_2AG_3)_4$ human telomere structure in Na^+ and K^+ solutions. Nucleic Acids Res 31:1605–1613
22. He Y, Neumann RD, Panyutin IG (2004) Intramolecular quadruplex conformation of human telomeric DNA assessed with ^{125}I-radioprobing. Nucleic Acids Res 32: 5339–5367
23. Xu Y, Noguchi Y, Sugiyama H (2006) The new models of the human telomere d[AGGG(TTaGGG)$_3$] in K^+ solution. Bioorg Med Chem 14:5584–5591
24. Vorlíčková M, Chládková J, Kejnovská I, Fialová M, Kypr J (2005) Guanine tetraplex topology of human telomere DNA is governed by the number of (TTAGGG) repeats. Nucleic Acids Res 33:5851–5860
25. Chang CC, Chien CW, Lin YH, Kang CC, Chang T-C (2007) Investigation of spectral conversion of d(TTAGGG)$_4$ and d(TTAGGG)$_{13}$ upon potassium titration by a G-quadruplex recognizer BMVC molecule. Nucleic Acids Res 35:2846–2860
26. Schaffitzel C, Berger I, Postberg J, Hanes J, Lipps HJ, Plückthun A (2001) *In vitro* generated antibodies specific for telomeric guanine-quadruplex DNA react with *Stylonychia lemnae* macronuclei. Proc Natl Acad Sci U S A 98:8572–8577
27. Siddiqui-Jain A, Grand CL, Bearss DJ, Hurley LH (2002) Direct evidence for a G-quadruplex in a promoter region and its targeting with small molecule to repress c-MYC transcription. Proc Natl Acad Sci U S A 99:11593–11598
28. Granotier C, Pennarun G, Riou L, Hoffschir F, Gauthier LR, Cian AD, Gomez D, Mandine E, Riou J-F, Mergny J-L, Mailliet P, Dutrillaux B, Boussin FD (2005) Preferential binding of a G-quadruplex ligand to human chromosome ends. Nucleic Acids Res 33:4182–4190
29. Chang CC, Kuo I-C, Ling I-F, Chen CT, Chen HC, Lou PJ, Lin JJ, Chang T-C (2004) Detection of quadruplex DNA structures in human telomeres by a fluorescent carbazole derivative. Anal Chem 76:4490–4494
30. Chang CC, Chu JF, Kao FJ, Chiu YC, Lou PJ, Chen HC, Chang T-C (2006) Verification of antiparallel G-quadruplex structure in human

telomeres by using two-photon excitation fluorescence lifetime imaging microscopy of the 3, 6-bis(1-methyl-4-vinylpyridinium)carbazole diiodide molecule. Anal Chem 78:2810–2815
31. Chang C-C, Wu J-Y, Chang T-C (2003) A carbazole derivative synthesis for stabilizing the quadruplex structure. J Chin Chem Soc 50:185–188
32. Chang CC, Wu JY, Chien CW, Wu WS, Liu H, Kang CC, Yu LJ, Chang T-C (2003) A fluorescent carbazole derivative: high sensitivity for quadruplex DNA. Anal Chem 75: 6177–6183
33. Yang DY, Chang T-C, Sheu SY (2007) Interaction between human telomere and a carbazole derivative: a molecular dynamics simulation of a quadruplex stabilizer and telomerase inhibitor. J Phys Chem A 111:9224–9232
34. Sun D, Thompson B, Cathers BE, Salazar M, Kerwin SM, Trent JO, Neidle S, Hurley LH (1997) Inhibition of human telomerase by a G-quadruplex-interactive compound. J Med Chem 40:2113–2116
35. Chen Q, Kuntz ID, Shafer RH (1996) Spectroscopic recognition of guanine dimeric hairpin quadruplexes by a carbocyanine dye. Proc Natl Acad Sci U S A 93:2635–2639
36. Cheng JY, Lin SH, Chang T-C (1998) Vibrational investigation of DODC cation for recognition of guanine dimeric hairpin quadruplex studied by satellite holes. J Phys Chem B 102:5542–5546
37. Han HY, Cliff CL, Hurley LH (1999) Accelerated assembly of G-quadruplex structures by a small molecule. Biochemistry 38:6981–6986
38. Kerwin SM, Chen G, Kern JT, Thomas PW (2002) Perylene diimide G-quadruplex DNA binding selectivity is mediated by ligand aggregation. Bioorg Med Chem Lett 12:447–450
39. Law KY (1980) Fluorescence probe for microenvironments: anomalous viscosity dependence of the fluorescence quantum yield of p-N, N-diakylaminobenzylidenemalononitrite in 1-alkanols. Chem Phys Lett 75:545–549
40. Kung CE, Reed JK (1986) Microviscosity measurements of phospholipid bilayers using fluorescent dyes that undergo torsional relaxation. Biochemistry 25:6114–6121
41. Haidekker MA, Ling T, Anglo M, Stevens HY, Frangos JA, Theodorakis EA (2001) New fluorescent probes for the measurement of cell membrane viscosity. Chem Biol 8:123–131
42. Allen BD, Bennistin AC, Harriman A, Rostron SA, Yu C (2005) The photophysical properties of a julolidene-based molecular rotor. Phys Chem Chem Phys 7:3035–3040
43. Oster G, Nishijima Y (1956) Fluorescence and internal rotation: their dependence on viscosity of the medium. J Am Chem Soc 78:1581–1584
44. Changenet P, Zhang H, van der Meer MJ, Glasbeck M, Plaza P, Martin MM (1998) Ultrafast twisting dynamics of photoexcited auramine in solution. J Phys Chem A 102:6716–6721
45. Castex MC, Olivero C, Pichler G, Adés D, Cloutet E, Siove A (2001) Photoluminescence of donor-acceptor carbazole chromophores. Synth Met 122:59–61
46. Chang CC, Chu JF, Kuo HH, Kang CC, Lin SH, Chang T-C (2006) Solvent effect on photophysical properties of a fluorescence probe: BMVC. J Lumin 119:84–90
47. Loutfy RO, Arnold BA (1982) Effect of viscosity and temperature on torsional relaxation of molecular rotors. J Phys Chem 86:4205–4211
48. Parkinson GN, Lee MPH, Neidle S (2002) Crystal structure of parallel quadruplexes from human telomeric DNA. Nature 417:876–880
49. Luu KN, Phan AT, Kuryavyi V, Lacroix L, Patel DJ (2006) Structure of the human telomere in K^+ solution: an intramolecular (3+1) G-quadruplex scaffold. J Am Chem Soc 128:9963–9970
50. Burge S, Parkinson GN, Hazel P, Todd AK, Neidle S (2006) Quadruplex DNA: sequence, topology and structure. Nucleic Acids Res 34:5402–5415
51. Balagurumoorthy P, Brahmachari SK, Mohanty D, Bansal M, Sasisekharan V (1992) Hairpin and parallel quartet structures for telomeric sequences. Nucleic Acids Res 20:4061–4067
52. Giraldo R, Suzuki M, Chapman L, Rhodes D (1994) Promotion of parallel DNA quadruplexes by a yeast telomere binding protein: a circular dichroism study. Proc Natl Acad Sci U S A 91:7658–7662
53. Wu JY, Chang CC, Yan CS, Chen KY, Kuo I-C, Mou CY, Chang T-C (2003) Structural isomers and binding sites of guanine-rich quadruplexes investigated by induced circular dichroism of thionin: loops and tails. J Biomol Struct Dyn 21:135–140
54. Hazel P, Huppert J, Balasubramanian S, Neidle S (2004) Loop-length-dependent folding of G-quadruplexes. J Am Chem Soc 126:16405–16415
55. Qi J, Shafer RH (2005) Covalent ligation studies on the human telomere quadruplex. Nucleic Acids Res 33:3185–3192

56. Chaires JB (2001) Analysis and interpretation of ligand-DNA binding isotherms. Methods Enzymol 340:3–22
57. Kang C, Zhang X, Ratliff R, Moyzis R, Rich A (1992) Crystal structure of four-stranded oxytricha telomeric DNA. Nature 356:126–131
58. Schultze P, Smith FW, Feigon J (1994) Refined solution structure of the dimeric quadruplex formed. Nature 356:164
59. Ambrus A, Chen D, Dai JX, Jones RA, Yang DZ (2005) Solution structure of the biologically relevant G-quadruplex element in the human c-MYC promoter: implications for G-quadruplex stabilization. Biochemistry 44:2048–2058
60. Phan AT, Modi YS, Patel DJ (2004) Propelle-type parallel stranded G-quadruplexes in the human c-myc promotor. J Am Chem Soc 31:2097–2107
61. Kypr J, Fialová M, Chládková J, Tůmová M, Vorlíčková M (2001) Conserved guanine-guanine stacking in tetraplex and duplex DNA. Eur Biophys J 30:555–558
62. Kypr J, Vorlíčková M (2001) Dimethylsulfoxide-stabilized conformer of guanine-adenine repeat strand of DNA. Biopolymers 62:81–84
63. Kypr J, Kejnovská I, Vorlíčková M (2003) DNA homoduplexes containing no pyrimidine nucleotide. Eur Biophys J 32:154–158
64. Carter T, Moerner WE (1987) Statistical fine structure in inhomogeneously broadened absorption lines. Phys Rev Lett 59: 2705–2708
65. Catasti P, Chen X, Moyzis RK, Bradbury EM, Gupta G (1996) Structure-function correlations of the insulin-linked polymorphic region. J Mol Biol 264:534–545
66. Phan AT, Modi YS, Patel DJ (2004) Propelle-type parallel stranded G-quadruplexes in the human c-myc promotor. J Am Chem Soc 126:8710–8716
67. Seenisamy J, Rezler EM, Powell TJ, Tye D, Gokhale V, Joshi CS, Siddiqui-Jain A, Hurley LH (2004) The dynamic character of the G-quadruplex element in the c-MYC promoter and modification by TMPyP4. J Am Chem Soc 126:8702–8709

Chapter 13

Isolation of G-Quadruplex DNA Using NMM-Sepharose Affinity Chromatography

Jasmine S. Smith and F. Brad Johnson

Abstract

DNA can adopt a variety of non-standard conformations, including structures known as G-quadruplexes (G4-DNA), which consist of stacked tetrads of guanines. There are growing indications that G4-DNA is of biological importance, including evidence that it plays roles in telomere function, DNA recombination and the regulation of transcription and translation. However, it has been difficult to obtain direct, physical evidence for the presence of G-quadruplex DNA *in vivo* due, in part, to a lack of tools for G4-DNA identification. Here, we describe a method for coupling the G4-DNA binding ligand *N*-methyl mesoporphyrin IX (NMM) to a Sepharose resin, and demonstrate the ability of the resin to bind tightly and selectively to DNA oligonucleotides with the capacity to form G4-DNA. This technique might also be extended to examine genomic distributions of G4-DNA isolated from *in vivo* sources.

Key words: G-quadruplex, G4-DNA, Affinity chromatography, *N*-methyl mesoporphyrin IX, Telomere

1. Introduction

DNA can adopt a variety of structures distinct from canonical Watson-Crick base-paired duplexes. Among these, G4-DNA structures, which consist of stacked tetrads of Hoogsteen hydrogen bonded guanines stabilized by small monovalent cations, are of interest due to growing evidence for biologically relevant functions including roles in telomere maintenance (1,2), transcriptional and translational regulation (2–6), and DNA recombination (7,8). Sequences with intramolecular G-quadruplex forming potential (QFP) are concentrated near promoters in a variety of divergent organisms including bacteria, yeast, chickens, and humans. Numerous G4-DNA interacting proteins that may mediate transcriptional regulatory effects of G4-DNA have been

P. Baumann (ed.), *G-Quadruplex DNA: Methods and Protocols*, Methods in Molecular Biology, vol. 608
DOI 10.1007/978-1-59745-363-9_13, © Humana Press, a part of Springer Science+Business Media, LLC 2010

characterized (2,9–14). There is also growing interest in the cancer therapeutic potential of G4-DNA stabilizing compounds, which inhibit telomerase activity (putatively, through stabilizing quadruplex structures that block telomerase access at the telomere), cause telomere uncapping via displacement of telomere binding proteins such as TRF2 and POT1, and preferentially induce apoptosis of tumor cells (15–20). Given these indications for their biological importance, selective isolation of G4-DNA species from populations of non-quadruplex nucleic acids may be of use for further assessing the biological function of these structures.

Thus far, few methodologies have been developed to assess the presence of G4-DNA *in vivo*. One approach has involved isolation of DNA from cells followed by its visualization using electron microscopy together with probes that selectively cleave or bind G4-DNA (8). This approach works well for small DNA species that can be easily isolated (e.g., plasmids), but is more difficult to apply on a genome-wide scale. Antibodies against G4-DNA have also been developed, and one, specific for the telomeric G4-DNA of the protist *Stylonychia lemnae*, has been employed to detect telomeric G4-DNA *in vivo* by immunofluorescent staining (1). However, this approach was facilitated by the hundreds of millions of telomeres in the macronucleus of this organism, and therefore it might not be successful in cells with fewer or less concentrated G4-DNA targets. Moreover, antibodies are highly specific for particular G4-DNA conformations, and thus may be of limited general use given the large variety of conformations in which G4-DNA can exist – from uni- to multi-stranded structures, with the potential for numerous loop and strand orientations and combinations of glycosidic bond angles (1,21,22). In contrast, small molecule G4-DNA binding ligands, although they bind with lower affinity than antibodies, have the potential to bind a broader spectrum of G4-DNA species. A good example is *N*-methyl mesoporphyrin IX (NMM), which binds a variety of G-quadruplex structures (albeit with varying affinities) (23) (Fig. 13.1). Some small molecule ligands display altered fluorescence when bound to G4-DNA, and might thus be used as *in vivo* probes. To date, the closest this approach has come to the study of G4-DNA *in vivo* has been the staining of telomeres in metaphase spreads of human chromosomes (24).

A critical test for any physical method that measures the presence of G4-DNA in or isolated from living cells is that it should reveal changes in G4-DNA following *in vivo* manipulations that are predicted to modulate G4-DNA levels (e.g., deletion or overexpression of a G4-DNA unwinding helicase). Otherwise, it is difficult to rule out the possibilities that the apparent G4-DNA is an artifact of isolation of the DNA from the cell or that it might have formed in response to stabilization by a small molecule or antibody probe. The antibody-based detection of *S. lemnae*

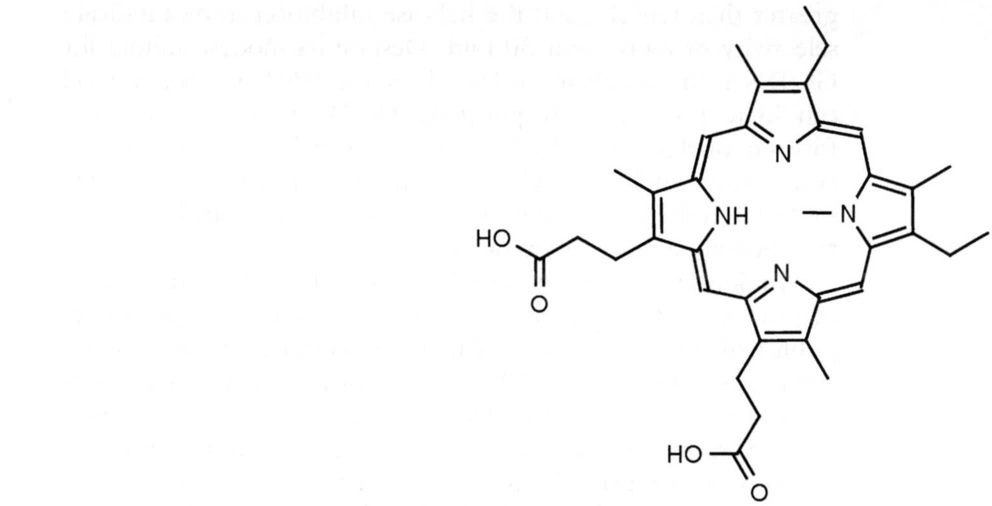

Fig. 13.1 Structure of NMM, an anionic porphyrin with high specificity for binding G4-DNA

telomeres, and the electron microscopic detection of G4-DNA in transcribed G-rich immunoglobulin loci have passed this test (1,8). Similarly, the ability of NMM to bind G4-DNA *in vivo* in *S. cerevisiae* was indicated by its selective modulation of gene expression at loci having QFP and by the preferential capacity of mutations in telomere maintenance factors to enhance or suppress the toxicity of NMM (2). Here, we describe our initial attempts to use NMM to selectively isolate G4-DNA species.

NMM is an anionic porphyrin that exhibits a high degree of selectivity for quadruplex DNA, having no significant binding activity for a variety of other nucleic acid species including single-stranded DNA, duplex DNA and RNA, duplex DNA–RNA hybrids, Z-DNA, and triplex DNA (25). NMM is more selective than several other porphyrins that have been studied (e.g., TMPyP4), probably in part due to the anionic nature of NMM, which should minimize non-specific ionic interactions with nucleic acids. Interestingly, NMM can bind both to a variety of G4-DNA species, albeit with differing affinities, as well as to the cytosine-rich i-motif that can be formed by the complement of sequences with intramolecular QFP (23,25,26). Based on equilibrium dialysis studies, free NMM has an apparent K_d for G4-DNA in the range of 10^{-4} to 10^{-5} M, but based on the selective inhibition of G4-DNA unwinding by the Sgs1 and BLM DNA helicases *in vitro*, it has an apparent K_i of 10^{-6} M (23,27,28). The reasons for these different affinity estimates are not clear, but may reflect the particular G4-DNA species used or possibly a higher affinity of NMM for helicase-G4-DNA complexes. The equilibrium dialysis studies indicate that the G4-DNA selectivity of NMM is much greater than tenfold, and the helicase inhibition studies indicate selectivity of more than 30-fold. Despite its modest affinity for G4-DNA, the specificity of NMM for G4-DNA makes it a good candidate for selectively purifying G4-DNA away from other forms of nucleic acids (23,25). Furthermore, it is possible that the cooperative binding to a G4-DNA target of more than one NMM molecule tethered to a solid support could substantially increase the apparent affinity of interaction.

In developing a method for isolating G4-DNA from mixtures of nucleic acids, we based our initial attempts on a previously published approach where NMM was coupled to an oxirane-acrylic resin, Eupergit C (29). However, we found the NMM was not stably coupled, and thus leached continuously from the resin. Moreover, the selectivity of NMM-Eupergit C resin for QFP species was not robust. We therefore switched to a different resin, EAH Sepharose 4B, which displays low inherent binding to nucleic acids, and which is ideal for efficient coupling to the carboxyl groups of NMM. Here, we provide a protocol for generating NMM-Sepharose resin, and describe its initial characterization using synthetic radiolabeled DNA substrates.

2. Materials

1. EAH Sepharose 4B resin (GE Healthcare Cat# 17-0569-01; store at 4°C)
2. Qiagen Buffer QG (Qiagen Catalog #: 19063)
3. Princeton Separation Centri-Spin 20 columns (Cat # CS-201)
4. *N*-methyl mesoporphyrin IX (NMM) (Frontier Scientific Cat# NMM580)
5. 1 M glycine
6. *N*-ethyl-*N*′-(3-dimethylaminopropyl) carbodiimide hydrochloride (EDC), Store airtight at −20°C (see Note 1)
7. 10% lithium dodecyl sulfate (LDS)
8. Phenol/chloroform/isoamyl alcohol (25:24:1), pH 8.0
9. T4 Polynucleotide Kinase (T4 PNK; New England Biolabs)
10. ^{32}P γ-ATP (3,000 Ci/mmol, 10 mCi/mL)
11. DNA oligonucleotides (Integrated DNA Technologies) dissolved in TE pH 7.5 at 100 μM
12. Coupling buffer (CB):100 mM 2-(*N*-morpholino) ethanesulfonic acid (MES),1 M NaCl; adjust pH to 4.5 with NaOH and store at 4°C

13. TE buffer: 10 mM Tris–HCl pH 7.5, 1 mM EDTA, pH 8.0
14. K-TE:100 mM KCl,10 mM Tris pH 7.5,1 mM EDTA pH 8.0
15. Sample Buffer (SB):100 mM Tris–acetate, pH 7.4, 750 mM Na-acetate,100 mM K-acetate, 10 mM Mg-acetate, 0.5% Triton-X 100, 5% DMSO, 0.5% SDS. At room temperature, SB will be opalescent, perhaps due to K-SDS precipitate, but can be cleared by warming to 37°C. Note that SDS concentrations may be varied (approximately between 0.3% and 0.5%) for different applications.

3. Methods

3.1. Generating a Stock Solution of N-Methyl Mesoporphyrin IX

NMM is deep red in color, has a molecular weight of 580.72, and an extinction coefficient of 1.03×10^5 L/mol at 382 nm.

1. For 25 mg of NMM, add 2.1 mL sterile 2 mM NaOH to solid in the supplier's bottle. Rotate at room temperature until dissolved. This may take around 10–15 min, and should make a solution of approximately 20 mM NMM, depending on the actual amount of NMM in the vial. If it is difficult to get NMM into solution, add more 2 mM NaOH until NMM is completely dissolved.
2. Verify NMM concentration by spectroscopy, by first diluting 4,000-fold to obtain a working solution, which should be approximately 5 μM. For example, add 3 μL of stock solution to 297 μL 2 mM NaOH, mix well, and then add 25 μL of this solution to 975 μL 2 mM NaOH.
3. Measure OD_{382} using a quartz cuvette (set blank with 2 mM NaOH). A true 5 μM solution made in 2 mM NaOH has an OD_{382} of 0.515. Use this information to calculate the true concentration of the NMM stock.
4. Store, protected from light, at −20°C.

3.2. Generating 0.5 mM NMM-coupled Sepharose Resin

3.2.1. Coupling NMM to EAH Sepharose 4B Beads

NMM may be covalently coupled to EAH Sepharose 4B using the carbodiimide method - addition of an *N*,*N*′-disubstituted carbodiimide to an acidic solution containing the resin and NMM promotes condensation between a free amino group of the resin and a free carboxyl group of NMM to yield a peptide bond by acid catalyzed removal of water (30). We used the water soluble carbodiimide *N*-ethyl-*N*′-(3-dimethylaminopropyl) carbodiimide hydrochloride (EDC) to couple NMM to Sepharose 4B in a pH 4.5 solution to promote an acid-catalyzed condensation reaction, described below.

1. Gently mix bottle of EAH Sepharose 4B to resuspend resin. A homogenous suspension will consist of ~50% resin.
2. Aliquot 4 mL of 50% resin mixture into a 15 mL polypropylene conical tube. Spin down in tabletop centrifuge at 450*g* for 2 min. Aspirate supernatant.
3. Wash resin by resuspending it in 12 mL dH_2O that has been adjusted to pH 4.5 with HCl. Mix by inverting or pipetting, not by vortexing. Spin, and aspirate supernatant as in **step 2**. Repeat 4× total.
4. Wash 2× with 12 mL Coupling Buffer ("CB"). Spin down and aspirate as in **step 2**.
5. Resuspend resin in 12 mL CB, and divide equally between two 15 mL tubes, one for a control resin and the other for the NMM resin. Spin and aspirate as in **step 2**.
6. In a fresh 15 mL conical tube, dilute NMM to a final concentration of 0.5 mM in CB at a final volume of 2 mL. Add this to one tube of resin, and add 2 mL of CB without NMM to the other.
7. In a 1.5 mL microcentrifuge tube, dissolve 125 mg of EDC in 1 mL of CB (*see* **Note 1**).
8. Add 250 μL of EDC solution to each of the resins and mix well. Rotate in the dark (cover in foil) at room temperature (RT) overnight.
9. Spin at 450*g* in a tabletop centrifuge for 2 min, and remove supernatant to a new tube to save for calculating coupling efficiency (*see* Subheading 3.2.2).
10. Wash 3× with 12 mL CB as in **step 4**.
11. Wash 1× with 12 mL 1 M glycine. The carbodiimide chemistry should not leave residual reactive species on the resin, but this step is included to block any unexpected reactive species.
12. Wash 3× with 12 mL CB. After removing final wash, add 2 mL CB and 1 mL 95% EtOH. Mix well. Spin briefly. Store, protected from light, at 4°C. The resin should be stable for weeks, and perhaps longer.

3.2.2. Calculating Coupling Efficiency

1. Add 50 μL of the supernatant from **step 9** above to 950 μL CB and measure OD_{382}. A 5 μM solution of NMM in CB has an OD_{382} of approximately 0.4. Use this to determine the NMM concentration in the supernatant.
2. To calculate: determine total volume in which coupling took place (volume resin + volume CB + volume EDC) and use this volume to determine the true concentration of NMM coupled (e.g., 2 mL 50% resin slurry + 2 mL CB + 0.25 mL

EDC = 4.25 mL, and thus the adjusted concentration of NMM from initial 0.5 mM in 2 mL CB = 0.235 mM in the 4.25 mL coupling reaction).

3. Determine the molarity of NMM in the supernatant. This is the uncoupled concentration of NMM.
4. Subtract uncoupled concentration from adjusted initial concentration of NMM (e.g., 0.235 mM – uncoupled concentration). This is the coupled concentration.
5. Divide the coupled concentration by the initial concentration to determine the coupling efficiency. In our hands, we generally get 60–90% coupling efficiency when generating NMM-Sepharose resin under the conditions of this protocol. At full efficiency, such resin would have an NMM concentration of 0.5 mM, using the macroscopic volume of the resin. However, because the surface of the resin, to which NMM is coupled, occupies only a fraction of the macroscopic volume, the actual NMM concentration on this surface will be higher.

3.3. QFP and Control DNA Molecules

Because our initial interest was to use the NMM-Sepharose resin to isolate potential G4-DNA at *S. cerevisiae* telomeres, we examined the capacity of the resin to bind a series of oligonucleotides based on the telomere sequence and with different predicted intramolecular QFP (Table 13.1). *S. cerevisiae* telomeres have imperfect repeats, following the consensus $(TG)_{0–6}$TGGGTGTG(G) (31), and thus different particular examples of the telomere sequence will have different QFP. The thermodynamic and structural details of G4-DNA formed by the varied *S. cerevisiae* telomere repeats have not been well studied (although at least some of the repeats form G4-DNA (32)), but based on the fact that GGG clusters generally form more stable G4-DNA than GG clusters (22), and that intramolecular QFP is generally inversely

Table 13.1
Oligonucleotides used for NMM-Sepharose binding studies

ScTel3	T**GGG**TGT*GG*TGT**GGG**TGT**GGG**TGT**GGG**
ScTel3con	TGAGTGT*GG*TGTGAGTGTGAGTGTGAG
ScTel4	T**GGG**TGT*GG*TGT**GGG**TGT**GGG**TGT*GG*TGT**GGG**
ScTel5	T**GGG**TGT*GG*TGT**GGG**TGT*GG*TGT**GGG**TGT*GG*TGT**GGG**
ScTel6	T**GGG**TGTGTGT**GGG**TGT**GGG**TGTGTGT**GGG**
ScTel7	T**GGG**TGTGTGT**GGG**TGTGTGT**GGG**TGTGTGT**GGG**
ScTel8	T**GGG**TGTGTGTGT**GGG**TGTGTGTGT**GGG**TGTGTGTGT**GGG**

Note: GG and GGG clusters are shown in italics and in bold, respectively, for emphasis related to the length of the loops between G clusters (33), the series of ScTel oligos are predicted to have intramolecular QFP that decreases from ScTel3 through ScTel8. Furthermore, because the intervening sequence between GGG clusters in natural *S. cerevisiae* telomeres has an average length of eight nucleotides and may contain GG clusters, it is probable that the intramolecular QFP of ScTel 5 and 6 are similar to that of a typical stretch of natural telomere repeat, whereas the QFPs of ScTel 3 and 4 are higher, and the QFPs of ScTel 7 and 8 are lower. For non-QFP controls, we used ScTel3con, in which the central G in each GGG cluster of ScTel3 was replaced by an A, and also the duplex form of ScTel3con.

ScTel3 bound the NMM-Sepharose resin quantitatively, and at least several hundred-fold more efficiently than either the single stranded or duplex forms of ScTel3con (Fig. 13.2), indicating preferential binding to QFP sequences. In contrast, no significant binding of any of the DNA molecules to the control resin was observed. Furthermore, ScTel4 through ScTel8 bound the NMM resin with decreasing efficiency, in proportion to their predicted intramolecular QFP (Fig. 13.3). Therefore, although we do not know the structures of the ScTel oligo species that are actually bound to the NMM (because NMM may itself stabilize particular G4-DNA folds), there is a remarkable correlation between predicted intramolecular QFP and binding to the NMM-Sepharose resin.

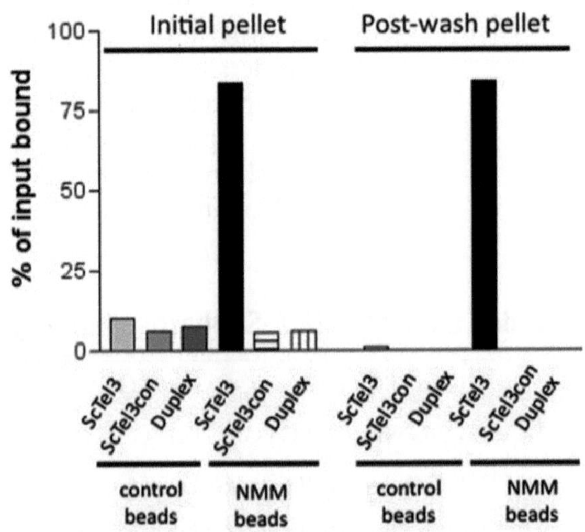

Fig. 13.2. Sample experiment showing ability of NMM to specifically retain G4-DNA and not single-stranded or duplex substrates. See Table 13.1 for sequences. *Duplex* refers to a duplex of ScTel3con and its cognate strand. The binding and wash were carried out in SB with 0.3% SDS. *Initial pellet* and *post-wash pellet* indicate counts on the NMM-Sepharose beads before and after one wash with SB.

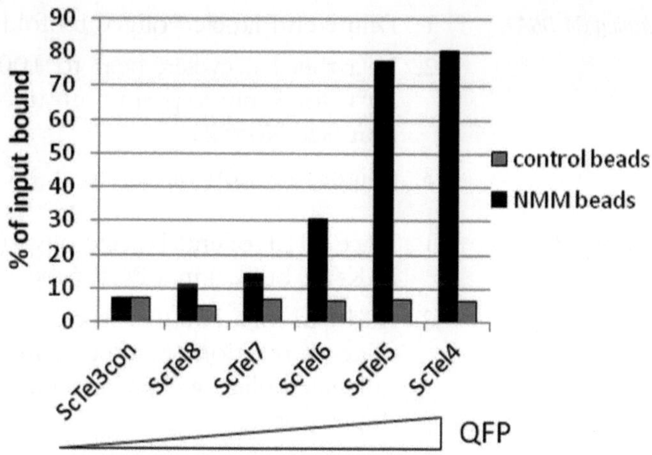

Fig. 13.3. Binding of ScTel control, or ScTel G4-DNA-forming oligonucleotides with increasing QFP, to NMM-coupled vs. control Sepharose resin. Binding was carried out in SB with 0.5% SDS. Note that beads were not washed after binding, and thus the level of residual counts in the control bead samples reflects residual sample volume, not actual binding.

3.3.1. Preparation of Labeled Oligonucleotides

1. Make a solution of 10 μM oligo in TE buffer.
2. End label oligos using T4 polynucleotide kinase (PNK), as outlined below:

   End labeling reaction (prepare components on ice)

   8 μL 10 μM oligonucleotide

   1.5 μL 10× PNK buffer

   4.5 μL ^{32}P γATP

   1 μL PNK (10 units)

   15 μL total
3. Incubate reaction at 37°C for 1 h.
4. During the end labeling reaction, prepare Princeton separation spin columns by swelling in 650 μL K-TE for at least 30 min. Spin down column at 750*g* for 2 min. Transfer columns to fresh 1.5 mL microcentrifuge tubes.
5. Add reaction to spin column and spin 1 min at 750*g*. Add an additional 50 μL K-TE to the column, spin for a further 2 min to ensure elution of DNA. The efficiency of incorporation should be at least 50%.
6. The samples can be stored at 4°C and used for at least a week after labeling.

3.3.2. Forming G4-DNA

1. Dilute end-labeled oligos tenfold in K-TE in PCR tubes.
2. In a thermal cycler, heat to 100°C for 3 min, then cool to 30°C for 5 min to permit intramolecular G-quadruplex formation (see Note 2).
3. Oligos may now be stored at 4°C if necessary.

3.3.3. Generating Control Duplex DNA

1. Place 10 μL of end-labeled oligo (final concentration of ~5 μM in K-TE buffer) in a PCR tube.
2. Add 1 μL of a 100 μM stock of the complementary unlabelled oligo to reaction (final concentration of 10 μM). The excess unlabeled oligo will ensure that most of the labeled oligo is in duplex form.
3. In a thermal cycler with a heated lid, heat to 100°C for 3 min, then incubate at 65°C for 3 h, then 60°C for at least 3 h, to permit duplex formation (optimal temperatures will depend on the melting point of the oligos used).

3.4. Binding Oligos to NMM-Sepharose Affinity Columns

1. Resuspend settled NMM resin in its CB-EtOH storage buffer, and aliquot enough for the present experiment into a 15 mL conical tube. We standardly use 25 μL of resin (i.e., 50 μL of 50% slurry) per binding reaction. To aliquot, cut off the distal ~3 mm of a pipet tip with a clean razor to allow the beads unimpeded passage through the tip.
2. Wash with 10–12 mL Sample Buffer (SB). Spin 450*g*×5 min, and pour off supernatant.
3. Add enough SB to make 50% slurry and then aliquot 50 μL/sample of this slurry into fresh 1.5 mL microcentrifuge tubes. Add 1 mL SB and invert to wash. Spin in microcentrifuge at 1,300*g* for 90 s, and aspirate supernatant. Store pellets protected from light until ready for use.
4. Determine the amount of DNA oligos to be added to the NMM-Sepharose. Approximately 10 μL of 1 μM oligo (10 pmol) is typical, but different amounts may be used depending on the specific radioactivity of the labeled DNA molecules (see Note 3).
5. Add 500 μL SB and the labeled DNA oligos to the tubes containing the NMM-Sepharose. Rotate the tubes for 1 h at RT in the dark (see Note 4).
6. Spin down binding reactions at 1,300*g* for 90 s in a microcentrifuge. Remove 50 μL of supernatant to fresh tube for counting, and discard the remainder of the supernatant. Count the pellet (around 50 μL volume itself given some small amount of residual binding buffer) in a portable Cerenkov counter or scintillation counter (see Note 5). At this point, the majority of counts should remain with the pellet if binding G4-DNA,

while the majority of counts for non-G4-DNA substrates should be removed.

7. Wash pellet by adding 1 mL SB, and mix well by inverting. Rotate for 15 min and spin at 1,300*g* for 90 s. Remove 50 μL of supernatant, and count radioactivity with a portable Cerenkov counter or scintillation counter, and discard the remaining supernatant. Also count the pellet.
8. To determine binding efficiency, plot the fraction of input oligo that is bound to the resin after washing. Equation: Number of counts in the pellet after wash #1/total counts added to the reaction (see Fig. 13.1 for results of a sample experiment).

3.5. Removing G4-DNA from NMM Columns

Some ScTel oligos proved very difficult to remove from the NMM resin. This might reflect cooperative binding of resin-bound NMM to G4-DNA. However, the following protocol is fairly efficient, and approximately 80–85% of ScTel4 could be removed. Because 100% of the bound DNA was not removed, this portion of the protocol is certainly open to further optimization (see Note 6 for list of wash conditions that were less efficient than the agents listed below for removing G4-DNA from the columns).

1. To remove residual SB, resuspend pellet from step 7 in 1 mL sterile dH_2O and rotate for 5 min. Spin down in microcentrifuge at 1,300*g* for 90 s, and discard supernatant into radioactive waste.
2. Resuspend the pellet in 250 μL of 10% lithium dodecyl sulfate (LDS) solution. Rotate for 30 min at RT to wash. This helps remove G4-DNA from NMM because lithium ions are too large a monovalent cation to stabilize quadruplex structures.
3. Spin down at 1,300*g* for 90 s. Remove supernatant to a fresh 15 mL conical tube.
4. Wash pellet with 350 μL of sterile TE pH 7.5 buffer for 5 min with rotation.
5. Spin down at 1,300*g* for 90 s and remove supernatant to tube with LDS supernatant.
6. Add 500 μL Buffer QG to pellet, resuspend, and rotate for 30 min (see Note 7).
7. Spin down at 1,300*g* for 90 s, and combine QG supernatant with LDS and TE washes. An LDS precipitate will form rapidly upon addition of the QG buffer.
8. To separate G4 DNA from the precipitate, perform a phenol/chloroform/isoamyl extraction by adding 1.1 mL phenol/chloroform isoamyl alcohol. Vortex for 1 min, and spin down at 1,300*g* for 5 min in a tabletop centrifuge. DNA should partition to the upper aqueous layer.

9. If desired, determine the volume of the aqueous layer, obtain counts for 50 μL of this volume, and use this number to determine total number of counts recovered in the aqueous layer.
10. The DNA may next be precipitated or purified by other methods depending on downstream applications.

3.6. Future Directions and Limitations

It is conceivable that the NMM-Sepharose resin could be used to isolate G4-DNA species derived from *in vivo* sources. Because NMM could stabilize G4-DNA species and thus artifactually force a G4-DNA conformation upon DNA not naturally in this state, great care needs to be taken with such approaches. It might be possible to use DNA and DNA-protein crosslinking to stabilize DNA in its native state prior to incubation with the NMM resin. Further, as noted above, it will be critical to show that the apparent level of G4-DNA derived from *in vivo* sources changes with genetic or pharmacologic manipulations in the source material that should alter G4-DNA stability.

Regardless of whether the NMM-Sepharose resin is used to isolate G4-DNA species from *in vivo* or *in vitro* sources, the imperfect sensitivity and selectivity of NMM for G4-DNA must also be considered. Although NMM appears to bind a broad array of G4-DNA conformations, it is probably not universal for all folds and is not perfectly selective for G4-DNA, having some ability to bind cytosine-rich i-motifs (**23**). In addition, for crosslinked chromatin isolated from *in vivo* sources, G4-DNA binding proteins may block access to the resin. Use of additional selective G4-DNA ligands that can be coupled to Sepharose, as well as control ligands (i.e., porphyrins without G4-DNA binding capacity) would be of clear utility in this regard. In addition, it is possible that a different concentration of NMM than the ~0.5 mM used in our protocol might yield higher selectivity, affinity, or elutability for particular G4-DNA species.

4. Notes

1. EDC decomposes rapidly in aqueous solution, and so the solution should be made fresh. Let the bottle warm to room temperature (RT) before opening to prevent condensation of moisture, reclose bottle quickly, and return to airtight storage in desiccant at –20°C.
2. This protocol is designed to promote the formation of intramolecular G-quadruplexes. However, some degree

of intermolecular G4-DNA may form depending on the oligo sequence and concentration. If generation of intermolecular quadruplex structures is desired, significantly longer incubation times and higher oligo concentrations may be necessary, and NaCl may be substituted for KCl.

3. Assuming a 1:1 ratio of NMM to G4-DNA binding, 25 μL of a 0.5 mM NMM-Sepharose resin should have enough binding sites to accommodate up to 12.5 nmol of G4-DNA. Thus, addition of 10 pmol of oligonucleotide is approximately three orders of magnitude below a saturating level for the resin.
4. Do not incubate Sepharose beads above 40°C because they will begin to melt and deform.
5. Counting a volume of 50 μL was an arbitrarily selected amount appropriate for the single-tube portable Cerenkov counter that we used. Because the detector sat below the microcentrifuge tube, a small volume concentrated at the bottom of the tube permitted accurate counts. Larger volumes would be quantifiable in a scintillation counter, if desired.
6. We tested several methods to remove G4-DNA from the NMM columns. The following methodologies were generally unsuccessful: washing the column with excess NMM (presumably because cooperative binding bind multiply tethered NMM molecules and could not be completed by single NMM molecules); organic solvents (dimethyl formamide, *N*-methylpyrrolidone). We also tried to break the peptide bond between NMM and the resin using a variety of proteases (pepsin, papain, pronase, subtilisin A); however, none were successful in separating NMM from the resin.
7. Buffer QG contains guanidine thiocyanate, a chaotropic salt. Other chaotropic salts tested included guanidine hydrochloride, sodium perchlorate, and sodium iodide. However, guanidine thiocyanate proved to be the most effective for removing G4-DNA from NMM-Sepharose beads.

Acknowledgments

We thank Robert M. Carlson, J. Brad Chaires, and Paul Ryvkin for advice and discussions. This work was supported by NIH grants R01-AG021521 and P01-AG031862.

References

1. Paeschke K, Simonsson T, Postberg J, Rhodes D, Lipps HJ (2005) Telomere end-binding proteins control the formation of G-quadruplex DNA structures *in vivo*. Nat Struct Mol Biol 12:847–854
2. Hershman SG, Chen Q, Lee JY, Kozak ML, Yue P, Wang L-S, Johnson FB (2008) Genomic distribution and functional analyses of potential G-quadruplex-forming sequences in Saccharomyces cerevisiae. Nucl Acids Res 36:144–156
3. Siddiqui-Jain A, Grand CL, Bearss DJ, Hurley LH (2002) Direct evidence for a G-quadruplex in a promoter region and its targeting with a small molecule to repress c-MYC transcription. Proc Natl Acad Sci USA 99:11593–11598
4. Cogoi S, Xodo LE (2006) G-quadruplex formation within the promoter of the KRAS proto-oncogene and its effect on transcription. Nucl Acids Res 34:2536–2549
5. Kumari S, Bugaut A, Huppert JL, Balasubramanian S (2007) An RNA G-quadruplex in the 5[prime] UTR of the NRAS proto-oncogene modulates translation. Nat Chem Biol 3:218–221
6. Khateb S, Weisman-Shomer P, Hershco-Shani I, Ludwig AL, Fry M (2007) The tetraplex (CGG)n destabilizing proteins hnRNP A2 and CBF-A enhance the *in vivo* translation of fragile X premutation mRNA. Nucleic Acids Res 35:5775–5788
7. Larson ED, Duquette ML, Cummings WJ, Streiff RJ, Maizels N (2005) MutS[alpha] binds to and promotes synapsis of transcriptionally activated immunoglobulin switch regions. Curr Biol 15:470–474
8. Duquette ML, Handa P, Vincent JA, Taylor AF, Maizels N (2004) Intracellular transcription of G-rich DNAs induces formation of G-loops, novel structures containing G4 DNA. Genes Dev 18:1618–1629
9. Huppert JL, Balasubramanian S (2007) G-quadruplexes in promoters throughout the human genome. Nucleic Acids Res 35:406–413
10. Du Z, Kong P, Gao Y, Li N (2007) Enrichment of G4 DNA motif in transcriptional regulatory region of chicken genome. Biochem Biophys Res Commun 354:1067–1070
11. Rawal P, Kummarasetti VB, Ravindran J, Kumar N, Halder K, Sharma R, Mukerji M, Das SK, Chowdhury S (2006) Genome-wide prediction of G4 DNA as regulatory motifs: Role in Escherichia coli global regulation. Genome Res 16:644–655
12. Johnson JE, Smith JS, Kozak ML, Johnson FB (2008) In vivo veritas: Using yeast to probe the biological functions of G-quadruplexes. Biochimie 90:1250–1263
13. Shklover J, Etzioni S, Weisman-Shomer P, Yafe A, Bengal E, Fry M (2007) MyoD uses overlapping but distinct elements to bind E-box and tetraplex structures of regulatory sequences of muscle-specific genes. Nucleic Acids Res 35:7087–7095
14. Fry M (2007) Tetraplex DNA and its interacting proteins. Front Biosci 12:4336–4351
15. Sun D, Thompson B, Cathers BE, Salazar M, Kerwin SM, Trent JO, Jenkins TC, Neidle S, Hurley LH (1997) Inhibition of human telomerase by a G-quadruplex-interactive compound. J Med Chem 40:2113–2116
16. Burger AM, Dai F, Schultes CM, Reszka AP, Moore MJ, Double JA, Neidle S (2005) The G-quadruplex-interactive molecule BRACO-19 inhibits tumor growth, consistent with telomere targeting and interference with telomerase function. Cancer Res 65:1489–1496
17. De Cian A, Lacroix L, Douarre C, Temime-Smaali N, Trentesaux C, Riou JF, Mergny JL (2008) Targeting telomeres and telomerase. Biochimie 90:131–155
18. Gomez D, O'Donohue MF, Wenner T, Douarre C, Macadre J, Koebel P, Giraud-Panis MJ, Kaplan H, Kolkes A, Shin-ya K, Riou JF (2006) The G-quadruplex ligand telomestatin inhibits POT1 binding to telomeric sequences *in vitro* and induces GFP-POT1 dissociation from telomeres in human cells. Cancer Res 66:6908–6912
19. Salvati E, Leonetti C, Rizzo A, Scarsella M, Mottolese M, Galati R, Sperduti I, Stevens MF, D'Incalci M, Blasco M, Chiorino G, Bauwens S, Horard B, Gilson E, Stoppacciaro A, Zupi G, Biroccio A (2007) Telomere damage induced by the G-quadruplex ligand RHPS4 has an antitumor effect. J Clin Invest 117:3236–3247
20. Tahara H, Shin-Ya K, Seimiya H, Yamada H, Tsuruo T, Ide T (2006) G-quadruplex stabilization by telomestatin induces TRF2 protein dissociation from telomeres and anaphase bridge formation accompanied by loss of the 3′ telomeric overhang in cancer cells. Oncogene 25:1955–1966
21. Fernando H, Rodriguez R, Balasubramanian S (2008) Selective recognition of a DNA G-quadruplex by an engineered antibody. Biochemistry 47:9365–9371

22. Lane AN, Chaires JB, Gray RD, Trent JO (2008) Stability and kinetics of G-quadruplex structures. Nucleic Acids Res 36:5482–5515
23. Ragazzon P, Chaires JB (2007) Use of competition dialysis in the discovery of G-quadruplex selective ligands. Methods 43:313–323
24. Chang CC, Chu JF, Kao FJ, Chiu YC, Lou PJ, Chen HC, Chang TC (2006) Verification of antiparallel G-quadruplex structure in human telomeres by using two-photon excitation fluorescence lifetime imaging microscopy of the 3, 6-Bis(1-methyl-4-vinylpyridinium) carbazole diiodide molecule. Anal Chem 78:2810–2815
25. Ren J, Chaires JB (1999) Sequence and structural selectivity of nucleic acid binding ligands. Biochemistry 38:16067–16075
26. Arthanari H, Basu S, Kawano TL, Bolton PH (1998) Fluorescent dyes specific for quadruplex DNA. Nucl Acids Res 26:3724–3728
27. Chaires JB (2005) Competition dialysis: An assay to measure the structural selectivity of drug-nucleic acid interactions. Curr Med Chem Anticancer Agents 5:339–352
28. Huber MD, Lee DC, Maizels N (2002) G4 DNA unwinding by BLM and Sgs1p: Substrate specificity and substrate-specific inhibition. Nucleic Acids Res 30:3954–3961
29. Li Y, Geyer R, Sen D (1996) Recognition of anionic porphyrins by DNA aptamers. Biochemistry 35:6911–6922
30. GEHealthcare (2006) EAH Sepharose 4B Instructions 71-7097-00AD
31. Forstemann K, Lingner J (2001) Molecular basis for telomere repeat divergence in budding yeast. Mol Cell Biol 21:7277–7286
32. Giraldo R, Rhodes D (1994) The yeast telomere-binding protein RAP1 binds to and promotes the formation of DNA quadruplexes in telomeric DNA. EMBO J 13:2411–2420
33. Bugaut A, Balasubramanian S (2008) A sequence-independent study of the influence of short loop lengths on the stability and topology of intramolecular DNA G-quadruplexes. Biochemistry 47:689–697

Chapter 14

Quantifying Interactions Between G-Quadruplex DNA and Transition-Metal Complexes

Roxanne Kieltyka, Pablo Englebienne, Nicolas Moitessier, and Hanadi Sleiman

Abstract

Telomerase inhibition through guanine quadruplex sequestration by small-molecule drugs is of great current interest as an anticancer strategy. G-quadruplexes (GQs) can be formed at the guanine-rich sequences found at the end of the telomere. They possess a large electron-rich π-surface which is favorable for the binding of electron-poor small molecules. Small molecules binding to GQs can sequester the telomere ends and inhibit the enzyme telomerase, which is expressed in cancer cells and absent in normal somatic cells. Transition-metal complexes present a myriad of geometries and numerous ligand coordination environments and allow for modular syntheses for development of compound libraries to target GQs. We have demonstrated the size of the π-surface, binding selectivity and affinity of phenanthroimidazole platinum (II) complexes $[PtPIX(en)]^{2+}2PF_6^-$ (X = naphthyl, phenyl and en = ethylenediamine) and metallosupramolecular complexes $[Pt(4,4'\text{-bpy})(en)]_4^{8+}8PF_6^-$ (where bpy = bipyridine) to GQs can be readily tuned and assayed through a number of biophysical techniques.

Key words: Transition-metal complexes, Supramolecular chemistry, G-quadruplexes, Antitumor therapy, Cancer, Circular dichroism, FRET, Docking, Molecular dynamics

1. Introduction

Targeting of G-quadruplexes (GQs) with small-molecule drugs has attracted significant attention as a therapeutic strategy for cancer. By stabilizing the folded GQ structures at the telomere ends, these molecules can inhibit the action of the enzyme telomerase, which is responsible for "immortalization" of cancer cells. Telomerase is expressed in ~85–90% of cancerous cells and is not present in normal somatic cells, thus constituting an attractive target for selective antitumor therapy (1). Based on this fact, there has been great interest in small molecules, which can assist in the

P. Baumann (ed.), *G-Quadruplex DNA: Methods and Protocols*, Methods in Molecular Biology, vol. 608
DOI 10.1007/978-1-59745-363-9_14, © Humana Press, a part of Springer Science+Business Media, LLC 2010

formation of guanine quadruplexes or sequester pre-existing ones for the purpose of telomerase inhibition. Thus far, many elegant studies on organic GQ binders (2–5) have appeared, but examples involving inorganic complexes are rare. Some inorganic GQ binders that have been reported include copper (6), manganese (7) and nickel porphyrins (8), nickel salens (9), and ruthenium bis-intercalating complexes (10). Implementation of transition metals in these scaffolds provides the opportunity to generate GQ binders of multiple geometries beyond those that carbon can access. Comparatively speaking, introduction of metals into these scaffolds can lead to simplified syntheses of these binders over their organic counterparts. Moreover, generation of compound libraries can be envisaged as a result of the modularity of both the coordination sphere and metal geometry, in addition to their facile syntheses. We have previously reported that both monometallic phenanthroimidazole platinum (II) complexes (11), ([Pt(PIX)(en)]$^{2+}$2PF_6^-, X = naphthyl, phenyl) and supramolecular platinum (II) complexes (12), [Pt(4,4′-bpy)(en)]$_4^{8+}$8PF_6^- (Fig. 14.1) can be synthesized in a facile and expeditious manner for the purpose of strong and selective binding to GQ DNA. In this chapter, we outline several biophysical techniques that can be used to probe transition-metal complex interactions with GQ DNA.

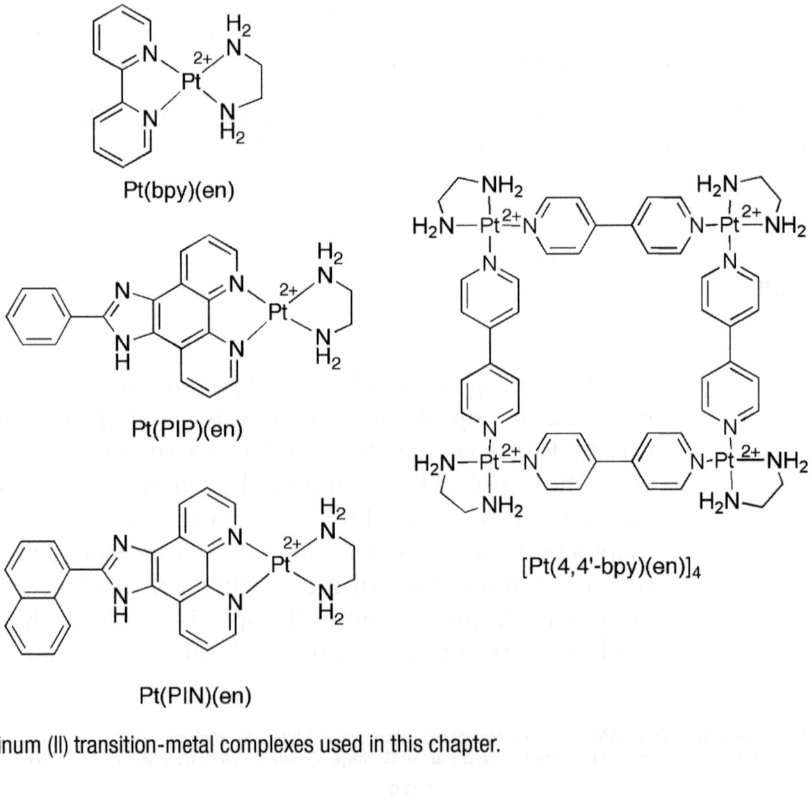

Fig. 14.1. Platinum (II) transition-metal complexes used in this chapter.

2. Materials

2.1. Wet Techniques

1. Phenanthroimidazole platinum (II) complexes ($[Pt(PIX)(en)]^{2+}2PF_6^-$, X = naphthyl, phenyl) and a platinum (II) supramolecular square complex $[Pt(4,4'\text{-bpy})(en)]_4^{8+}8PF_6^-$ are synthesized as previously reported (11,12) (*see* Note 1).
2. Phenanthroimidazole platinum (II) complexes are dissolved in dimethylsulfoxide (DMSO) to obtain a 10 mM concentration. From these original stock solutions, 1 mM stock solutions in DMSO are prepared. These platinum (II) complex solutions are stored at 4°C.
3. The platinum (II) molecular square complex $[Pt(4,4'\text{-bpy})(en)]_4^{8+}8PF_6^-$ is prepared as a 10 mM stock solution in water. This supramolecular platinum (II) complex solution is stored at 4°C.
4. Several buffers are prepared for the various biophysical studies, such as (a) 10 mM K_2HPO_4/KH_2PO_4 buffer with 49 mM KCl (pH 7.2), (b) 10 mM sodium cacodylate buffer with 100 mM LiCl (pH 7.4), (c) 10 mM Na_2HPO_4/NaH_2PO_4 with 100 mM NaCl (pH 7.2), and (d) 1 mM NaH_2PO_4/Na_2HPO_4 and 2 mM NaCl buffer (pH 7.2). These buffers are stored at room temperature.
5. DNA oligonucleotides are purchased from SigmaGenosys. The intermolecular quadruplex-forming sequence $(T_4G_4T_4)_4$ is dissolved in 100 μL of a 10 mM K_2HPO_4/KH_2PO_4 buffer with 49 mM KCl resulting in the formation of the GQ (6) (*see* Note 2). Both fluorescently labeled oligonucleotides 5′-FAM-$G_3(T_2AG_3)_3$-TAMRA-3′ (F21T) (*see* Note 3) and 5′-FAM-TATAGCTATA-HEG-TATAGCTATA-TAMRA-3′ (*see* Note 3) (where FAM = fluorescein, TAMRA = tetracarboxylrhodamine, and HEG = hexaethyleneglycol) are dissolved in 100 μL deionized water.
6. Calf thymus DNA (SigmaAlrich) is dissolved in 2 mL of 1 mM NaH_2PO_4/Na_2HPO_4 and 2 mM NaCl buffer (pH 7.2) to provide a stock solution concentration between 2,000 and 3,000 μM in base pair (*see* Note 5).

2.2. Molecular Modeling

Software required:

1. Jaguar (Schrödinger, Inc.) (*see* Note 6).
2. FITTED docking program, including the PROCESS and SMART modules (13,14) (*see* Note 7).
3. AMBER molecular dynamics package: *sander*, *pmemd*, *mm_pbsa.pl* modules (15) (*see* Note 8).

4. AmberTools set: *leap*, *antechamber*, *respgen*, *resp* modules (16).
5. Maestro (Schrödinger, Inc.) (*see* Note 9).

3. Methods

3.1. Spectroscopic Binding Titrations using UV–Vis Spectroscopy

Spectroscopic binding titrations of metal-based GQ binders can be performed to determine binding affinity to the quadruplex DNA motif. This experiment typically requires that the binding molecule of interest possesses a maximum in its absorbance spectrum over 320 nm and exhibits optical changes when binding to nucleic acids. Through the addition of several aliquots of GQ DNA to a platinum (II) complex solution of known concentration, a decrease in absorbance maximum and red shift are observed in a UV–vis spectrophotometer. From the absorbance values, the bound and free ligand concentrations can be determined mathematically and extrapolated to obtain a binding constant from a reciprocal plot of the change in the apparent extinction coefficient of the ligand with respect to DNA concentration. We have previously applied this method to determine the binding affinity and preference of the $[Pt(PIX)(en)]^{2+}2PF_6^-$ series of binders for the intermolecular GQ-forming sequence $(T_4G_4T_4)_4$ over duplex DNA (11). Below is a sample UV–vis binding titration for $[Pt(PIN)(en)]^{2+}2PF_6^-$.

A 1-mL solution of 10 μM $[Pt(PIN)(en)]^{2+}2PF_6^-$ is prepared from the 1 mM DMSO stock solution in the same buffer used for the GQ $(T_4G_4T_4)_4$ in a 1-cm quartz cuvette. This solution can be gently vortexed to ensure sample homogeneity.

Aliquots (1 μL) of a 1.5 mM solution of $(T_4G_4T_4)_4$ (*see* Note 10), as prepared in Subheading 14.2, are added sequentially to the $[Pt(PIN)(en)]^{2+}2PF_6^-$ solution and vortexed gently with each addition. A UV-vis spectrum on a Cary 300 Bio (Varian) is then recorded from 500 to 200 nm after the addition of each aliquot following the maximum at 302 nm. With each aliquot, a decrease in the absorbance and red shift at this maximum are observed (Fig. 14.2).

The titration of GQ DNA is continued until a plateau in the decrease of the peak at 302 nm is achieved. This titration should be performed in triplicate to verify the results obtained, and the cuvettes should be cleaned thoroughly between trials (*see* Note 11).

GQ DNA is then titrated to a cuvette containing the aforementioned buffer following steps 2 and 3 as a blank (*see* Note 12).

Absorbance data for both experiments collected in triplicate from steps 2 and 4 are imported into Microsoft Excel (or an

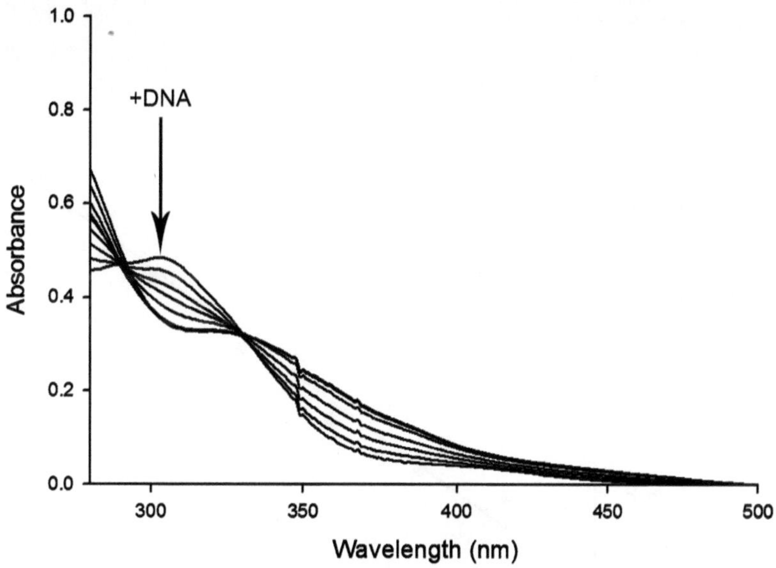

Fig. 14.2. Decrease in absorbance upon the addition of G-quadruplex DNA, monitoring the peak at 302 nm (11). Copyright Wiley-VCH Verlag GmbH & Co. KGaA. Reproduced with permission.

equivalent spreadsheet program). The absorbance traces obtained at 302 nm for the GQ in step 4 are subtracted from titration of the GQ to the platinum (II) complex in step 2.

Using a previously reported binding model (17), the binding constant for these complexes can be determined. From the initial absorbance (A) of the complex itself before any addition of nucleic acid permits the concentration (C)of the free complex in solution to be calculated using Beer's law $C = A/\varepsilon_f l$, with the molar extinction coefficient (ε_f) of 20,134 M^{-1} cm^{-1} for $[Pt(PIN)(en)]^{2+}2PF_6^-$, and a path length ($l$) of 1 cm.

The apparent extinction coefficient of the complex upon the addition of each aliquot is then calculated by dividing the observed absorbance by the concentration of platinum (II) complex calculated in step 6, $\varepsilon_A = A/Cl$ (*see* Note 13).

The apparent molar extinction coefficient of the bound, $\Delta\varepsilon_{ap}$, is then determined by subtracting the absolute value of the molar extinction coefficient of the free complex $\varepsilon_f([Pt(PIN)(en)]^{2+}2PF_6^-)$ from the apparent extinction coefficient calculated with each addition of DNA (from step 7) ε_A, to give $\Delta\varepsilon_{ap} = |\varepsilon_A - \varepsilon_f|$. This calculation is repeated for each aliquot addition.

The concentration of GQ (D) added to the platinum (II) solution with every aliquot is divided by 4 to give a concentration in base tetrad (*see* Note 14).

Each concentration of GQ DNA calculated in step 9, D, is divided by the apparent molar extinction coefficient $\Delta\varepsilon_{ap}$ of the bound species to give $D/\Delta\varepsilon_{ap}$.

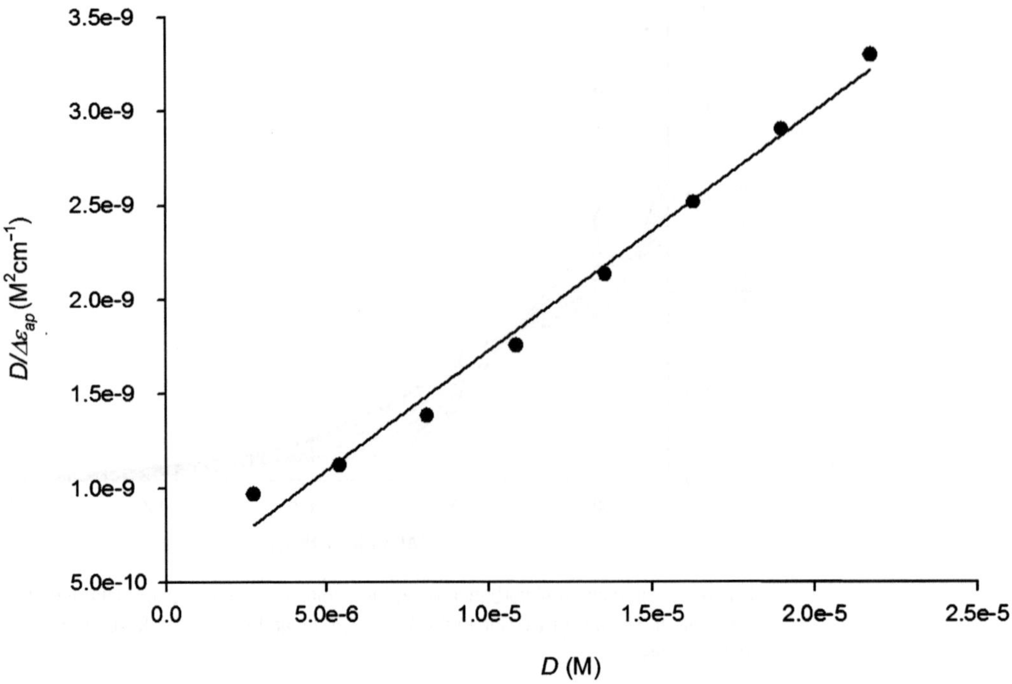

Fig. 14.3. Plot of $D/\Delta\varepsilon_{ap}$ vs. D for $[Pt(PIN)(en)]^{2+}2PF_6^{-}$. Copyright Wiley-VCH Verlag GmbH & Co. KGaA. Reproduced with permission.

The values obtained for $D/\Delta\varepsilon_{ap}$ are then plotted on the *y*-axis (within Microsoft Excel) against the concentration of each aliquot of GQ DNA added, D (Fig. 14.3).

Binding constants K can then be determined from the fitted line of the reciprocal plot of $D/\Delta\varepsilon_{ap}$ versus D using the equation below, where the *y*-intercept is equal to $1/(\Delta\varepsilon \times K)$ and the slope is $1/\Delta\varepsilon$. The slope is then divided by the *y*-intercept to obtain K (*see* Notes 15, 16).

$$D/\Delta\varepsilon_{ap} = D/\Delta\varepsilon + 1/(\Delta\varepsilon \times K)$$

Steps 1 through 13 can be repeated for calf thymus DNA against the platinum (II) complex in 1 mM Na_2HPO_4/NaH_2PO_4 and 20 mM NaCl buffer (*see* Note 10).

3.2. Continuous Variation Method (Job Plot)

The continuous variation method or Job plot (18) can be used to determine the binding stoichiometry of transition-metal complexes to GQ DNA. In this method, the total concentration of the phenanthroimidazole platinum (II) complexes and GQ DNA are held constant but their mole fractions relative to one another are varied. With each addition of GQ DNA or platinum (II) complex, a UV–vis spectrum is recorded. From the obtained absorbance values, a graph of absorbance versus mole fraction is plotted, and the intersection of the data points or their maximum defines the binding stoichiometry of the platinum (II) complex to the

G-quadruplex. We previously used the continuous variation method to determine the binding stoichiometry of the $[Pt(PIX)(en)]^{2+}2PF_6^-$ series of complexes (11).

1. Prepare 5 mL of a 5 μM solution of $[Pt(PIP)(en)]^{2+}2PF_6^-$ from the 1 mM DMSO stock solution in 10 mM K_2HPO_4/KH_2PO_4 buffer with 49 mM KCl (*see* Note 17).
2. Prepare 5 mL of a 5 μM solution of the GQ $(T_4G_4T_4)_4$ in the same buffer used in step 1 from the stock solution of this nucleic acid.
3. Add 700 μL of the 5 μM GQ DNA solution from step 2, $(T_4G_4T_4)_4$, to a 1-cm path length quartz cuvette. A UV–vis spectrum is recorded on a Cary 300 Bio (Varian) spectrometer from 500 to 200 nm prior to adding the $[Pt(PIN)(en)]^{2+}2PF_6^-$. Add 100 μL aliquots of the $[Pt(PIP)(en)]^{2+}2PF_6^-$, from step 1, recording an absorbance spectrum with each addition (*see* Note 18).
4. The cuvette should be cleaned thoroughly from the previous trial. Pipette 700 μL of a 5 μM solution of $[Pt(PIP)(en)]^{2+}2PF_6^-$ from step 1. Add 25 μL aliquots of the 5 μM GQ DNA solution from step 2, $(T_4G_4T_4)_4$, until a plateau in the absorbance increase is achieved at 305 nm.
5. The cuvettes should be cleaned thoroughly from the previous trials. Steps 1–4 should be repeated using the aforementioned buffer in place of the platinum complex with the GQ sequence listed above (*see* Note 19).
6. Import raw absorbance data for steps 3–5 into Microsoft Excel. Subtract the absorbances of GQ blanks from both sets of titration data obtained in steps 3 and 4.
7. Normalize each set of titration data from steps 3 and 4 with respect to the highest absorbance value within each respective step.
8. The mole fraction of $[Pt(PIP)(en)]^{2+}2PF_6^-$ for the titration performed in step 3 is calculated using the equation $X_{PIP} = V_{PIP}/(V_{PIP} + V_G)$, where V_{PIP} is the volume of $[Pt(PIP)(en)]^{2+}2PF_6^-$ added with each aliquot and V_G is the amount of GQ in the cuvette at the start.
9. The mole fraction of $[Pt(PIP)(en)]^{2+}2PF_6^-$ for the titration performed in step 4 is calculated using the equation $X_{PIP} = 1 - [V_G/(V_G + V_{PIP})]$, where V_G is the volume of GQ added to the cuvette with each aliquot and V_{PIP} is the volume of $[Pt(PIP)(en)]^{2+}2PF_6^-$ at the start.
10. This data is then plotted in SigmaPlot, placing the normalized absorbances on the *y*-axis and the mole fraction of platinum complex on the *x*-axis (*see* Note 20, Fig. 14.4). From the intersection of the data, the binding stoichiometry of the platinum (II) complex can be determined.

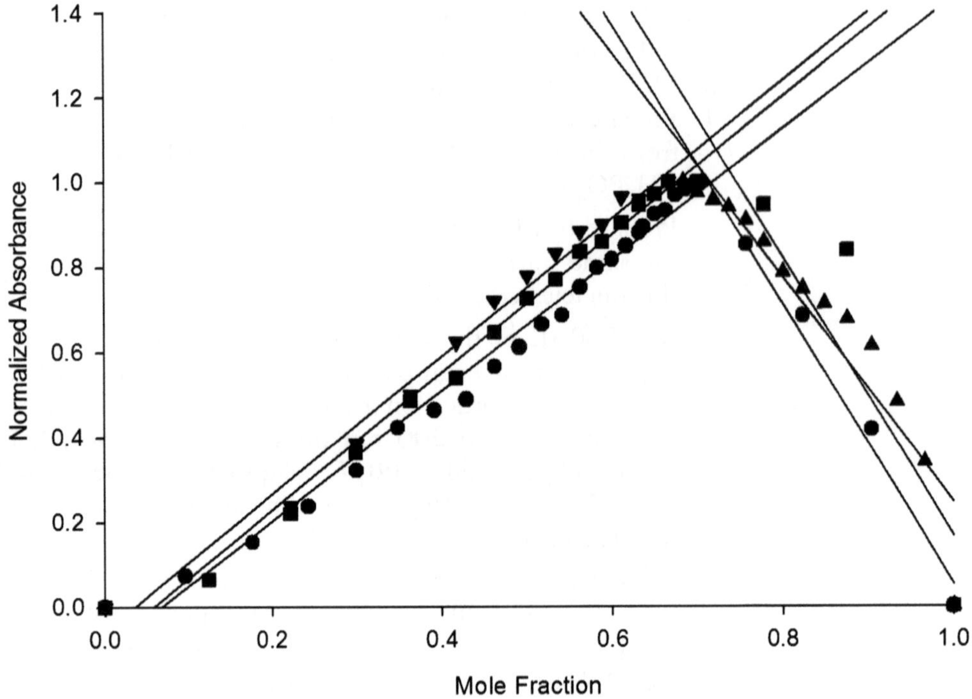

Fig. 14.4. Job plot of Pt(II) intercalators with quadruplex DNA, total concentration maintained between 5 and10 μM: complex $[Pt(bpy)(en)]^{2+}$ (●); $[Pt(PIP)(en)]^{2+}$ (▲); and $[Pt(PIN)(en)]^{2+}$ (■) (11). Copyright Wiley-VCH Verlag GmbH & Co. KGaA. Reproduced with permission.

3.3. Temperature-Dependent Circular Dichroism

Circular dichroism (CD) can be used to estimate the increased stability of GQ DNA upon small molecule binding. The CD signature of the GQ DNA is followed before and after heating to 90°C in the presence of a transition-metal binder. Hence, relative degrees of stability can be implied amongst several binders due to their ability to maintain the original quadruplex signature observed within the CD. The success of this experiment relies on the slow reassociation kinetics of the intermolecular $(T_4G_4T_4)_4$ quadruplex once denatured, allowing relative degrees of stability to be monitored after cooling to room temperature (19). Previously we have reported the stabilization of the intermolecular GQ $(T_4G_4T_4)_4$ by several platinum (II) complexes after heating to 90°C (11). Here we describe a sample trial for $[Pt(PIP)(en)]^{2+}2PF_6^-$ with the intermolecular GQ.

1. A sample of $(T_4G_4T_4)_4$ is diluted from its stock solution in 10 mM K_2HPO_4/KH_2PO_4 buffer and 49 mM KCl, to obtain a 34.8 μM solution of the GQ in base tetrad in 250 μL buffer. This solution should be vortexed gently to ensure homogeneity (*see* Note 21). This sample is prepared in duplicate.
2. A sample of $[Pt(PIP)(en)]^{2+}2PF_6^-$ is diluted from its 10 mM DMSO stock solution in 10 mM K_2HPO_4/KH_2PO_4 buffer and

49 mM KCl, to obtain a 34.8 μM solution of the metal complex in 250 μL buffer. Included in this 250-μL volume, enough $(T_4G_4T_4)_4$ to obtain a 34.8 μM concentration of nucleic acid in base tetrad is added. This sample should be vortexed gently to ensure homogeneity and is also prepared in duplicate.

3. A sample of $[Pt(PIP)(en)]^{2+}2PF_6^-$ is diluted from its 10 mM DMSO stock solution in 10 mM K_2HPO_4/KH_2PO_4 buffer and 49 mM KCl, to obtain a 34.8 μM solution of the metal complex in 250-μL buffer. This solution should be vortexed gently to ensure homogeneity (*see* Note 21). This sample is prepared in duplicate.
4. One of each pair of samples prepared in steps 1 (*see* Note 22), 2 (*see* Note 23), and 3 (*see* Note 24) are first assayed individually at room temperature using CD spectroscopy. Sequentially, each sample is transferred to a 0.1-cm path length quartz cuvette and a CD spectrum is recorded on a JASCO J-810 spectropolarimeter from 350 to 200 nm at a scan speed of 100 nm/min (Fig. 14.5). The quartz cuvette should be washed and dried thoroughly between samples.
5. The three other prepared samples are then heated to 95°C for 10 min and cooled to room temperature over 2–3 h within a Techne Flexigene Thermal Cycler.

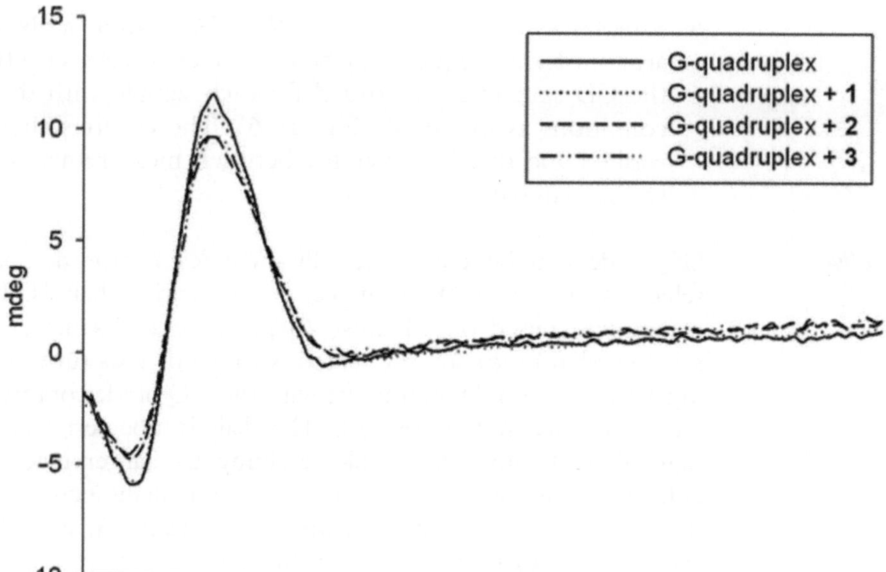

Fig. 14.5. CD spectra of intermolecular G-quadruplex/platinum (II) complexes prior to heating (**1** = $[Pt(bpy)(en)]^{2+}$, **2** = $[Pt(PIP)(en)]^{2+}$, **3** = $[Pt(PIN)(en)]^{2+}$) (11). Copyright Wiley-VCH Verlag GmbH & Co. KGaA. Reproduced with permission.

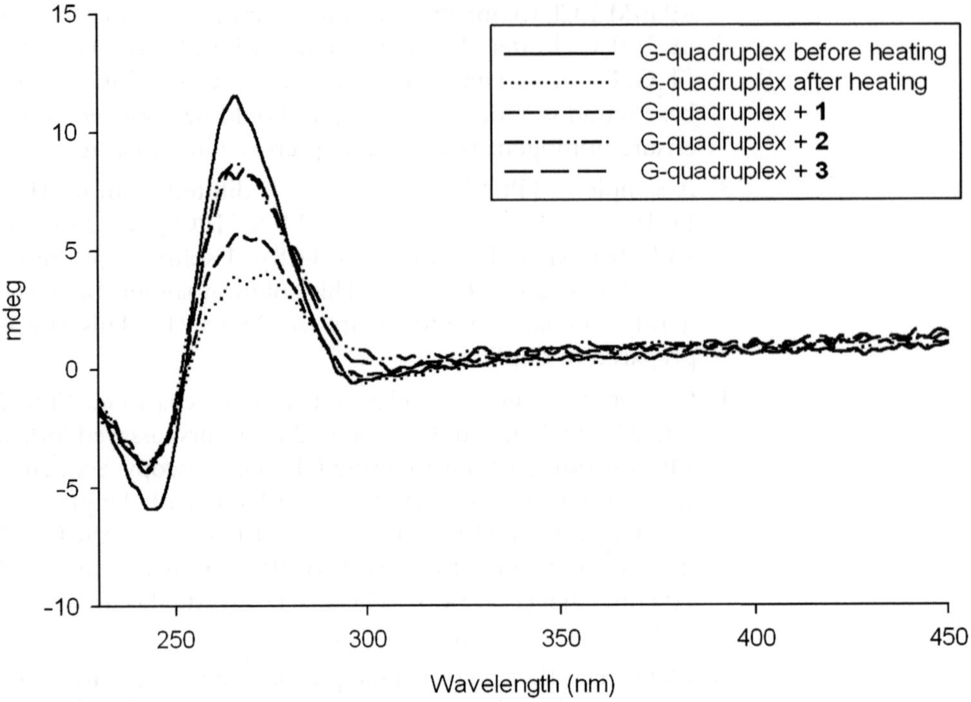

Fig. 14.6. CD spectra of intermolecular G-quadruplex before heating and G-quadruplex with platinum (II) complexes after heating (**1** = $[Pt(bpy)(en)]^{2+}$, **2** = $[Pt(PIP)(en)]^{2+}$, **3** = $[Pt(PIN)(en)]^{2+}$) (11). Copyright Wiley-VCH Verlag GmbH & Co. KGaA. Reproduced with permission.

6. Samples 1 (*see* Note 25), 2 (*see* Note 26), and 3 (*see* Note 27) are transferred sequentially to the 0.1-cm quartz cuvette and the CD spectra are recorded for each sample with the same conditions as in step 4 (Fig. 14.6). The cuvettes should be washed and dried thoroughly between measurements of the various samples.

3.4. Competitive Dialysis

GQ binders can be screened rapidly and efficiently for their binding selectivity to the GQ motif using competitive dialysis. In this experiment besed on Chaires (20), nucleic acid samples at the same concentration are incubated within a dialysis tube containing a semipermeable membrane with the GQ binder of interest at a certain concentration for 24 h. This dialysis tube permits the passage of small molecules while retaining the larger macromolecules, allowing for the equilibration of the small molecule between the dialysis tube and the surrounding solution. At equilibrium, the dialysis tubes are assayed using UV–vis spectroscopy using peaks above 320 nm to determine the concentration of GQ binder with each nucleic acid sequence. Using the $[Pt(PIX)(en)]^{2+}2PF_6^-$ series of complexes, binding selectivity and affinity of these complexes to calf thymus DNA and $(T_4G_4T_4)_4$ can be assayed (11). Described is the competitive dialysis of $[Pt(PIP)(en)]^{2+}2PF_6^-$ against the intermolecular GQ and calf thymus DNA.

1. An 80-mL volume of a 5 μM solution (*see* Note 28) of $[Pt(PIP)(en)]^{2+}2PF_6^-$ is prepared in 1 mM NaH_2PO_4/Na_2HPO_4 and 2 mM NaCl buffer (pH 7.0) from 1 mM stock solution of the complex in DMSO. This solution is poured into a 100-mL beaker.
2. Dialysis bags (Pierce slide-a-lyzer mini-dialysis kits, MWCO 3500 Da) filled with 200 μL of 75 μM solution of each nucleic acid measured in base pair or base quartet are suspended using a metal wire slightly above the dialysis solution with the base of the dialysis units in contact with the solution in the 100-mL beaker.
3. A magnetic stir bar is added to the beaker and the setup is then covered with parafilm and aluminum foil. This setup is then allowed to equilibrate for 24 h with stirring at room temperature.
4. At equilibrium, the solutions within the dialysis bags (180 μL) are transferred to microcentrifuge tubes along with a 10% (w/v) sodium dodecyl sulfate (SDS) solution (20 μL) to give final concentrations of 1% SDS (*see* Note 29). At this point, these samples were incubated with SDS for 10 min prior to any further manipulation.
5. These SDS-treated solutions are first diluted one-fifth of the total volume from step 4 in the aforementioned buffer in a sub-microcuvette with a 1-cm path length cell and 160-μL volume (*see* Note 30). This solution is then analyzed by UV–vis spectroscopy on a Cary 300 Bio (Varian) spectrometer following the metal complex absorbance at 305 nm. This UV–vis experiment is run in single-beam mode from 500 to 200 nm.
6. Another sample is prepared where the GQ and duplex DNA are at the same concentration as in step 2 and subjected to the same SDS treatment as in step 4 and the dilution in step 5. UV–vis spectra are acquired for both of these samples under the same conditions mentioned in step 5 (*see* Note 31) and subtracted from the spectra collected in this step.
7. The platinum (II) complex concentration is assayed using UV–vis spectroscopy under the same conditions listed in step 5. Two hundred microliters of the bulk solution is transferred into the clean sub-microcuvette and the absorbance maximum at 305 nm is followed.
8. Since the UV–vis experiment is run in single-beam mode, a blank of the buffer used in the experiment is run under the conditions listed in step 5 and subtracted from the absorbance spectra of the bulk solution collected in step 7.
9. To calculate the total concentration bound, the absorbances (step 5) obtained within the dialysis tubes at 305 nm are first multiplied by the dilution factors 1.11 (200 μL/180 μL) and subsequently by 5 (*see* Note 32) before substitution into $C_t = A/\varepsilon l$.

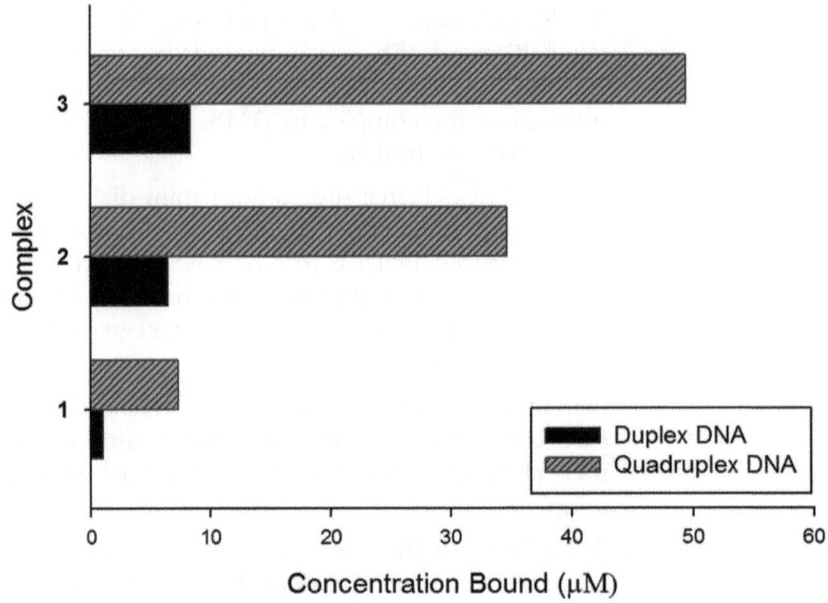

Fig. 14.7. Data obtained from the competitive dialysis experiment for phenanthroimidazole platinum (II) G-quadruplex DNA binders compared with a smaller π-surface Pt^{II} complex (**1** = $[Pt(bpy)(en)]^{2+}$, **2** = $[Pt(PIP)(en)]^{2+}$, **3** = $[Pt(PIN)(en)]^{2+}$). Data was scaled to 1 μM for comparison between complexes (11). Copyright Wiley-VCH Verlag GmbH & Co. KGaA. Reproduced with permission.

In this equation, A is equal to the observed absorbance, ε is the molar extinction coefficient of the platinum (II) complex (11), $\varepsilon = 25{,}626\ M^{-1}\ cm^{-1}$, and l is the path length of the cuvette.

10. The concentration of the free ligand (C_f) is then determined from the absorbance measured of the bulk solution containing the platinum (II) metal complex using the same equation above (*see* Note 33).
11. The amount of GQ binder bound to each nucleic acid sequence is then determined by subtracting the concentration of the free ligand from the total ligand obtained spectrophotometrically in the experiment: $C_b = C_t - C_f$. The concentrations of the platinum (II) complexes bound are then plotted with Microsoft Excel (Fig. 14.7).
12. Binding constants can be further extrapolated from this data using the equation $K_{app} = C_b / C_f \times (S_{tot} - C_b)$, where C_b is the amount of platinum (II) complex bound, S_{tot} is the total concentration of nucleic acid in either base pair or tetrad, and C_f is the concentration of the unbound platinum (II) complex (*see* Note 34).

3.5. Fluorescence Resonance Energy Transfer (FRET) Experiment

The ability of a small molecule to stabilize a GQ structure can be evaluated by performing a thermal denaturation assay of the quadruplex structure that relies on communication between two fluorophores (fluorescence resonance energy transfer – FRET)

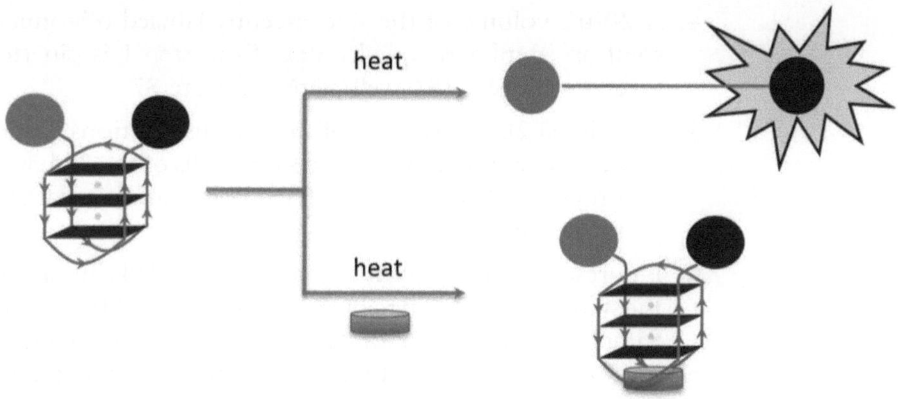

Fig. 14.8. FRET melting assay experiment: addition of the G-quadruplex binder retards the denaturation of the oligonucleotide and its onset of fluorescence.

covalently bound at opposite ends of the oligonucleotide strand (21,22). Upon heating this ordered quadruplex structure without any binder present, the distance between the FRET pair increases because of the conversion of the quadruplex structure into a random coil, resulting in a detectable fluorescence signal, which can be ascribed to a melting temperature transition (Fig. 14.8).

In this assay, metal complexes that stabilize the GQ to a greater degree will increase its melting temperature more significantly at low concentrations of ligand than if there was no binder present. Stronger GQ binders typically require lower concentrations to achieve substantial changes in melting temperature from the GQ itself. We have reported this method for the molecular square complex $[Pt(4,4'\text{-bpy})(en)]_4^{8+}8PF_6^-$, with F21T = 5′-FAM-GGG$(TTAGGG)_3$-TAMRA-3′ and 5′FAM-TATAGCTATA-HEG-TATAGCTATA-TAMRA-3′ (12).

1. Fluorescently labeled duplex (5′FAM-TATAGCTATA-HEG-TATAGCTATA-TAMRA-3′ where FAM = fluorescein, TAMRA = tetracarboxylrhodamine, and HEG = hexaethyleneglycol) and quadruplex (F21T = 5′-FAM-GGG$(TTAGGG)_3$-TAMRA-3′) forming oligonucleotides dissolved in deionized water were diluted to 400 nM in 10 mM sodium cacodylate buffer with 100 mM LiCl (pH 7.4).
2. These oligonucleotide solutions are heated to 90°C in a Cary 300 Bio (Varian) instrument for 5 min and then allowed to cool for 2–3 h (*see* Note 35).
3. For the molecular square complex $[Pt(4,4'\text{-bpy})(en)]_4^{8+}8PF_6^-$, several solutions are prepared in the buffer in step 1 with concentrations of 200 nM, 500 nM, 1 μM, 1.5 μM, 2 μM, and 2.5 μM (*see* Note 36). These solutions are prepared from the 1 mM stock solution of the supramolecular complex in deionized water.

4. A 20-μL volume of the fluorescently labeled oligonucleotide solution (duplex or quadruplex) from step 1 is pipetted into several wells of a 384-well plate (*see* Note 37).
5. Additional 20 μL volumes of the various solutions of the complex are then pipetted into the same wells of the 384-well plate to a total of 40 μL per well. Each concentration data point is performed in triplicate.
6. Fluorescence measurements were recorded in an Applied Biosystems Real-Time PCR (ABI HT 7900) instrument. Samples are first equilibrated within the instrument at 25°C for 5 min, prior to heating to 95°C in 71 cycles at 1°C/min. Fluorescence readings are recorded at 0.5°C/min intervals. The fluorescence of FAM is followed by excitation at 494 nm and emission at 522 nm.
7. After the experiment is completed, raw FAM emission data is exported to Microsoft Excel, examined, and then normalized with respect to the highest fluorescence intensity value (*see* Note 38). Normalized fluorescence data is graphed with respect to temperature (Fig. 14.9). The triplicate datasets are averaged and the thermal denaturation temperature ($\Delta T_{1/2}$) is taken as the temperature at which the normalized fluorescence is equal to 0.5 (*see* Note 39).

3.6. Molecular Modeling

Computational studies are directed at the prediction of the binding mode and the relative binding affinity of the transition-metal complexes. We have applied a hybrid docking/molecular dynamics (MD)

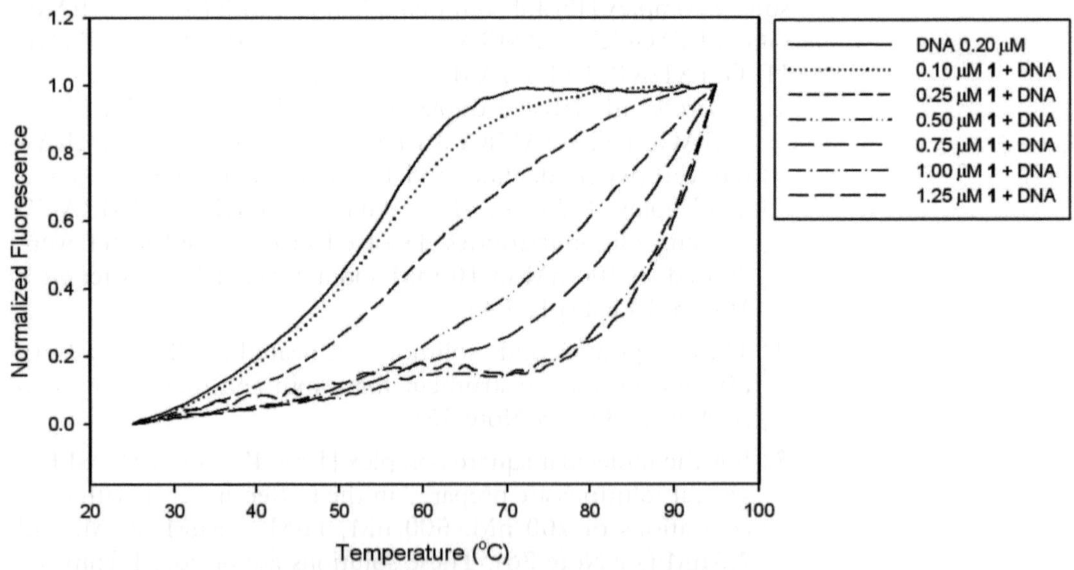

Fig. 14.9. An example of processed FRET data for the molecular square with the intramolecular G-quadruplex. Reprinted with permission from (12). Copyright 2008 American Chemical Society.

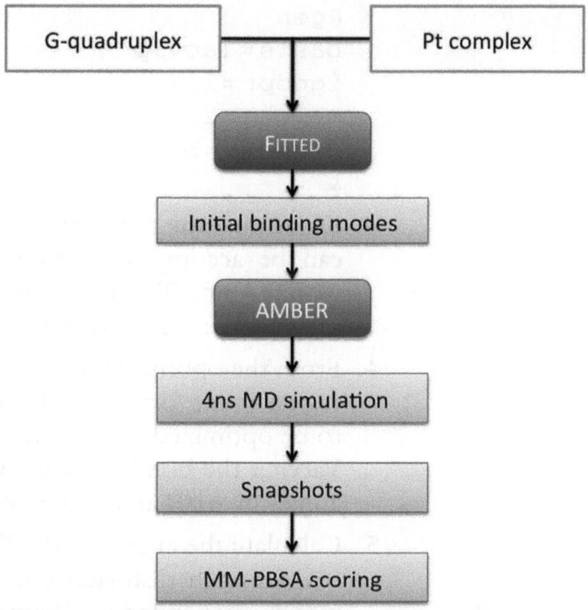

Fig. 14.10. Workflow for the molecular modeling of G-quadruplex binders.

protocol to this effect (12), which we describe in the following sections (*see* Fig. 14.10).

The docking and the MD simulations require the prior development of force field parameters for the platinum (II) complexes, which can be accomplished by fitting the parameters to *ab initio* (e.g., density functional theory – DFT) calculations. Subheading 14.3.6.1 outlines the process for developing these parameters, Subheadings 14.3.6.2–14.3.6.4 describe the procedure for docking the platinum (II) complexes to the GQs, and Subheadings 14.3.6.5 and 14.3.6.6 deal with the MD simulations and MM-PBSA (molecular mechanics Poisson–Boltzmann surface area) scoring.

3.6.1. Development of Force Field Parameters for the Pt(II) Complexes

The following protocol outlines the steps needed for the development of additional force field parameters (GAFF (23)) for the simulations with FITTED and AMBER.

1. Determine a suitable minimal model for which to determine the parameters that need to be calculated; in our case, Pt(en)(bpy)$_2$ proved to be appropriate. Obtain (e.g., draw in a visualization tool) a three-dimensional description of the minimal model; save it as a MOL2 file (*ligand.mol2*) and a Jaguar input file (*ligand.in*).
2. Optimize the minimal model at the B3LYP/LACV3P** level of theory (*see* Note 40). This can be accomplished in Jaguar by prepending the following commands to the *z*-matrix portion of an input file:

```
&gen
basis=lacv3p**
igeopt=1
molchg=2
dftname=b3lyp
&
```

3. Determine the parameters missing from the force field. This can be accomplished with the *parmchk* module included within AmberTools (*see* Note 41):

```
% parmchk -i ligand.mol2 -f mol2 -o frcmod.Pt
```

4. From the optimized model, generate distorted structures (in Jaguar input format) with values for the parameters needing to be optimized around the equilibrium point (*see* Note 42). Varying the bond distance by ±0.1 Å at 0.01 Å intervals and angles by ±10° at 1° intervals is usually satisfactory.
5. Calculate the energy at the B3LYP/LACV3P** level of theory for each distorted conformation. The following commands, prepended to a Jaguar input file (*optimize.in*), can be used for the single-point energy calculations:

```
&gen
basis=lacv3p**
molchg=2
dftname=b3lyp
&
```

Run the *jaguar* calculation:

```
% jaguar -run optimize.in
```

6. For each distorted structure, convert the Jaguar file to MOL2 format with the following command:

```
% jaguar -babel -ijagin ligand.in -omol2
ligand.mol2
```

7. Assign RESP (Restrained ElectroStatic Potential) charges to the ligand (*see* Subheading 14.3.6.3 and also Note 43).
8. Generate topology and coordinate files for each conformation using *leap*; the following is the content of a text file (*leap.in*) containing the *leap* commands (*frcmod.Pt* is the file generated by *parmchk* in step 3):

```
Source leaprc.gaff
loadamberparams frcmod.Pt
model=loadmol2 "ligand.mol2"
saveamberparm complex ligand.top ligand.
crd
quit
```

leap is run with the following command:

```
% tleap -s -f LEAP.IN
```

9. Calculate the AMBER energy for each distorted conformation generated in the previous step, without considering the

contribution from the parameters being optimized. The following *sander* input file (*sander.in*) can be used for this purpose:

```
&cntrl
imin=1, maxcyc=0
cut=16,
igb=0,
ntb=0,
ntpr=1,
/
&end
```

The calculation is run with the following command:

```
% sander -O -i sander.in -p ligand.top -c
ligand.crd
```

10. Calculate the difference between the AMBER and the DFT energies (*see* Note 44) for each coordinate being scanned, and fit the energy difference to a quadratic term of the form $K(x-x_0)^2$ (*see* Note 45).
11. Repeat steps 4–10 for all the parameters that needed to be defined as found in step 3.
12. Add the optimized parameters (K, x_0) to an AMBER additional force field parameter file (in our case, *frcmod.Pt*) and to the FITTED force field file (*fitted.txt*).

3.6.2. Preparation of the DNA GQ for Docking

The following steps start from a PDB (protein data bank) crystal structure and generate the receptor files necessary for FITTED.

1. Obtain a suitable three-dimensional structure of a DNA GQ, e.g., from the RCSB PDB (24,25), in PDB format. In our case, we used 1KF1 (*see* Note 46).
2. Add hydrogens and fix bond orders (*see* Note 47); save the model in MOL2 format (*1KF1.mol2*).
3. Prepare a keyword file for PROCESS (*g4.txt*):

```
Protein 1
1KF1.mol2
Output  1KF1
Binding_Site_Cav    1KF1_BS
Interaction_Sites   1KF1_IS
Num_of_IS      1000
Active_Site   0
AutoFind_Site Yes
AutoFind_Center     No
Find_Residues Number
Renumber_Residues   1
Assign_Gyes
Truncateno
Cutoff  11
```

```
United  yes
Grid_Sphere_Size   40
Grid_Center   0 0 0
Grid_Size     50   50   50
Ligand_Cutoff 30
```

4. Run PROCESS to prepare receptor input files for FITTED.
```
% process g4.txt
```

3.6.3. Preparation of the Platinum Complex for Docking

The following protocol outlines the assignment of RESP charges (26) and FITTED atom types to the platinum complexes.

1. Run an ESP (electrostatic potential) calculation on the DFT-optimized ligand (B3LYP/LACV3P** level of theory). Using Jaguar, this can be accomplished by adding the following lines to the input file (*ligand.in*):
```
&gen
basis=LACV3P**
dftname=B3LYP
ip172=2
icfit=1
&
```
This will output a *ligand.resp* file, which will be used in the following step.
2. Convert the Jaguar file to MOL2 format with the following command:
```
% jaguar -babel -ijagin ligand.in -omol2
ligand.mol2
```
3. Run the two-stage RESP calculation using the *antechamber*, *respgen*, and *resp* modules of AmberTools. The following commands convert the MOL2-formatted ligand file to AC format, generate the *resp* input files, and fit the ESP to atomic point charges:
```
% antechamber -i ligand.mol2 -fi mol2 -o
ligand.ac -fo ac
% respgen -i ligand.ac -o ligand.respin1 -f
resp1
% respgen -i ligand.ac -o ligand.respin2 -f
resp2
% resp -i ligand.respin1 -o ligand.respout1
-e ligand.resp -t qout_1 -O
% resp -i ligand.respin2 -o ligand.respout2
-e ligand.resp -t qout_2 \
-q qout_1 -O
% antechamber -i ligand.ac -fi ac -o ligand_
resp.mol2 -fo mol2 \
-c rc -cf qout_2
```
4. Assign FITTED atom types to the charged ligand by using the SMART module:
```
% smart -nocharge ligand_resp.mol2
```

3.6.4. Docking of the Platinum Complex Onto the GQ DNA

1. Place the SMART-prepared ligand and PROCESS-prepared protein files in the appropriate FITTED directory.
2. Prepare a keyword file for FITTED (*g4_ligand.txt*):

```
Protein 1
1KF1
Ligand   ligand_resp.mol2
Output   1KF1-ligand
Forcefield       fitted_ff.txt
Parameters       Auto
Binding_Site_Cav      1KF1_BS.mol2
Interaction_Sites 1KF1_IS_4.mol2
Displaceable_Waters off
Number_of_Runs        10
Matching_Algorithm  off
Min_MatchScore        1
GA_Max_Iter 20
GI_Max_Iter    40
GI_Initial_E  1000
GI_Minimized_E        200
Mode      Dock
Flex_Type        Rigid
Pop_Size100
Max_Tx   20
Max_Ty   20
Max_Tz   20
Anchor_Coor     0 0 0
```

3. Run the docking simulation

```
% fitted g4_ligand.txt
```

   Out of the found poses, select distinct binding modes to evaluate the energy with MD/MM-PBSA (*see* Note 48). *See* Fig. 14.11 flowchart of the steps described in this section. The file names inside the boxes depict the coordinates of the system used as input/output for each step. On top of the arrows (in italics) is the AMBER module used to carry out the step; below the arrow is the name of the input file used (as described in the corresponding step). Files with ".rst" extension contain a single structure of the system, while the ones with a ".crd" extension contain multiple snapshots of a MD trajectory.
4. See figure 14.12 for an example.

3.6.5. MD Simulations

The following steps use the docked poses as starting points for MD simulations (Fig. 14.11). The PDB structure for the GQ is assumed to be in *1KF1.pdb*; the docked ligand in *ligand.mol2*. To set the restraints, the DNA is assumed to be a 22-mer oligonucleotide possessing two internal cations, and the platinum (II) complex is a single residue (*see* Note 49). We refer the readers to the

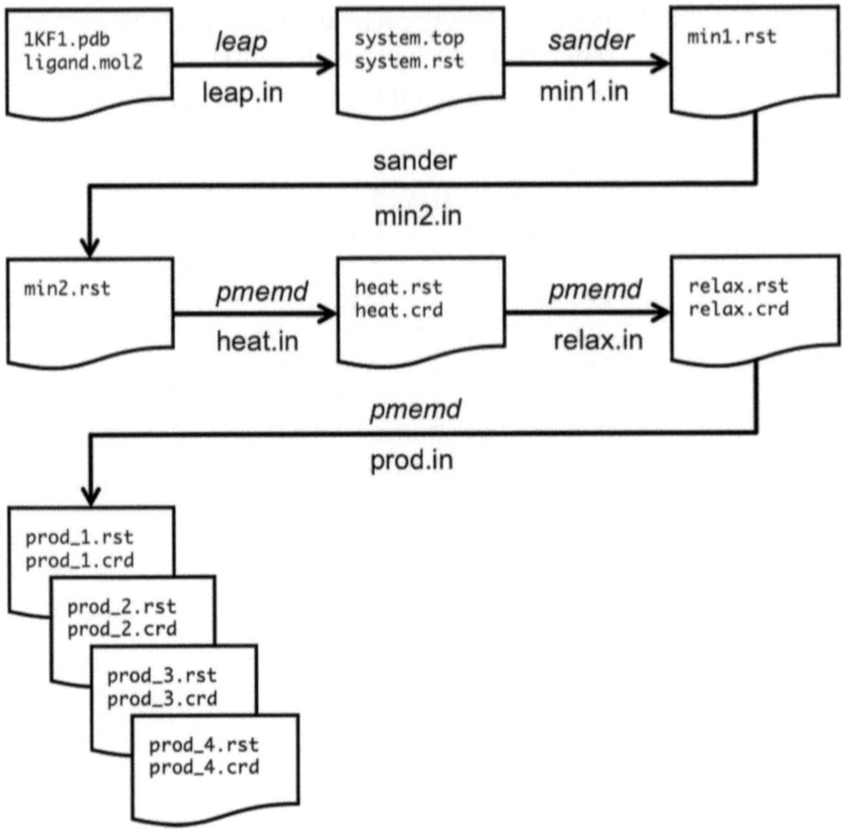

Fig. 14.11. Flowchart of the steps described in this section. The filenames inside the boxes depict the coordinates of the system, used as input/output for each step. On top of the arrows (*in italics*) is the AMBER module used to carry out the step; below the arrow is the name of the input file used (as described in the corresponding step). Files with ".rst" extension contain a single structure of the system, while the ones with a ".crd" extension contain multiple snapshots of a MD trajectory.

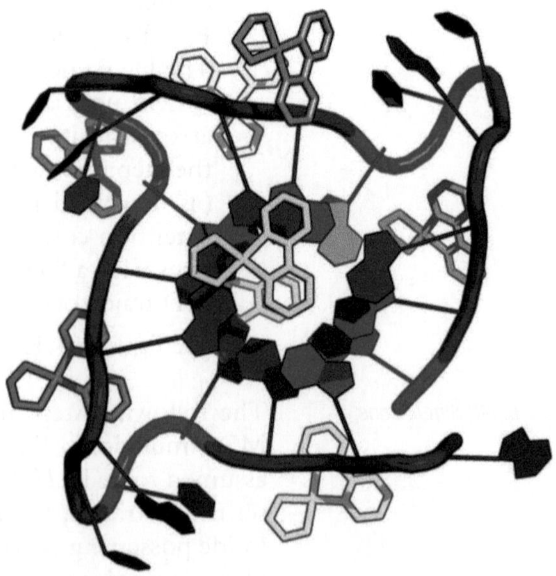

Fig. 14.12. Poses of $[Pt(bpy)(en)]^{2+}$ docked to X-ray crystal structure of human telomeric sequence (PDB code 1KF1). This picture was generated with PyMol (44).

AMBER DNA MD simulation tutorial for more details (27), and to the review by Sponer et al. for interesting discussions on the topic (28).

1. Using the *tleap* program, create topology and coordinate files for the solvated (10 Å truncated octahedron of TIP3P water), potassium-neutralized system, as well as for the system in vacuo (for the MM-PBSA calculation). The DNA is parmeterized with the *parmbsc0* (29) set of parameters for DNA in combination with the *parm99* force field (30,31), while the Pt complex is described with GAFF (23) and the *ad hoc* generated parameters derived as above. This can be accomplished with the following commands in a *leap.in* file:
```
source leaprc.ff99bsc0
dna=loadpdb "1KF1.pdb"
source leaprc.gaff
loadamberparams frcmod.Pt
complex=loadmol2 "ligand.mol2"
model=combine {dna complex}
saveamberparm dna gquart-vac.top gquart-vac.rst
saveamberparm complex complex-vac.top complex-vac.rst
saveamberparm model system-vac.top system-vac.rst
solvateOct model TIP3PBOX 10
addions model K+0
saveamberparm model system.top system.rst
quit
```
tleap can be run with the following command:
```
% tleap -s -f leap.in
```
2. Minimize the position of the water molecules keeping the DNA and platinum (II) complex in place with strong restraints. Assuming that the oligonucleotide, structural cations, and platinum (II) complex are the first 25 residues in the topology file (*see* Note 49), this can be achieved with the following *sander* input file (*min1.in*):
```
&cntrl
imin=1,
maxcyc=2000,
ncyc=750,
ntb=1,
ntr=1,
restraint_wt=500,
restraintmask=':1-25',
cut=10
/
```
sander is run with the following command:

```
% sander -O -i min1.in -p system.top -c system.
rst -ref system.rst -r min1.rst -o min1.out
```

3. Using the optimized coordinates from the previous step, minimize the whole system without restraints with the following *sander* input file (*min2.in*):

```
&cntrl
imin=1,
maxcyc=2500,
ncyc=1000,
ntb=1,
ntr=0,
cut=10
/
```

sander is run with the following command:

```
% sander -O -i min2.in -p system.top -c
min1.rst -r min2.rst -o min2.out
```

4. With the fully minimized coordinates from the last step as input, heat the system from 0 to 300 K keeping the DNA and the ligand fixed with weak restraints (*see* Note 49). A 20-ps simulation with a timestep of 1.6 fs (using the SHAKE algorithm to decouple the highest frequency vibrational modes) is shown in the following *pmemd* input file (*heat.in*, *see* Note 50):

```
&cntrl
imin=0,
irest=0,
ntx=1,
ntb=1,
cut=10,
ntr=1,
ntc=2,
ntf=2,
tempi=0.0,
temp0=300.0,
ntt=3,
gamma_ln=1.0,
nstlim=12500, dt=0.0016
ntpr=125, ntwx=125, ntwr=1250
/
10.0
RES 1 25
END
END
```

pmemd is run with the following command:

```
% pmemd -O -i heat.in -p system.top -c min2.
rst -ref min2.rst -r heat.rst -x heat.crd -o
heat.out
```

5. Relax the system through 100 ps of unrestrained MD (*relax.in*):
```
&cntrl
imin=0, irest=1, ntx=7,
ntb=2, pres0=1.0, ntp=1,
taup=2.0,
cut=10, ntr=0,
ntc=2, ntf=2,
tempi=300.0, temp0=300.0,
ntt=3, gamma_ln=1.0,
nstlim=62500, dt=0.0016,
ntpr=6250, ntwx=6250, ntwr=6250
/
```
pmemd is run with the following command:
```
% pmemd -O -i relax.in -p system.top -c heat.
rst -r relax.rst -x relax.crd -o relax.out
```
6. Run 1 ns of production MD, with 1.6 fs timestep and collecting snapshots every 10 ps (*prod.in*):
```
&cntrl
imin=0, irest=1, ntx=7,
ntb=2, pres0=1.0, ntp=1,
taup=2.0,
cut=10, ntr=0,
ntc=2, ntf=2,
tempi=300.0, temp0=300.0,
ntt=3, gamma_ln=1.0,
nstlim=625000, dt=0.0016,
ntpr=6250, ntwx=6250, ntwr=6250
/
```
pmemd is run with the following command:
```
% pmemd -O -i prod.in -p system.top -c relax.
rst -r prod_1.rst -x prod_1.crd -o prod_1.out
```
7. Repeat step 6 three additional times (saving the coordinates as successive *prod_X.rst* and *prod_X.crd* (*X*=2, 3, 4), and using the successive *prod_X.rst* as input for the next step), for a final simulation time of 4 ns.
8. At this point, the snapshots of the production run should be analyzed for convergence and stability (*see* Note 51).

3.6.6. MM-PBSA Scoring

The following steps take snapshots from the MD production run to calculate the binding affinity by the MM-PBSA formalism (32). Within AMBER, a Perl script (*mm_pbsa.pl*) is included to set up and run the calculations. The procedure consists of two steps, involving the extraction of the snapshots from the MD simulation and the subsequent calculations. The reader is referred to the AMBER tutorial on MM-PBSA for further details (33).

1. Prepare an *extract_coords.mmpbsa* file. Replace the placeholders (marked as *#XXXXX* in the following section) by the corresponding value in the system (*see* Note 52):

```
@GENERAL
PREFIX system
PATH ./
COMPLEX 1
RECEPTOR 1
LIGAND 1
COMPT system-vac.top
RECPT gquart-vac.top
LIGPT complex-vac.top
GC 1
AS 0
DC 0
MM 0
GB 0
PB 0
MS 0
NM 0
@MAKECRD
BOX YES
NTOTAL #NTOTAL
NSTART 1
NSTOP 200
NFREQ 1
NUMBER_LIG_GROUPS 1
LSTART #LSTART
LSTOP #LSTOP
NUMBER_REC_GROUPS 1
RSTART #RSTART
RSTOP #RSTOP
TRAJECTORY system_prod_1.crd.gz
TRAJECTORY system_prod_2.crd.gz
TRAJECTORY system_prod_3.crd.gz
TRAJECTORY system_prod_4.crd.gz
```

2. Run the coordinate extraction:

```
% mm_pbsa.pl extract_coords.mmpbsa
```

3. Prepare a *binding_energy.mmpsa* file for energy evaluation:

```
@GENERAL
PREFIX system
PATH ./
COMPLEX 1
RECEPTOR 1
LIGAND 1
COMPT system-vac.top
RECPT gquart-vac.top
```

```
LIGPT complex-vac.top
GC 0
AS 0
DC 0
MM 1
GB 1
PB 1
MS 1
NM 0
@PB
PROC 2
REFE 0
INDI 1.0
EXDI 80.0
SCALE 2
LINIT 1000
PRBRAD 1.4
ISTRNG 0.0
RADIOPT 0
NPOPT 1
CAVITY_SURFTEN 0.0072
CAVITY_OFFSET 0.00
SURFTEN 0.0072
SURFOFF 0.00
@MM
DIELC 1.0
@GB
IGB 2
GBSA 1
SALTCON 0.00
EXTDIEL 80.0
INTDIEL 1.0
SURFTEN 0.0072
SURFOFF 0.00
@MS
PROBE 0.0
```

4. Run the binding energy calculations:
   `% mm_pbsa.pl binding_energy.mmpbsa`
5. Extract the MM-PBSA binding affinities from the *system_statistics.out* file, in the "PBTOT" line of the "DELTA" table (*see* Note 53).

4. Notes

1. Caution should be taken when handling DNA intercalators as they may be toxic to humans.

2. Solutions of the intermolecular quadruplex sequence $(T_4G_4T_4)_4$ are quantified by UV–vis spectroscopy prior to use, monitoring the absorbance at 260 nm using an experimentally determined molar extinction coefficient, 63,197 M^{-1} cm^{-1} per quadruplex (11).
3. Solutions of fluorescently labeled oligonucleotides are quantified using UV–vis spectroscopy at 260 nm using molar extinction coefficients of 267,300 M^{-1} cm^{-1} for 5′-FAM-TATAGCTATA-HEG-TATAGCTATA-TAMRA-3′ and 272,000 M^{-1} cm^{-1} for 5′-FAM-$G_3(T_2AG_3)_3$-TAMRA-3′. These molar extinction coefficients are calculated using the IDT DNA Oligo Analyzer 3.1 (34) for the oligonucleotide sequences and adding molar extinction coefficients for FAM and TAMRA at 260 nm (35).
4. The calf thymus DNA is quantified by UV–vis spectroscopy using a molar extinction coefficient of 13,200 M^{-1} cm^{-1} per base pair (36).
5. Other suitable *ab initio* software packages could also be used (such as Gaussian, GAMESS).
6. Other suitable docking programs could be used, as long as they are able to handle transition-metal complexes as ligands and nucleic acids as targets.
7. Other MD programs could be used as well (e.g., CHARMM, GROMOS, NAMD, TINKER).
8. Other visualization programs (e.g., Sybyl, PyMol, Discovery Studio) allowing for the building of three-dimensional models could also be used, although the interface of Jaguar and Maestro makes it easier for the nonexpert to use.
9. The starting concentration of the oligonucleotide can vary from sample to sample depending on the yield obtained from a 1 μmol synthesis of the desired sequence.
10. The titration of the platinum (II) complex is performed with both calf thymus DNA and the intermolecular GQ in order to compare selectivity and binding affinity of these complexes for both these nucleic acid motifs.
11. Due to the very weak metal-to-ligand charge transfer (MLCT) bands of these phenanthroimidazole platinum (II) complexes around 400 nm, the shoulders of the peaks near 300 nm were followed during the titrations. At these wavelengths, there is a very small amount of absorbance that occurs in the later titration points from calf thymus and quadruplex DNA. Therefore, one needs to account for this absorbance and perform titrations of both calf thymus and quadruplex DNA to buffer. The obtained values are then subtracted from the absorbance traces of the platinum (II) metal complex/DNA

mixture to better monitor the hypochromic effects that are occurring to the complex. This step does not need to be performed if the metal complex provides a reasonable absorbance maximum above 320 nm.

12. The variable *C* in the equation provided is the concentration of the free platinum (II) complex calculated in step 7. This concentration is approximated to be constant throughout the titration because of the relatively insignificant volume changes that occur upon the addition of the nucleic acid motif to the overall volume of the solution (1 mL).
13. This factor can change depending on the number of tetrads present within the quadruplex sequence one is using. If one is measuring the binding constant per tetramolecular quadruplex, the DNA concentration is as calculated previously.
14. There are multiple mathematical treatments that can be applied to determined UV–vis binding constants, such as the McGhee–von Hippel (37) approach. However, we selected the method previously used by Meehan (17) and developed by Schmehel and Crothers (38) because of the better fit that we obtained for our data.
15. Due to the application of a duplex DNA binding treatment to this system, caution needs to be taken when considering these binding constant values as absolute for these platinum (II) complexes to the GQ. Although several research groups have previously applied these treatments for GQ ligand binding (6,39), these systems may actually violate some of the conditions that these binding models are based on (17,37,38).
16. Job plot concentrations can be changed to observe binding; we used 5 μM for some trials and 10 μM for others, mainly due to the weak absorbance signal that we observed for the some of the platinum (II) complexes above 300 nm.
17. Since it is not known where the intersection of the data points from the titration of platinum (II) complex to GQ and vice versa will be, one may need to graph their data before knowing where to end the outlined titrations. In addition, this process may help in determining the amount that will need to be titrated per aliquot for either nucleic acid or platinum (II) complex to achieve the necessary information to determine binding stoichiometry.
18. Due to the very slight overlap in absorbance of quadruplex DNA and the platinum (II) complexes at 305 nm, a titration of the nucleic acid to buffer needs to be performed to remove any increases in absorbances pertaining to the quadruplex. Therefore, the titrations using the GQ alone as a blank are performed in the same method as steps 3 and 4 using the

same concentrations of quadruplex without any platinum (II) complex present.

19. Data was plotted in SigmaPlot for cosmetic reasons. One can keep plots in Excel or transfer the data to any other graphing program of their choice.
20. A minimum volume of 250 μL is required for this experiment, so that the beam of light from the CD instrument passes through the solution only. This amount may differ on an individual basis for each cuvette and spacer used.
21. The CD signature of the intermolecular GQ possesses a maximum at 260 nm and a minimum at 240 nm (control sample prepared in step 1).
22. Upon the addition of platinum (II) complex to the GQ DNA solution, the CD signature is slightly decreased at 260 nm and increased at 240 nm, which is most likely due to perturbation of the GQ structure (sample prepared in step 2, this effect is more pronounced for larger π-surface binders).
23. The sample containing the platinum (II) complex shows a CD signal (control sample prepared in step 3) due to lack of chirality.
24. The GQ DNA control (step 1) shows a substantial change in CD signature, namely a significant decrease with a red shift of the peak at 260 nm to 273 nm, and an increase in the negative peak with a slight blue shift from 243 nm to 249 nm.
25. The sample with the GQ DNA/platinum (II) intercalator (step 2) solution shows retention of the circular dichroic signal and is closest in appearance to the GQ DNA prior to heating. This result indicates significant stabilization and inhibition of denaturation.
26. The platinum (II) control solution (step 3) shows no change in CD signal after heating and cooling.
27. Occasionally, the amount of GQ binder in the bulk buffer solution for the competitive dialysis experiment requires adjustment depending on the maximum absorbance that can be obtained at the wavelength one is monitoring. This problem occurs for some of the platinum (II) complexes that possess lower molar extinction coefficients for their absorbances over 300 nm.
28. The SDS solution is used to dissociate bound ligands from DNA. Since we initially reported this experiment (11), a contribution by Chaires (40) suggests the use of detergents other than SDS, such as Triton X-100 or Tween 80, when using phosphate buffers to prevent precipitation.
29. Solutions within the dialysis bags are diluted for purposes of analysis by UV–vis spectroscopy, because of large absor-

bance values encountered for DNA containing solutions that were out of range. The absorbances monitored at a certain wavelength can then be multiplied by the dilution factor. This dilution, or the need for one, can be decided on an individual basis.

30. Due to running the experiment in single-beam mode and overlapping absorbances of GQ and duplex DNA with $[PtPIP(en)]^{2+}2PF_6^-$ at 305 nm, a blank of the GQ or duplex DNA in buffer (at concentration in step 2), subjected to the SDS treatment (step 4) and dilution (step 5), needs to be acquired for subtraction from the metal complex/DNA mixture. This step is performed only if the complex has an absorbance below 320 nm in order to accurately determine the amount of metal complex present within the dialysis bags.
31. The dilution factor of 1.11 arises from the addition of the 1% SDS solution, and the dilution factor of 5 arises from dilution of the complex solution with the nucleic acids.
32. The obtained concentration of the bulk solution after 24 h should be similar to the concentration that is prepared prior to dialysis.
33. Although absolute binding constants can be derived from competitive dialysis data, caution should be taken, as it has been previously reported that the numerical values obtained are not always reproducible (41). Hence relative binding constants, rather than absolute values, can be determined from this experiment.
34. This annealing step is performed to form the duplex DNA or the GQ from the sequences reported.
35. Concentrations in the FRET experiment can be modified to observe stabilization on an individual basis. Weaker binders will show small alterations to the melting temperature of the GQ being assayed even at higher concentrations (10 μM) and vice versa. Therefore, various concentrations may need to be tested before obtaining a window of thermal denaturation values which can be evaluated.
36. In this case, 21 wells are filled because triplicates are run for each concentration listed within the text, in addition to 3 wells that are assigned to the intramolecular GQ forming sequence.
37. For FRET experiments, caution should be taken when normalizing data because of the potential for artifacts. It is recommended to carefully examine the raw data before proceeding. In addition, this experiment should be repeated several times on different well plates in order to obtain confidence in triplicate data obtained, because this experiment is sensitive to small changes in pipetting.

38. FRET melting data should be validated using another method where the oligonucleotide has no dye molecules attached, such as a thermal denaturation experiment by CD or UV–vis to eliminate the possibility of π-stacking of the GQ ligand to the dye molecules.
39. This denotes a triple-ζ basis set with polarization functions on all atoms except hydrogens.
40. Consider all the lines stating "ATTN, need revision" in the output *frcmod.Pt* file as missing parameters.
41. The distorted structures can be generated through the *scan* function in Jaguar, or via a Python script within Maestro.
42. In practice, a single set of charges could be used (i.e., assuming the charges are conformation-independent), and they can be copied from a template file with a Pyhon or awk script.
43. Pay special attention to the units of the calculated energies; Jaguar's (and most *ab initio* programs') output is in Hartrees while AMBER's is in kcal/mol. 1 Hartree = 627.509 kcal/mol.
44. This can be accomplished, for example, with the Solver module in Microsoft Excel.
45. 1KF1 is the only X-ray crystallographic structure of the intramolecular human GQ-forming sequence AGGG(TTAGGG)$_3$. Nuclear magnetic resonance (NMR) structures could also be used (e.g., 143D, 2JSL, 2JSM), but in these cases the internal structural cations should be placed manually, as their placement is not defined in the NMR experiments (42). There are multiple NMR and X-ray crystallographic structures of intermolecular GQs deposited in the PDB.
46. This step is best done with a visual interface such as Maestro or Sybyl, but the *leap* module of AmberTools can also be used. The PDB format only includes the atomic coordinates and the atom connectivity, but does not provide any information about the bond orders (e.g., if a bond single, double, or aromatic). For X-ray crystal structures, only the coordinates of heavy atoms are reported; NMR structures include hydrogens but not the structural cations. When the PDB structure contains only standard nucleotides, the visualization program should automatically assign bond orders, although it is a good habit to verify the assignment.
47. This can be done either by visual inspection or using clustering techniques (e.g., with the XCluster module within the Schrödinger Suite). The goal is to select binding modes that are diverse and that represent the total population of possible poses found by docking, for further evaluation by MD.
48. If this is not the case, modify the *restraintmask* in step 2 and the *RES* statement in step 4 accordingly.

49. The pmemd program is heavily optimized for parallel execution of particle mesh Ewald (PME) MD simulations with respect to *sander*. It requires a separate compilation from the remainder of the AMBER package, and provides a significant speedup if multiple procesors are available (e.g., in a cluster environment or a multicore/multiprocessor workstation).
50. The Perl script *process_mdout.perl* by Ross Walker is very helpful in this task (43). The output from *process_mdout.perl* should be inspected, for example, by graphing the different parameters (temperature, pressure, components of the energy) against simulation time.
51. #NTOTAL is the total number of atoms (including counterions and solvent molecules) in each snapshot; it is the first number under the %FLAG POINTERS header in the *system.top* file. #RSTART and #RSTOP are the first and last atoms of the receptor in each snapshot; when building the system with *leap*, the DNA was the first object combined, hence #RSTART = 1 and #RSTOP is the first number under the %FLAG POINTERS header in the *gquart-vac.top* file generated before. The ligand will appear directly after the DNA, hence #LSTART = (#RSTOP + 1) and #LSTOP = (#LSTART + N), where N is the first number under the %FLAG POINTERS header in the *complex-vac.top* file generated in step 1 of Subheading 14.3.6.5.
52. The *system_statistics.out* file will report the individual energy contributions from ligand, receptor, and complex, and then the "DELTA" section reports the difference such that DELTA = COMPLEX − LIGAND − RECEPTOR. Notice that there is a "PB" section and a "PBTOT"; the former is only the contribution from the continuum electrostatics, while the latter is the actual MM-PBSA binding energy. Besides MM-PBSA, one might be interested in considering MM-GBSA, which uses the generalized Born continuum solvent model instead of solving the Poisson–Boltzmann equation; this is likewise reported in the "GBTOT" line.

References

1. Kim NW, Piatyszek MA, Prowse KR, Harley CB, West MD, Ho PLC, Coviello GM, Wright WE, Weinrich SL, Shay JW (1994) Specific association of human telomerase activity with immortal cells and cancer. Science 266:2011–2015
2. Harrison RJ, Gowan SM, Kelland LR, Neidle S (1999) Human telomerase inhibition by substituted acridine derivatives. Bioorg Med Chem Lett 9:2463–2468
3. Sun D, Thompson B, Cathers BE, Salazar M, Kerwin SM, Trent JO, Jenkins TC, Neidle S, Hurley LH (1997) Inhibition of human telomerase by a G-quadruplex-interactive compound. J Med Chem 40:2113–2116
4. Wheelhouse RT, Sun D, Han H, Han FX, Hurley LH (1998) Cationic porphyrins as telomerase inhibitors: the interaction of tetra-(N-methyl-4-pyridyl)porphine with quadruplex DNA. J Am Chem Soc 120:3261–3262
5. Read M, Harrison RJ, Romagnoli B, Tanious FA, Gowan SH, Reszka AP, Wilson WD, Kelland LR, Neidle S (2001) Structure-based design of selective and potent G quadruplex-

mediated telomerase inhibitors. Proc Natl Acad Sci U S A 98:4844–4849

6. Keating LR, Szalai VA (2004) Parallel-stranded guanine quadruplex interactions with a copper cationic porphyrin. Biochemistry 43:15891–15900
7. Dixon IM, Lopez F, Tejera AM, Esteve JP, Blasco MA, Pratviel G, Meunier B (2007) A G-quadruplex ligand with 10000-fold selectivity over duplex DNA. J Am Chem Soc 129:1502–1503
8. Dixon IM, Lopez F, Esteve JP, Tejera AM, Blasco MA, Pratviel G, Meunier B (2005) Porphyrin derivatives for telomere binding and telomerase inhibition. ChemBioChem 6:123–132
9. Reed JE, Arnal AA, Neidle S, Vilar R (2006) Stabilization of G-quadruplex DNA and inhibition of telomerase activity by square–planar nickel(II) complexes. J Am Chem Soc 128:5992–5993
10. Rajput C, Rutkaite R, Swanson L, Haq I, Thomas JA (2006) Dinuclear monointercalating RuII complexes that display high affinity binding to duplex and quadruplex DNA. Chemistry 12:4611–4619
11. Kieltyka R, Fakhoury J, Moitessier N, Sleiman HF (2008) Platinum phenanthroimidazole complexes as G-quadruplex DNA selective binders. Chemistry 14:1145–1154
12. Kieltyka R, Englebienne P, Fakhoury J, Autexier C, Moitessier N, Sleiman HF (2008) A platinum supramolecular square as an effective G-quadruplex binder and telomerase inhibitor. J Am Chem Soc 130:10040–10041
13. Corbeil CR, Englebienne P, Moitessier N (2007) Docking ligands into flexible and solvated macromolecules. 1. Development and validation of FITTED 1.0. J Chem Inf Model 47:435–449
14. Corbeil CR, Englebienne P, Yannopoulos CG, Chan L, Das SK, Bilimoria D, L'Heureux L, Bedard J, Moitessier N (2008) Docking ligands into flexible and solvated macromolecules. 2. Development and application of FITTED 1.5 to the virtual screening of potential HCV polymerase inhibitors. J Chem Inf Model 48:902–909
15. Case DA, Cheatham TE III, Darden T, Gohlke H, Luo R, Merz KM Jr, Onufriev A, Simmerling C, Wang B, Woods RJ (2005) The Amber biomolecular simulation programs. J Comput Chem 26:1668–1688
16. Rester U (2006) Dock around the clock – current status of small molecule docking and scoring. QSAR Comb Sci 25:605–615
17. Wolfe A, Shimer GH, Meehan T (1987) Polycyclic aromatic hydrocarbons physically intercalate into duplex regions of denatured DNA. Biochemistry 26:6392–6396
18. Huang CY, Daniel LP (1982) Methods enzymol., vol 87, Academic Press, pp 509–525
19. Mergny J-L, De Cian A, Ghelab A, Saccà B, Lacroix L (2005) Kinetics of tetramolecular quadruplexes. Nucleic Acids Res 33:81–94
20. Chaires JB (2003) Current protocols in nucleic acid chemistry. In: Beaucage SL, Bergstrom DE, Herdewijn P, Matusda A (eds) Wiley, Hoboken
21. De Cian A, Guittat L, Kaiser M, Saccà B, Amrane S, Bourdoncle A, Alberti P, Teulade-Fichou MP, Lacroix L, Mergny JL (2007) Fluorescence-based melting assays for studying quadruplex ligands. Methods 42:183–195
22. Guyen B, Schultes CM, Hazel P, Mann J, Neidle S (2004) Synthesis and evaluation of analogues of 10H-indolo[3, 2-b]-quinoline as G-quadruplex stabilising ligands and potential inhibitors of the enzyme telomerase. Org Biomol Chem 2:981–988
23. Wang J, Wolf RM, Caldwell JW, Kollman PA, Case DA (2004) Development and testing of a general amber force field. J Comput Chem 25:1157–1174
24. Berman HM, Westbrook J, Feng Z, Gilliland G, Bhat TN, Weissig H, Shindyalov IN, Bourne PE (2000) The Protein Data Bank. Nucleic Acids Res 28:235–242
25. Davis AM, Teague SJ, Kleywegt GJ (2003) Application and limitations of X-ray crystallographic data in structure-based ligand and drug design. Angew Chem Int Ed Engl 42:2718–2736
26. Bayly CI, Cieplak P, Cornell WD, Kollman PA (1993) A well-behaved electrostatic potential based method using charge restraints for deriving atomic charges: the RESP model. J Phys Chem 97:10269–10280
27. Walker R. Simulating a polyA-polyT DNA decamer. http://ambermd.org/tutorials/basic/tutorial1. Accessed 30 Nov 2008
28. Sponer J, Spackova N (2007) Molecular dynamics simulations and their application to four-stranded DNA. Methods 43:278–290
29. Pérez A, Marchán I, Svozil D, Sponer J, Cheatham TE III, Laughton CA, Orozco M (2007) Refinement of the AMBER force field for nucleic acids: improving the description of α/γ conformers. Biophys J 92:3817–3829
30. Cheatham TE III, Cieplak P, Kollman PA (1999) A modified version of the Cornell et al. force field with improved sugar pucker

phases and helical repeat. J Biomol Struct Dyn 16:845–862

31. Cornell WD, Cieplak P, Bayly CI, Gould IR, Merz KM Jr, Ferguson DM, Spellmeyer DC, Fox T, Caldwell JW, Kollman PA (1995) A second generation force field for the simulation of proteins, nucleic acids, and organic molecules. J Am Chem Soc 117:5179–5197
32. Kollman PA, Massova I, Reyes C, Kuhn B, Huo S, Chong L, Lee M, Lee T, Duan Y, Wang W, Donini O (2000) Calculating structures and free energies of complex molecules: combining molecular mechanics and continuum models. Acc Chem Res 33:889–897
33. Walker R, Steinbrecher T. MM-PBSA AMBER tutorial. http://ambermd.org/tutorials/advanced/tutorial3. Accessed 30 Nov 2008
34. IDT Sci Tools Oligo Analyzer 3.1 http://www.idtdna.com/analyzer/Applications/OligoAnalyzer/(accesed November 27, 2008)
35. Dragan AI, Liu Y, Mekeyeva EN, Privalov PL (2004) DNA-binding domain of GCN4 induces bending of both the ATF-CREB and AP-1 binding sites of DNA. Nucleic Acids Res 32:5192–5197
36. Cosa G, Focsaneanu KS, McLean JRN, McNamee JP, Scaiano JC (2001) Photophysical properties of fluorescent DNA-dyes bound to single- and double-stranded DNA in aqueous buffered solution. Photochem Photobiol 73:585–599
37. McGhee JD, von Hippel PH (1974) Theoretical aspects of DNA-protein interactions: co-operative and non-co-operative binding of large ligands to a one-dimensional homogeneous lattice. J Mol Biol **86**: 469–489
38. Schmechel DE, Crothers DM (1971) Kinetic and hydrodynamic studies of the complex of proflavine with poly A-poly U. Biopolymers 10:465–480
39. Goncalves DPN, Rodriguez R, Balasubramanian S, Sanders JKM (2006) Tetramethylpyridiniumporphyrazines – a new class of G-quadruplex inducing and stabilising ligands. Chem Commun, 4685–4687
40. Ragazzon PA, Garbett NC, Chaires JB (2007) Competition dialysis: a method for the study of structural selective nucleic acid binding. Methods 42:173–182
41. Rosu F, De Pauw E, Guittat L, Alberti P, Lacroix L, Mailliet P, Riou J-F, Mergny J-L (2003) Selective interaction of ethidium derivatives with quadruplexes: an equilibrium dialysis and electrospray ionization mass spectrometry analysis. Biochemistry 42: 10361–10371
42. Neidle S, Parkinson GN (2003) The structure of telomeric DNA. Curr Opin Struct Biol 13:275–283
43. Walker R. Perl script to process mdout AMBER files. http://ambermd.org/tutorials/basic/tutorial1/files/process_mdout.perl. Accessed 30 Nov 2008
44. DeLano W (2008) The PyMOL Molecular Graphics System

Chapter 15

G4-FID: A Fluorescent DNA Probe Displacement Assay for Rapid Evaluation of Quadruplex Ligands

David Monchaud and Marie-Paule Teulade-Fichou

Abstract

Currently, small molecules able to interact specifically with G-quadruplex DNA (G-quadruplex ligands) are intensively studied since they appear to inhibit the growth of cancer cells via an unprecedented mode of action based on structural perturbation of telomeres. It is thus of crucial interest to develop methods that enable easy, rapid and reliable detection of valuable candidates. Herein, we would like to report on the G4-FID assay, a simple fluorescence screening method based on the displacement of a DNA light-up probe (thiazole orange) from both quadruplex and duplex DNA, therefore allowing a concomitant evaluation of the quadruplex-affinity and quadruplex- over duplex-selectivity of the tested candidates.

Key words: G-quadruplex DNA, FID assay, Ligands, Thiazole orange, Fluorescence, G-quadruplex ligands

1. Introduction

G-quadruplex DNA (1) is a peculiar DNA structure that is becoming increasingly suspected to play a critical role in the regulation of genomic functions at both telomeric (2) and intra-chromosomal localizations (3). This DNA structure is comprised of four guanine-rich DNA strands, held together via the formation of G-quartets i.e., coplanar associations of four guanine residues via Hoogsteen-type hydrogen bounds (Fig. 15.1) (4). G-quartets thus provide a broad aromatic surface highly prone to welcome and interact with small aromatic molecules, which has been demonstrated as readily accessible for drug targeting (5). Interestingly, since G-quadruplex DNA is structurally dynamic in nature and resolved in cells by specific enzymes (helicases, resolvases), the artificial stabilization of this structure by small molecules may perturb several crucial processes related to DNA,

P. Baumann (ed.), *G-Quadruplex DNA: Methods and Protocols*, Methods in Molecular Biology, vol. 608
DOI 10.1007/978-1-59745-363-9_15, © Humana Press, a part of Springer Science+Business Media, LLC 2010

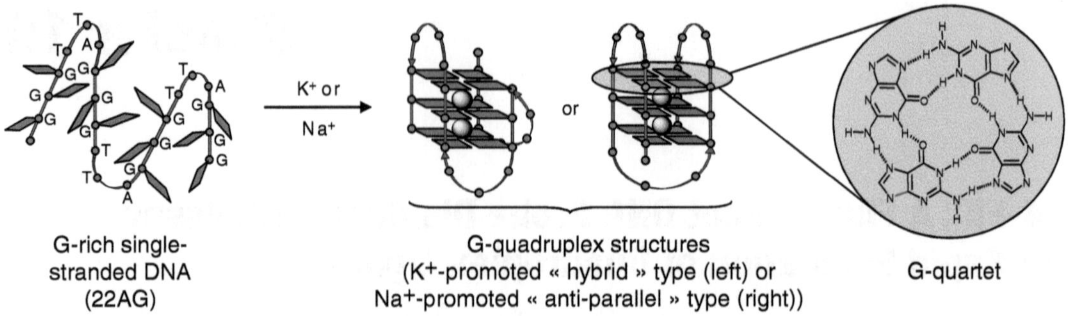

Fig. 15.1. Schematic representation of the folding of a G-rich oligonucleotide (22AG, *left*) into quadruplex-structures (whose structure depends on the nature of the cation) and detail of the G-quartet architecture (*right*).

ultimately leading to cellular senescence or apoptosis (2). Indeed, small molecules able to stabilize G-quadruplex DNA induce both in vitro and in vivo effects such as telomere-mediated chromosomal instability, reduction in activity of telomerase, a cancer-specific enzyme, and transcriptional inhibition of oncogenes (2). In addition, several quadruplex-binding molecules have shown exceptional growth inhibitory activity in cancer cells (2,5). Although their mechanism of action remains to be fully demonstrated, it is now agreed that small molecules able to interact efficiently and selectively with G-quadruplex DNA (termed hereafter "G-quadruplex ligand") can be considered as putative anticancer agents.

In this context, it is of crucial interest to develop easy-to-use and reliable methods to detect G-quadruplex ligands. Several elegant methods are currently routinely used, among which FRET-melting (6), equilibrium dialysis (7), SPR (8), ESI-MS (9), UV-melting, (10) and multifluorescent probes assays (11) can be cited. All of these methods present advantages (including high-throughput screening, access to thermodynamic parameters of ligand/DNA interactions, etc.) and drawbacks (immobilized DNA, nonisothermal conditions, etc.). These have already been thoroughly discussed and consequently will not be presented herein (12). However, since most of them require specific equipment or modified oligonucleotides, we decided to develop a simple alternative assay, G4-FID, (12) derived and adapted from assays used with duplex DNA (13). The G4-FID (for G-quadruplex Fluorescent Intercalator Displacement) assay relies on the labelling of DNA matrices (both duplex and quadruplex DNA) with a light-up fluorescent probe, thiazole orange (TO) (the principle of G4-FID assay is schematically presented in Fig. 15.2). This dye is strongly fluorescent when bound to DNA with a quantum yield comprised between 0.1 and 0.4, whereas it is virtually nonfluorescent in the free state (14). Therefore its displacement from DNA can be readily monitored via the decrease of its fluorescence.

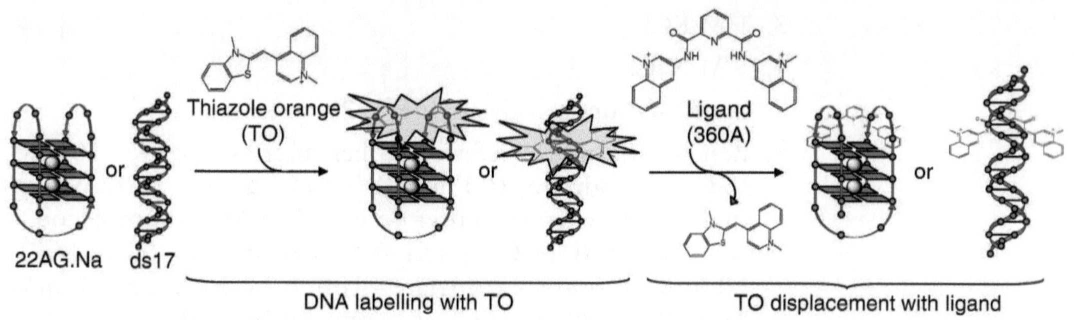

Fig. 15.2. Schematic representation of the two main steps of the G4-FID assay, i.e., (**a**) labelling of the DNA matrices (quadruplex- or duplex DNA) by thiazole orange (TO) and (**b**) displacing the fluorescent probe from the DNA matrices by a small molecule candidate (360A, see text).

Thus, the quadruplex-affinity of a candidate compound can be evaluated through its ability to displace TO from quadruplex DNA. Furthermore, on the basis that quadruplex and duplex DNA are labelled with TO with a comparable efficiency, insights into the quadruplex over duplex DNA selectivity of a given candidate can be gained by comparing its TO displacement ability from quadruplex and duplex DNA.

The validity of the G4-FID assay has been assessed by comparing G4-FID results obtained with a representative array of ligands with FRET-melting and ESI-MS results (12). This comparison program highlighted advantages (broad variety of unmodified DNA matrices, biologically relevant conditions, etc.) and limitations (indirect competition, incompatible spectroscopic properties, etc.). Nonetheless, most importantly, when G4-FID is performed in parallel with other assays, it actively participates in the quest for ligands, since the use of multiple tests minimizes the probability of studying false positives and helps to eliminate false negatives.

As previously mentioned, G4-FID is a readily realizable test that once successfully adopted, enables the evaluation of several ligands per day. In the following sections, essential key steps will be revealed in order to rapidly and efficiently setup this assay.

2. Materials

1. Spectrophotometer: In our laboratory, the G4-FID assay is performed on a FluoroMax-3 spectrophotometer (Horiba Jobin-Yvon), but equivalent instruments from other manufacturers serve equally well. Water-bathed thermostated cell holders are essential to keep samples at a constant 20°C.
2. Quartz cuvettes (3 mL, Hellma) (*see* Note 1).

3. 1 M KCl.
4. 1 M NaCl.
5. 0.1 M sodium cacodylate (pH 7.2).
6. Benchmark buffer (referred to hereafter as Caco.K): 10 mM sodium cacodylate, 100 mM KCl, pH 7.2. It is worth noting that other types of buffer (e.g., 10 mM sodium cacodylate + 100 mM NaCl (referred to hereafter as "Caco.Na"), 10 mM lithium cacodylate + 100 mM KCl, or 10 mM Tris-HCl + 100 mM KCl, etc.) can also be used.
7. DNA. All oligonucleotides are synthesized in ~200 nmol scale, at OligoGold purity grade (Eurogentec), further purified by Reverse-Phase High Performance Liquid Chromatography (RP-HPLC). The sequences of the oligonucleotides used are as follows:
   - Quadruplex DNA: 22AG is (5′-$AG_3T_2AG_3T_2AG_3T_2AG_3$-3′).
   - Duplex DNA: ds17 and ds26 are both duplex DNA, comprised of 17 and 26 base pairs respectively; ds17 is comprised of two complementary strands that are (5′-$C_2AGT_2CGTAGTA_2C_3$-3′) and (5′-$G_3T_2ACTACGA_2CTG_2$-3′), while ds26 is the self-complementary sequence (5′-$CA_2TCGGATCGA_2T_2CGATC_2GAT_2G$-3′).
8. 2 mM Thiazole orange (TO, Sigma-Aldrich) in DMSO. Store at −20°C for up to 2 months (*see* **Note 2**).
9. 2 mM ligand stock solutions in DMSO. Store at −20°C for up to 4–6 months (*see* **Note 3**).

3. Methods

3.1. Preparing the DNA Matrices

The G4-FID assay is based on the use of both quadruplex and duplex DNA matrices; the preparation (also termed "folding") of these DNA structures is as follows.

3.1.1. Preparation of DNA Aliquots

Oligonucleotides are first solubilized in deionized water (18.2 Ωcm^{-1} resistivity, simply referred to hereafter as "water") to prepare 500 μM solutions, referred to hereafter as "mother solutions," which are kept at −20°C (for a period of time not exceeding ~4–5 months). Small aliquots (80 μL for quadruplexes, 160 μL for duplexes) of diluted folded DNA (~250 μM for quadruplexes, ~125 μM for duplexes) are prepared from mother solutions. One of the advantages of the G4-FID assay is that it can be performed in near-physiological conditions. Thus we use preferentially the photochemically inert sodium cacodylate buffer ($(CH_3)_2AsO_2Na$) (pH 7.2) with cellular-mimicking high ionic strength (100 mM K^+).

Quadruplex DNA 22AG

This sequence was chosen since it corresponds to a fragment of the human telomeric 3′-overhang which is constituted of hexa-nucleotidic repeats 5′-TTAGGG-3′ (~30–40 repeats). This overhang is of high biological interest since it is a cellular-stable single-stranded DNA, able to form G-quadruplex folds which is well-documented in vitro (15). In this regard, 22AG is a model mimicking a single telomeric quadruplex unit. Given that the formation of quadruplexes within the telomeric region is highly suspected to concomitantly increase the chromosomal instability and indirectly inhibit telomerase activity, stabilization of telomeric quadruplexes by small molecules is currently under investigation as a novel anticancer strategy (2). Therefore, we decided to firstly focus on this particular quadruplex, but the possibility of using other biologically relevant quadruplex-structures will be discussed in the Notes section.

- 22AG in potassium-rich conditions: the preparation of the quadruplex-structure derived from 22AG in the presence of potassium (called hereafter "22AG.K," whose structure is schematized in Fig. 15.1) is prepared by mixing 40 μL of DNA mother solution (500 μM), 8 μL 1 M KCl, 8 μL 100 mM sodium cacodylate buffer, pH 7.2 and 24 μL water.
- 22AG in sodium-rich conditions: the preparation of the quadruplex-structure derived from 22AG in the presence of sodium (called hereafter "22AG.Na," whose structure is schematized in Figs. 15.1 and 15.2) is prepared by mixing 40 μL DNA mother solution (500 μM), 8 μL 1 M NaCl, 8 μL 100 mM sodium cacodylate buffer, pH 7.2 and 24 μL water.

It is worth noting that 22AG adopts a distinct topology in each cationic condition as demonstrated by NMR studies (15). The Na^+ fold and the K^+ fold differ by both strand orientations and loop nature, therefore offering two distinct matrices for small molecule targeting.

Duplex DNA ds26 and ds17

These two random sequences were selected since they have previously been studied and used in other projects (6,12,16).

- ds26: the preparation of ds26 is prepared by mixing 80 μL DNA mother solution (500 μM), 16 μL 1 M KCl, 16 μL 100 mM sodium cacodylate buffer pH 7.2 and 48 μL water.
- ds17: the preparation of ds17 is prepared by mixing 40 μL of both complementary strands (500 μM), 16 μL 1 M KCl, 16 μL 100 mM sodium cacodylate buffer pH 7.2 and 48 μL water (the structure of ds17 is schematized in Fig. 15.2).

3.1.2. Folding the DNA

Quadruplex structures, i.e., 22AG.K and 22AG.Na, are prepared by heating the corresponding and previously prepared 80 μL aliquots at 90°C for 5 min and cooling on ice to favor the intramolecular folding by kinetic trapping.

The preparation of duplex DNAs, i.e., ds17 and ds26, is carried out by heating the two corresponding, previously prepared 160 μL aliquots at 90°C for 5 min followed by a slow and gradual cooling to 20°C over 6–7 h by leaving the samples in the heating block and gradually decreasing the temperature.

Quadruplex and duplex DNA aliquots are subsequently stored at least overnight at 4°C before use, for a period of time not exceeding 2–3 months.

3.1.3. Calculating the Concentration of the DNA Aliquots

The concentration of the resulting DNA samples is evaluated via UV-absorbance spectra analysis at 260 nm, using the molar extinction coefficient value provided by the manufacturer (17). To this end, 8 μL of the previously prepared aliquots are added to 992 μL water and after thorough mixing transferred to 1 mL quartz cuvettes (125-fold dilution). These solutions are heated at 80°C for 5 min (to unfold both quadruplex and duplex structures), and UV-vis spectra are recorded in the 200–400 nm wavelength range (at least in duplicate for a more accurate determination of the concentration); concentrations are evaluated on the basis of optical density (OD) at 260 nm, following the equation: $OD = \varepsilon.l.c$, i.e., $c = OD/(\varepsilon.l)$, in which ε (the coefficient of molar extinction, expressed in $M^{-1} cm^{-1}$) is an inherent characteristic of the studied oligonucleotide (herein ε = 228,500, 253,200, 160,900 and 167,400 $M^{-1} cm^{-1}$ for 22AG, ds26 and both complementary strands of ds17 respectively, provided by the manufacturer) and l is the length of the optical path (expressed in cm, frequently l = 1 cm).

3.1.4. Confirming the Structure of DNA Aliquots

The structure of the quadruplex DNA can be checked by circular dichroism (CD) studies, given that quadruplex-structures are characterized by typical CD signatures (18): indeed, the 22AG.K structure, a "mixed-hybrid" structure (Fig. 15.1), presents a CD signature with two positive ellipticity maxima at 295 and 268 nm, whereas the CD spectrum of 22AG.Na, an "antiparallel" structures (Figs. 15.1 and 15.2), has a maxima at 295 (positive) and 264 nm (negative). Insights gained with CD signals are very informative, since CD spectra of quadruplex-structures drastically differ from spectra of random-coiled oligonucleotides (characterized by a positive ellipticity peak at 257 nm and a weak negative maximum at ~280 nm).

The structure of both quadruplex and duplex DNA can also be checked by UV-melting experiments (17). The thermal denaturation (from 10 to 90°C) of folded oligonucleotides followed by the increase of absorbance at 260 nm for duplex DNA and the decrease of absorbance at 295 nm for quadruplex DNA (17) leads to typical sigmoid-shaped curves that enable the determination of melting temperature (Tm) that has to be compared to those reported in the literature (e.g., 63 and 56°C for 22AG.K and

22AG.Na respectively, Caco.K and Caco.Na respectively) (17). It is worth noting that the observation of a sigmoid-shaped curve implies that the studied oligonucleotide is unfolded upon heating and the analysis of the obtained melting-curve confirms that the quadruplex-structure is the main species under the conditions of the G4-FID assay (20°C).

3.2. Performing G4-FID Assay

G4-FID evaluation of a given sample is readily performed according to the following four steps:

1. Buffer: 3 mL of buffer (Caco.K for 22AG.K, ds17 and ds26, Caco.Na for 22AG.Na) is dispensed in a 3 mL quartz cuvette (*see* Note 1), a fluorescence spectrum is recorded upon excitation (λ_{ex}) at 501 nm, and data is collected (λ_{em}) from 510 to 750 nm (excitation slits: 3 nm) (*see* Note 4). This spectrum will be used as a reference (or blank) and will be systematically subtracted from the following spectra. Given that both λ_{ex} and λ_{em} will be identical in all following experiments, it will be simply referred to hereafter as "and a fluorescence spectrum is recorded."
2. DNA: DNA is added to afford a 0.25 μM solution (in the ideal case of a 250 μM DNA sample this represents an addition of 3 μL) and a fluorescence spectrum is recorded (*see* Note 5).
3. Thiazole orange (TO): Two or three molar equivalents of TO are added, depending on the nature of the DNA used (2 equivalents with 22AG.K, 22AG.Na and ds17, representing an addition of 0.75 μL 2 mM TO solution; 3 equivalents with ds26, representing an addition of 1.12 μL 2 mM TO solution). After thorough mixing by hand, a fluorescence spectrum is recorded immediately ($t=0$) and again after a 5 min equilibration period ($t=5$ min). This double spectrum recording being performed systematically hereafter will not be further indicated and only the spectrum recorded at $t=5$ min will be used for results interpretation.
4. Candidate (*see* Note 3): increasing concentrations (from 0 to 10 molar equivalents, i.e., from 0 to 2.5 μM) of compounds to be tested are added to displace TO from DNA matrices, following a 11-step gradual addition (0.5, 1.0, 1.5; 2.0, 2.5, 3.0, 4.0, 5.0, 6.0, 8.0 and 10.0 molar equivalents) that corresponds to the successive addition of 0.75, 0.75, 0.75, 0.75, 0.75, 0.75, 0.75, 1.50, 1.50, 1.50, 3.0 and 3.0 μL of a 500 μM solution of ligand in DMSO. Fluorescence spectra are recorded after each addition.

3.3. Treating the Data

The TO displacement is evaluated by measuring the fluorescence area (FA, from 510 to 750 nm) of all 12 experimentally determined spectra (from FA_1 (the fluorescence area of the spectrum recorded after the addition of TO) to FA_{12} (the fluorescence area of the spectrum recorded after the addition of 10 equivalents of compounds). The percentage of TO displacement (^{TO}D) is calculated for each of the 12 previously determined FA using the following equation: $^{TO}D_x = 100-((FA_x/FA_1)\times 100)$, with $1<x<12$. The percentage of TO displacement is then plotted as a function of the concentration of added compound, $^{TO}D = f$((compound)).

3.4. Quantifying the Results

The affinity of the studied molecule is quantified via DC_{50} values, i.e., the concentration required to displace TO from the DNA matrix by 50%, that is to say to decrease the TO fluorescence by 50%. The DC_{50} values are calculated with both quadruplex ($^{G4}DC_{50}$) and duplex matrices ($^{ds}DC_{50}$). The selectivity for quadruplex over duplex DNA (^{G4}S) is calculated via the following equation: $^{G4}S = {}^{ds}DC_{50}/{}^{G4}DC_{50}$ (with $^{G4}S \rightarrow 1$ (and not 0) for unselective molecules and $^{G4}S \rightarrow +\infty$ for selective molecules). The example of the quinacridine MMQ_{16} (further developed in reference 12) was illustrative of such a situation, with $^{G4}DC_{50} = 0.11$ μM (from 22AG.Na) and $^{ds}DC_{50} = 1.89$ μM (from ds26), leading to a selectivity factor of $^{G4}S = 17$.

When the interaction with duplex DNA is very weak, no $^{ds}DC_{50}$ values can be determined, for example this is the case with the bisquinolinium 360A, (*see* Subheading 3.5). In this particular situation, ^{G4}S cannot be calculated but has to be estimated (Est.^{G4}S). To this end, the TO displacement obtained with 2.5 μM (10 molar equivalents) of ligand is determined ($^{TO}D_{2.5\,\mu M}$, expressed in %); the concentration of ligand required for obtaining the same displacement from quadruplex DNA is thus determined (^{G4}C). The Est.^{G4}S value can thus be calculated on the basis of a concentration ratio, using the following equation: Est.$^{G4}S = 2.5/{}^{G4}C$.

The comparison of DC_{50} (along with ^{G4}S and/or Est.^{G4}S) determined with the various DNA matrices (22AG.K, 22AG.Na, ds17 and ds26) offers a global overview ("mapping") of the interaction of the studied molecule with DNA.

3.5. Example: 360A

The bisquinolinium pyridodicarboxamide compound known as 360A (Figs. 15.2 and 15.3) is one of the most interesting G-quadruplex ligands recently reported in the literature, thanks to its very high affinity and selectivity for quadruplex DNA, along with a rapid and efficient synthetic access (19–23). Although not yet structurally characterized, the interaction of 360A with quadruplex DNA is probably mostly oriented by G-tetrad recognition, given that 1) it interacts very efficiently with various types of quadruplexes, including "loopless" tetramolecular quadruplexes (21) and 2) its structural simplicity does not enable a concomitant recognition of various structural elements of a quadruplex-structure (tetrad, loop and/or groove). 360A is therefore considered a very representative example of a

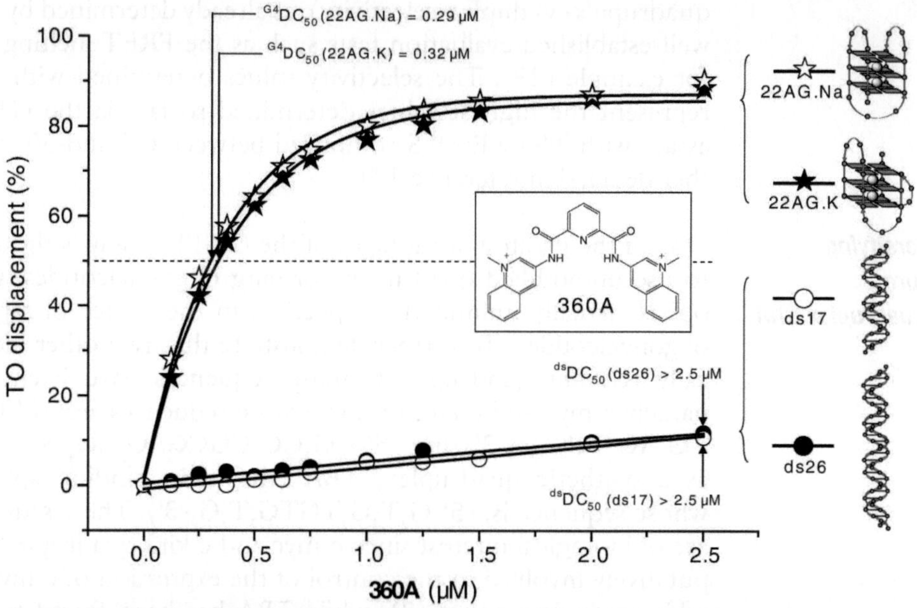

Fig. 15.3. G4-FID assay performed with 360A and both quadruplex (22AG.Na (*white star*) and 22AG.K (*black star*)) and duplex DNA (ds17 (*white circle*) and ds26 (*black circle*)).

G-quadruplex ligand, which must be unambiguously detected to validate the assay.

When evaluated through the G4-FID assay (Fig. 15.3), 360A is indeed found to be a very powerful TO displacing molecule in the case of quadruplex matrices, since it is characterized by low $^{G4}DC_{50}$ values with 22AG.K and 22AG.Na ($^{G4}DC_{50}=0.29$ and 0.32 μM for 22AG.Na and 22AG.K respectively). These values indicate the strong ability of 360A to interact with quadruplex DNA, implying a strong quadruplex-affinity.

When performed with duplex DNA (both ds17 and ds26), the TO displacing ability of 360A is very weak; indeed, even 2.5 μM (10 molar equivalents) of 360A appears unable to reach the 50% TO displacement level that enables the determination of $^{ds}DC_{50}$ values, meaning that 360A has a poor affinity for duplex DNA while exhibiting a high quadruplex- over duplex-selectivity. However, given that no $^{ds}DC_{50}$ values are available, the ^{G4}S cannot be determined, but Est.^{G4}S has to be calculated. In the present case, the presence of 2.5 μM of 360A leads to a TO displacement ($^{TO}D_{2.5\,\mu M}$) of 12 and 14% for ds17 and ds26 respectively; the concentration of 360A required to displace 12 and 14% of TO are 0.06 and 0.07 μM respectively from 22AG.K and 0.05 and 0.07 μM respectively from 22AG.Na, leading to Est.^{G4}S values of 42, 36, 50 and 42 respectively.

360A has thus been found via the G4-FID assay to be a very potent displacer of TO from quadruplex DNA (i.e., quadruplex-affinity) with a low interaction with duplex DNA (i.e., high quadruplex- vs duplex-selectivity), as already determined by other well-established evaluation tests such as the FRET-melting assay for example (23). The selectivity values determined with 360A represent the highest values determined so far via the G4-FID assay (with ^{G4}S or Est.^{G4}S comprised between 0.3 and 50, as further detailed in reference 12).

3.6. Diversifying the Nature of the Quadruplex DNA

One of the greatest advantages of the G4-FID assay is the ability to use unmodified quadruplex-forming oligonucleotides, which opens virtually unlimited perspectives in the choice of targeted oligonucleotides. To further demonstrate this, two other biologically relevant quadruplex-forming sequences have been used, namely c-myc and c-kit2 (3,24), whose sequences are: (5′-$GAG_3TG_4AG_3TG_4A_2G$-3′) and (5′-$CG_3CG_3CGCGAG_3AG_4$-3′) as well as a synthetic quadruplex, TBA (thrombin binding aptamer) whose sequence is: (5′-$G_2T_2G_2TGTG_2T_2G_2$-3′). These sequences are of biological interest since c-myc and c-kit2 quadruplexes are putatively involved in the control of the expression of c-myc and c-kit proto-oncogenes (3), while TBA has been found to be a very potent inhibitor of α-thrombin, a serine protease involved in neuronal calcium homeostasis (25). It was thus of importance to evaluate them as quadruplex-matrices for the G4-FID assay (*see* Note 6).

3.6.1. Preparing and Folding the Quadruplex DNA

Since c-myc, c-kit2 and TBA sequences can form intramolecular quadruplex-structures, their folding is performed according to the procedure described for 22AG (*see* Subheading 3.1).

1. Mother solutions (500 μM) of the three purchased oligonucleotides are prepared in water.
2. Small aliquots (80 μL) of the oligonucleotides are prepared in Caco.K by mixing 40 μL of DNA mother solutions (500 μM), 8 μL 1 M KCl solution, 8 μL 100 mM sodium cacodylate buffer pH 7.2 and 24 μL water.
3. Quadruplex-structures from c-myc, c-kit2 and TBA are prepared (folded) by heating the corresponding, previously prepared 80 μL aliquots at 90°C for 5 min then cooling on ice to favor the intramolecular folding by kinetic trapping.

3.6.2. Calculating the Concentration and Confirming the Structure of the DNA

As performed with 22AG (*see* Subheading 3.1), the concentration of the obtained DNA samples is calculated via UV-absorbance spectra analysis, using the molar extinction coefficient of the unfolded species at 260 nm (herein ε =228,700, 199,100 and 144,700 $M^{-1}cm^{-1}$ for c-myc, c-kit2 and TBA respectively).

The folded topology is checked by CD measurements, as described with 22AG (*see* Subheading 3.1). Given that c-myc and c-kit2 quadruplexes have been described as "parallel" structures, they are therefore characterized by a typical CD signature comprising of a strong positive maximum at 265 and 262 nm for c-myc and c-kit2 respectively and a negative one at 242 and 240 nm for c-myc and c-kit2 respectively. TBA folds into 22AG-like "antiparallel" structures characterized by maxima at 295 nm (positive) and 264 nm (negative).

3.6.3. Using c-myc, c-kit2 and TBA as Matrices for G4-FID

Given the ability of 360A to interact efficiently with various G-quadruplexes irrespective of their structure (19–23), this ligand was used as a test for evaluating the possibility of using various quadruplex-forming oligonucleotides to perform G4-FID experiments. As seen in Fig. 15.4, 360A was able to displace TO from c-myc, c-kit2 and TBA quadruplexes with an efficiency in a range comparable with that of 22AG.K, the TO displacement from TBA being the most difficult ($^{G4}DC_{50}$ = 0.44 μM), and the displacement from c-myc and c-kit2 being the most efficient ($^{G4}DC_{50}$ = 0.29 μM in both cases). This series of results confirms that other quadruplex-forming sequences can be valuably employed in the G4-FID assay, the differences obtained between the various determined $^{G4}DC_{50}$ originating in, and reflecting, a different affinity of 360A for the different quadruplex-architectures.

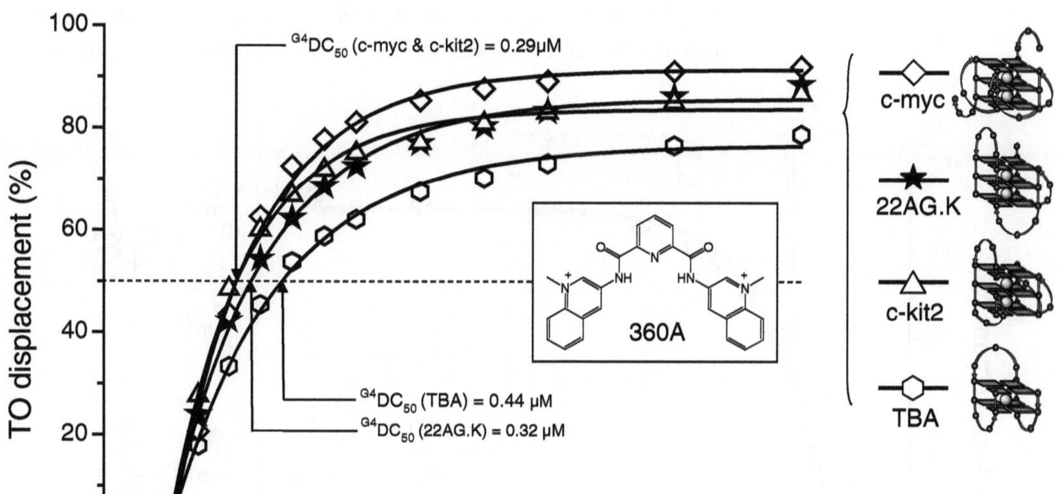

Fig. 15.4. G4-FID assay performed with 360A and 22AG.K (*black star*), c-myc (*white diamond*), c-kit2 (*white triangle*) and TBA (*white hexagon*).

4. Notes

1. The original protocol of the G4-FID assay is based on the use of 3 mL (10 × 10 mm) quartz cuvettes; however, it can also be applicable to 1 mL (10 × 10 mm) cuvettes, thus allowing minimization of both DNA and ligand consumption, which is especially useful for compounds difficult to access (e.g., extraction from natural sources or challenging multiscale synthesis). In this case, we used more diluted stock solutions of both TO and ligands (500 and 250 μM respectively) and the protocol described in Subheading 3.1 is modified as follows: 1) 1 mL of buffer (Caco.K) is dispensed in a 1 mL quartz cuvette and a fluorescence spectrum is recorded; 2) DNA is thus added to afford a 0.25 μM solution (in the ideal case of a 250 μM DNA sample this represents an addition of 1.0 μL) and a fluorescence spectrum is recorded; 3) 2 or 3 molar equivalents of TO are thus added depending on the nature of the used DNA (2 equivalents with c-myc and 3 equivalents with ds26, representing additions of 1.0 μL and 1.5 μL of a 500 μM TO solution respectively, Fig. 15.5) and a fluorescence spectrum is recorded; 4) increasing concentrations (from 0 to 2.5 μM) of compounds to be tested are thus added to displace TO from DNA matrices, following a 11-step gradual addition (0.5, 1.0, 1.5; 2.0, 2.5, 3.0, 4.0, 5.0, 6.0, 8.0, and 10.0 molar equivalents) that corresponds to the successive addition of

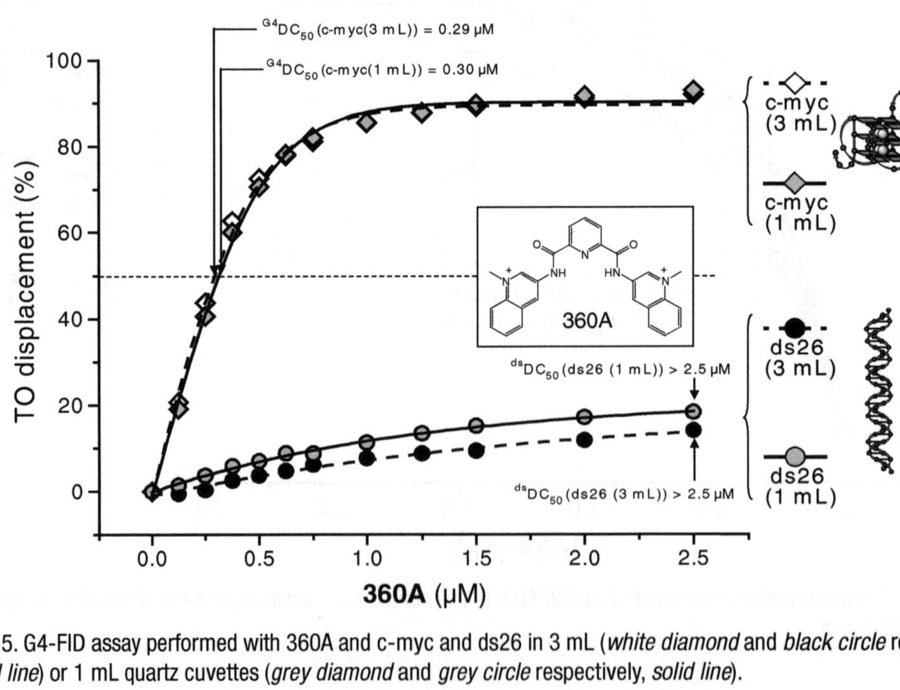

Fig. 15.5. G4-FID assay performed with 360A and c-myc and ds26 in 3 mL (*white diamond* and *black circle* respectively, *dashed line*) or 1 mL quartz cuvettes (*grey diamond* and *grey circle* respectively, *solid line*).

6×0.50, 3×1.0, and 2×2.0 μL of a 250 μM DMSO solution; fluorescence spectra are recorded after each addition. As seen in Fig. 15.5, the obtained results are quite similar to those obtained using 3 mL quartz cuvettes.

2. Thiazole orange, as well as its interaction with DNA, have been thoroughly studied by Kubista and colleagues (14). Since TO is relatively unstable in water and prone to form dimers or other high-order aggregates, it is prepared in DMSO (2 mM). In order to operate in optimal conditions, freezing/heating cycles of the TO/DMSO solution should be avoided, and the use of small aliquots (~20 μL) is preferred.

3. In order to compare the broadest library of putative ligands in strictly identical conditions, all candidates are used in DMSO as 500 μM solutions (even if they are soluble in water or buffered solutions). In order to avoid freezing/heating cycles, the preparation of small aliquots (~50 μL) is preferred (from mother solutions generally 2 mM in DMSO). These solutions are kept at −20°C, for a period of time not exceeding 4–6 months (depending on the stability of the chemicals, which can be regularly checked by recording their UV-vis spectrum).

   Some limitations from the ligand point of view: even if this situation is rare, the spectroscopic characteristics of a given candidate can hamper its evaluation via the G4-FID assay. Indeed, the overlap of the absorption or the emission spectra of the TO with the absorption characteristics of the studied molecule must be avoided to prevent inner filter effects and re-absorption. Such a situation has been observed with the well-known ligand TMPyP4 (26–29). This tetracationic porphyrin could not be properly evaluated since it presents a significant absorbance between 480 and 540 nm ($\varepsilon_{501\ nm=}$ 19,200 M^{-1} cm^{-1}) leading to a strong overlap with that of TO (510–700 nm) resulting in a biased fluorescence decrease. Thus, the absorption spectrum of candidates must be recorded before starting their evaluation via the G4-FID assay.

4. Depending on the software used, each acquired spectrum can be an average of several scans (it is for example the case with DataMAX for Windows® software); even if this possibility has not been exploited herein, it has to be taken into account when possible to increase the accuracy of obtained results.

5. Given the low concentrations used for the G4-FID assay (both DNA and chemicals), experiments have to be at least duplicated, although triplicates are preferred.

6. The use of various quadruplex structures is directly dependent on the quantum yield of the probe once bound to DNA.

For instance when TO was associated with tetramolecular quadruplexes such as d(TG_4T)$_4$ or d(TG_3T)$_4$, a much weaker fluorescence enhancement was observed (~7-fold lower than with 22AG), preventing the use of these substrates for ligand screening (12). Herein, while the fluorescence of TO bound to 22AG, c-myc and TBA stands in a comparable range, it is of lower intensity with c-kit2 (~2-fold lower than with 22AG) in the conditions of the assay.

Acknowledgement

The authors would like to deeply acknowledge Clémence Allain for the preliminary work on the G4-FID assay, Hélène Bertrand, Elsa de Lemos, Candide Hounsou and Patrick Mailliet for synthesis of ligands used to calibrate this assay, Anne De Cian, Aurore Guedin, Jean-Louis Mergny for the FRET-melting measurements and Valérie Gabélica, Nicolas Smargiasso and Frédéric Rosu for ESI-MS evaluation of the compounds used to calibrate this assay.

References

1. Neidle S, Balasubramanian S (2006) Quadruplex nucleic acid. RSC, Cambridge
2. De Cian A, Lacroix L, Douarre C, Temine-Smaali N, Trentesaux C, Riou J-F, Mergny J-L (2008) Targeting telomeres and telomerase. Biochimie 90:131–155
3. Qin Y, Hurley LH (2008) Structures, folding patterns, and functions of intramolecular DNA G-quadruplexes found in eukaryotic promoter regions. Biochimie 90:1149–1171
4. Patel DJ, Phan AT, Kuryavyi V (2007) Human telomere, oncogenic promoter and 5′-UTR G-quadruplexes: diverse higher order DNA and RNA targets for cancer therapeutics. Nucleic Acids Res 35:7429–7455
5. Monchaud D, Teulade-Fichou M-P (2008) A hitchhiker's guide to G-quadruplex ligands. Org Biomol Chem 6:627–636
6. De Cian A, Guittat L, Kaiser M, Saccà B, Amrane S, Bourdoncle A, Alberti P, Teulade-Fichou M-P, Lacroix L, Mergny J-L (2007) Fluorescence-based melting assays for studying quadruplex ligands. Methods 42:183–195
7. Ragazzon PA, Chaires JB (2007) Use of competition dialysis in the discovery of G-quadruplex selective ligands. Methods 43:313–323
8. White EW, Tanious F, Ismail MA, Reszka AP, Neidle S, Boykin DW, Wilson WD (2007) Structure-specific recognition of quadruplex DNA by organic cations: influence of shape, substituents and charge. Biophys Chem 126:140–153
9. Rosu F, De Pauw E, Gabelica V (2008) Electrospray mass spectrometry to study drug-nucleic acids interactions. Biochimie 90:1074–1087
10. Rachwal PA, Fox KR (2007) Quadruplex melting. Methods 43:291–301
11. Paramasivan S, Bolton PH (2008) Mix and measure fluorescence screening for selective quadruplex binders. Nucleic Acids Res 36:e106
12. Monchaud D, Allain C, Bertrand H, Smargiasso N, Rosu F, Gabelica V, De Cian A, Mergny J-L, Teulade-Fichou M-P (2008) Ligands playing musical chairs with G-quadruplex DNA: a rapid and simple displacement assay for identifying selective G-quadruplex binders. Biochimie 90:1207–1223
13. Tse WC, Boger DL (2004) A fluorescent intercalator displacement assay for establishing DNA binding selectivity and affinity. Acc Chem Res 37:61–69

14. Nygren J, Svanvik N, Kubista M (1998) The interactions between the fluorescent dye thiazole orange and DNA. Biopolymers 46:39–51
15. Dai J, Carver M, Yang D (2008) Polymorphism of human telomeric quadruplex structures. Biochimie 90:1172–1183
16. Teulade-Fichou M-P, Carrasco C, Guittat L, Bailly C, Alberti P, Mergny J-L, David A, Lehn J-M, Wilson WD (2003) Selective recognition of G-quadruplex telomeric DNA by a bis(quinacridine) macrocycle. J Am Chem Soc 125:4732–4740
17. Mergny J-L, Phan A-T, Lacroix L (1998) Following G-quartet formation by UV-spectroscopy. FEBS Lett 435:74–78
18. Paramasivan S, Rujan I, Bolton PH (2007) Circular dichroism of quadruplex DNAs: applications to structure, cation effects and ligand binding. Methods 43:324–331
19. Lemarteleur T, Gomez D, Paterski R, Mandine E, Mailliet P, Riou J-F (2004) Stabilization of the c-myc gene promoter quadruplex by specific ligands' inhibitors of telomerase. Biochem Biophys Res Commun 323:802–808
20. Pennarun G, Granotier C, Gauthier LR, Gomez D, Hoffschir F, Mandine E, Riou J-F, Mergny J-L, Mailliet P, Boussin FD (2005) Apoptosis related to telomere instability and cell cycle alterations in human glioma cells treated by new highly selective G-quadruplex ligands. Oncogene 24:2917–2928
21. De Cian A, Mergny J-L (2007) Quadruplex ligands may act as molecular chaperones for tetramolecular quadruplex formation. Nucleic Acids Res 35:2483–2493
22. De Cian A, Cristofari G, Reichenbach P, De Lemos E, Monchaud D, Teulade-Fichou M-P, Shin-ya K, Lacroix L, Lingner J, Mergny J-L (2007) Reevaluation of telomerase inhibition by quadruplex ligands and their mechanisms of action. Proc Natl Acad Sci U S A 104:17347–17352
23. Monchaud D, Yang P, Lacroix L, Teulade-Fichou M-P, Mergny J-L (2008) A metal-mediated conformational switch controls G-quadruplex binding affinity. Angew Chem Int ed 47:4858–4861
24. Dash J, Shirude PS, Balasubramanian S (2008) G-quadruplex recognition by bis-indole carboxamides. Chem Commun (26):3055–3057
25. Pagano B, Martino L, Randazzo A, Giancola C (2008) Stability and binding properties of a modified thrombin binding aptamer. Biophys J 94:562–569
26. Han FX, Wheelhouse RT, Hurley LH (1999) Interactions of TMPyP4 and TMPyP2 with quadruplex DNA. Structural basis for the differential effects on telomerase inhibition. J Am Chem Soc 121:3561–3570
27. Shi D-F, Wheelhouse RT, Sun D, Hurley LH (2001) Quadruplex-interactive agents as telomerase inhibitors: synthesis of porphyrins and structure-activity relationship for the inhibition of telomerase. J Med Chem 44:4509–4523
28. Kim M-Y, Gleason-Guzman M, Izbicka E, Nishioka D, Hurley LH (2003) The different biological effects of telomestatin and TMPyP4 can be attributed to their selectivity for interaction with intramolecular or intermolecular G-quadruplex structures. Cancer Res 63:3247–3256
29. Freyer MW, Buscaglia R, Kaplan K, Cashman D, Hurley LH, Lewis EA (2007) Biophysical studies of the c-MYC NHEIII$_1$ promoter: model quadruplex interactions with a cationic porphyrin. Biophys J 92:2007–2015

Index

U

V

X

MIX
Papier aus verantwortungsvollen Quellen
Paper from responsible sources
FSC® C105338

If you have any concerns about our products,
you can contact us on
ProductSafety@springernature.com

In case Publisher is established outside the EU,
the EU authorized representative is:
Springer Nature Customer Service Center GmbH
Europaplatz 3, 69115 Heidelberg, Germany

Printed by Libri Plureos GmbH
in Hamburg, Germany